Joining Processes

Joining Processes
An Introduction

David Brandon and Wayne D. Kaplan
Technion, Israel Institute of Technology, Israel

JOHN WILEY AND SONS
Chichester · New York · Weinheim · Brisbane · Singapore · Toronto

Baffins Lane, Chichester,
West Sussex PO19 1UD, England
National 01243 779777
International (+44) 1243 779777
e-mail (for orders and customer service enquiries): cs-books@wiley.co.uk
Visit our Home Page on http://www.wiley.co.uk
or http://www.wiley.com

Other Wiley Editorial Offices
John Wiley & Sons, Inc., 605 Third Avenue,
New York, NY 10158-0012, USA

VCH Verlagsgesellschaft mbH, Pappelallee 3,
D-69469 Weinheim, Germany

Jacaranda Wiley Ltd, 33 Park Road, Milton,
Queensland 4064, Australia

John Wiley & Sons (Asia) Pte Ltd, 2 Clementi Loop #02-01,
Jin Xing Distripark, Singapore 129809

John Wiley & Sons (Canada) Ltd, 22 Worcester Road,
Rexdale, Ontario MW 1L1, Canada

Library of Congress Cataloging-in-Publication Data

Brandon, D. G.
Joining processes : an introduction / David Brandon and Wayne D. Kaplan.
p. cm.
Includes bibliographical references and index.
ISBN 0-471-96487-5 (hbk : alk. paper). – ISBN 0-471-96488-3 (pbk.)
1. Surfaces (Technology) 2. Sealing (Technology) 3. Welding. 4. Joints (Engineering) I. Kaplan, Wayne D.
TA418.7.B67 1997
620′.44–dc21 96-43512
CIP

British Library Cataloguing in Publication Data

A catalogue record for this book is available from the British Library

ISBN 0-471-96487 5 Hbk
ISBN 0 471 96488 3 Pbk

Contents

Preface

The distillation of any body of human knowledge into written forms is a curious process. We talk of an *area of expertise*, or a *body of knowledge*, but neither two nor three dimensions are really quite adequate as a framework to describe any subject. *Multifaceted* is also a common term, but *multidimensional* would be more accurate. This complexity must then be reduced to a *single dimension*, the written word, and sentence follows sentence in what we hope is a logical order. We seek to *represent* our topic: selecting the textual content from the plethora of options, presenting a few *factual data*, defining some *concepts* which we consider significant, and exploring those *consequences* which we decide are important. All this is done in a totally artificial textual framework of our own choosing, and with the conscious understanding that, by the time the reader opens the page, *our* state of knowledge will have developed and moved on.

This present text is no exception. The *joining of components* and their *assembly* into engineering systems covers a diverse range of technical knowledge and experience which has accumulated in *all* branches of the engineering profession. What do the problems associated with the bonding of plaster to cement and brick have in common with the welding of alloy steels? What common principles can be distinguished in the selection of orthopaedic implants and the packaging of a microelectronic circuit? Is it even worthwhile *attempting* to bring together what can, quite justifiably, be described as a complete *hotchpotch* of technical skills? Do the topics to be covered in this text constitute a suitable subject for academic study within the framework of a credit course in *Materials Science* or *Engineering*?

The authors' answers to these last two questions are clearly *yes*, but for two reasons they may not convince either the materials educator or the professional expert who earns his living in this field.

In the first place, it is a common defect in most materials education that, while strong emphasis is given to the interrelation between the *bulk properties* of engineering materials and their *microstructural features*, very little attention is paid to the ways in which the components of an engineering system are *assembled*, or the ways in which the components may *interact* at the contact surfaces and interfaces between them.

Secondly, this educational neglect of *interfacial properties* and *interface interactions* is often only partial. Students are commonly offered elective courses in *Welding Technology*, or *Adhesives*, or *Surface Science*. Depending on the aims and inclinations of the lecturer, these courses cover, to varying degrees and at various levels, the *science* behind the technology and the *engineering methodology* which is associated with the processes and their applications. What they generally do *not* do is to give the student a comprehensive understanding of what is best termed *interface science and technology*.

For the past several years the authors have been teaching third year Materials Engineering students an elective course in *Joining Processes*. This had replaced an earlier course on *Welding Science*. The new course has been well-received by the students, who seem to have appreciated the blend of *science* and *technology*, as well as the conscious effort which has been made to include topics which are commonly regarded as unrelated. In particular, they have commented favourably on the combination of *physical* principles, derived from *surface science*, with the *mechanical* aspects of joining, derived from concepts in *strength of materials*.

The level of presentation is *elementary*, but we have assumed that the reader is familiar with the mechanical properties of materials, their microstructure and the relation between the two. The book is divided into *three* sections. The first three chapters deal with those topics which are common to all joining processes. The next six chapters describe the engineering options currently available for the joining and assembly of the components of an engineering system. The last two chapters take two areas of special interest, *vacuum systems* and *microelectronics*, and survey the range of solutions to joining problems which have been developed for these specific applications.

The reader will probably discover omissions and appreciable repetition, together with many faults, but it is our hope that he will also find the text *readable* and *informative*. Perhaps he will conclude that the authors have succeeded in their aim, and that this artificially selected sequence of sections and chapters does indeed comprise a *body of knowledge*. If so, the omissions, the repetitions and the faults may perhaps be forgiven. As in the old days of accounting, the statement from the ledger may close with the term *E&OE—errors and omissions excepted*.

David Brandon   Wayne D. Kaplan
Haifa, March 1997

Part I

PRINCIPLES OF JOINING

1

Introduction

The joining and assembly of engineering compounds is not usually treated in a single, introductory text intended for engineering students. Instead the various technologies associated with joining and assembly are normally presented as specialist topics in textbooks which are intended for elective courses. Such courses may be on, for example, welding and brazing, or adhesives and glues. Mechanical joining methods and the mechanical aspects of joining are often treated in a somewhat cursory fashion. These mechanical aspects include geometrical problems which are associated with joint design, as well as the methods available for the mechanical testing of joints. A reasoned discussion of the applications and limitations of mechanical joining methods, with one or two exceptions, is commonly found only in the professional handbooks.

The nature of solid surfaces, on the other hand, is generally regarded as a topic in physics. The discipline of *surface science* is then divorced from any technological problems in the joining and assembly of engineering components, instead of being integrated into what ought to be a single branch of engineering science. For the purposes of this text we attempt to combine these varied specialist topics and define our subject as: *the joining and assembly of components in the construction of engineering systems*.

1.1 TEXT OBJECTIVES

In this book an elementary treatment of surface science is integrated with a description of the common engineering methods for the joining and assembly of engineering components. These engineering technologies include both *metallurgical* processes, based on welding, brazing, soldering or diffusion bonding, as well as the methods of *polymer science*, which employ adhesives and glues to join components. *Mechanical* methods of attachment are also included, and are contrasted with alternative methods of assembly.

The *mechanical performance* of joints and the evaluation of their *mechanical properties* are also treated. This elementary presentation includes the basic concepts of stress concentration, residual stress and fracture mechanics. The intention is to provide the student with a *teaching text* which presents the scientific background to joining and assembly and which includes most of the commonly available joining technologies.

1.1.1 Prerequisite Knowledge

We assume that the reader is familiar with basic concepts in *materials science*. In particular, he is expected to understand elementary concepts in *strength of materials* (stress, strain and elastic modulus), *crystallography* (symmetry concepts and the nature of the crystal lattice) and the interpretation of *phase diagrams* (phase stability and phase transitions). The text further assumes that students have been exposed to introductory courses on engineering materials, and that they are aware of the significance of *microstructure* and its relation to the common *engineering properties* of materials (such as hardness, ductility and yield strength). The reader is also expected to be familiar with the basic differences in microstructure and properties of the major classes of materials which are used in the construction of engineering systems: *metals* and *alloys*, *polymers* and *plastics*, and *ceramics* and *semiconductors*.

1.2 CLASSIFICATION OF JOINING METHODS

The three basic options available for the assembly and joining of engineering components may be designated *mechanical*, *chemical* or *physical*. Some familiar examples:

1. Nailing two pieces of wood together relies on the mechanical *frictional forces* between the wood and the nail to keep the two pieces of wood in contact at the point of attachment. The pieces are held in place by a balance of mechanical forces, *tensile* in the nail and *compressive* in the wood (Fig. 1.1).

2. A flour and water paste will stick sheets of paper together because the wet flour (starch) swells and penetrates the cellulose fibres of the paper, to form a stiff joint when the excess water evaporates. Hydration of the starch (a *chemical* reaction) combines with mechanical interlocking of the hardened starch with the cellulose fibres to ensure the mechanical integrity of the bond.

3. An electrical copper contact can be soldered because the flux in the flux-cored solder dissolves the protective oxide film on the copper, allowing molten solder to *wet* the copper. The solder provides a strong joint because of the strength of the metallic bond which is formed between the (clean) copper substrate and the solder alloy. Copper and the common electrical solders may react to form intermetallic

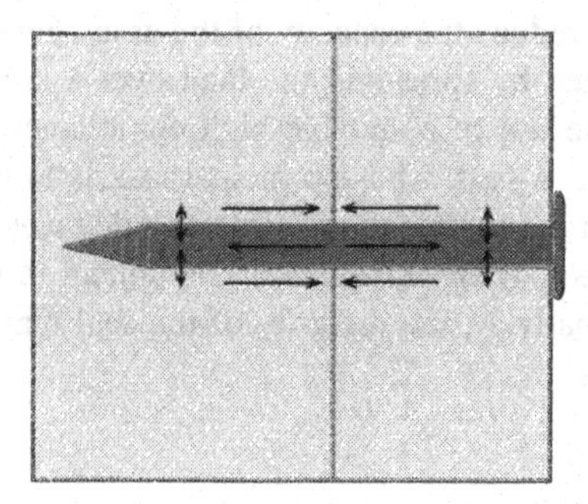

Fig. 1.1. Nailing together two pieces of wood places them in compression, leaving the nail in tension.

compounds, but these reactions are usually slow. The *wetting* of the copper and the *spreading* of the solder are *physical* processes.

1.2.1 Mechanical Joining Methods

Mechanical joining methods are often based on localized, *point-attachment processes*, in which the join is provided by a nail, a rivet, a screw or a bolt. All such joints depend on residual *tensile* stresses in the attachment to hold the components in *compression*. The joint is usually formed by an ordered array of point-attachments, as in the equally spaced rivets at the edge of a ship's plate, or the uniformly spaced bolts around a pressure vessel flange.

Mechanical joints are also made along a *line* of attachment, such as that formed when a piece of sheet is bent to form a cylinder (a paint can, for example) and the two edges are joined with an interlock seam. The complicated sequence of mechanical operations required to form such a seam is illustrated in Fig. 1.2. Here too the *residual stresses* (tensile around the circumference of the can, compressive along the join) ensure the integrity of the joint.

Many mechanical joints are designed for ease of assembly and disassembly (for example, bolted joints). A *demountable* joint often depends on *frictional forces* which exist over a predetermined contact *area*. A good example would be the ground conical joints which are used to assemble glassware in a chemical laboratory (Fig. 1.3). In this case the presence of a viscous layer of grease between the two surfaces both prevents leakage and aids assembly or disassembly.

1.2.2 Role of Residual Stress

Mechanical joints use *compressive* residual stresses across the join in order to maintain the components in contact. They therefore require a balancing *tensile* stress elsewhere in the system. These tensile stresses may either be in the fastening (nails, bolts or rivets), or in the components themselves (the interlock seam). Residual tensile stresses generally reduce the capacity of the assembly to carry a tensile load, but they can be minimized by spreading the load over a larger area. This is often accomplished through the use of a suitable high compliance (low rigidity) *gasket*. Such a gasket also acts as a seal, whose principal task is to ensure that the joint is leak-tight, especially in a point-attachment system. Adequate sealing often depends on the extreme compliance of an *elastomeric* component (a rubber gasket or elastomeric adhesive). Alternatively, the integrity of the seal may be assured by a *wax*

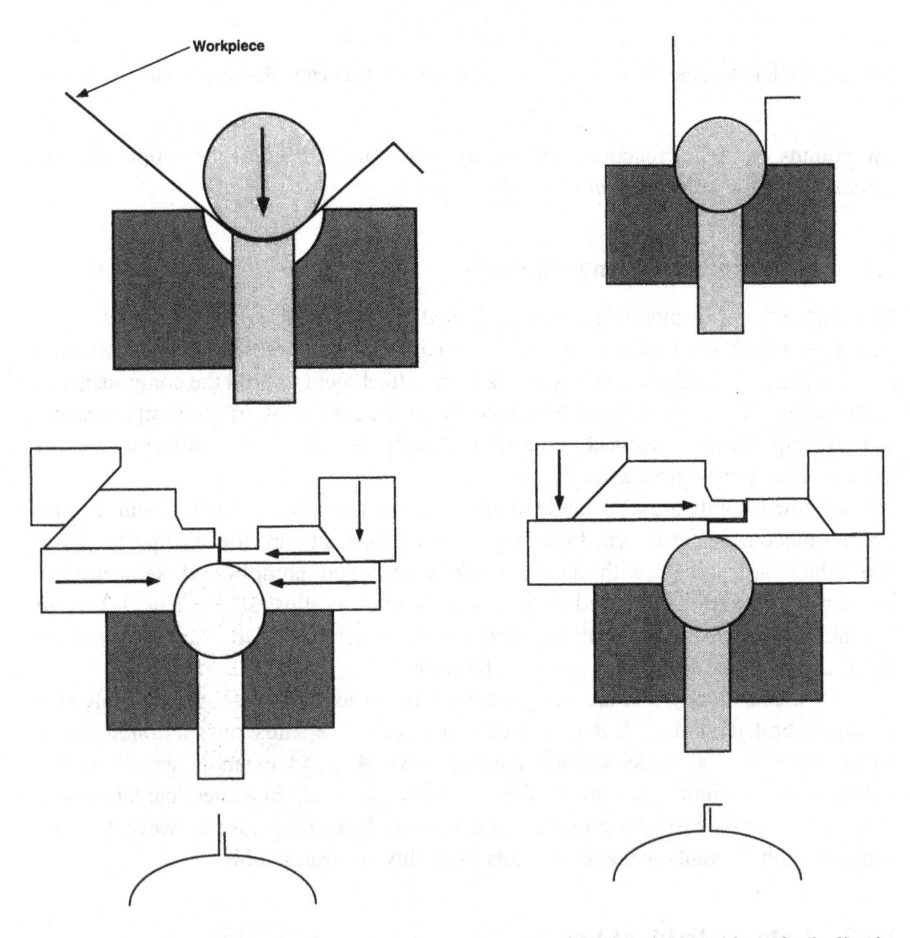

Fig. 1.2. Stages in the formation of an interlock seam. (a) The sheet, with a single lip, is formed into a half-cylinder. (b) The half-cylinder is clamped in the die. (c) The complete cylinder is formed together with the second lip. (d) The second lip is turned over the first by 90°. (e) Bending of the second lip over the first is completed. (f) The first lip is compressed by the second. (g) The seam is bent parallel to the wall of the can. (h) The seam is flattened against the can. (i) The finished seam.

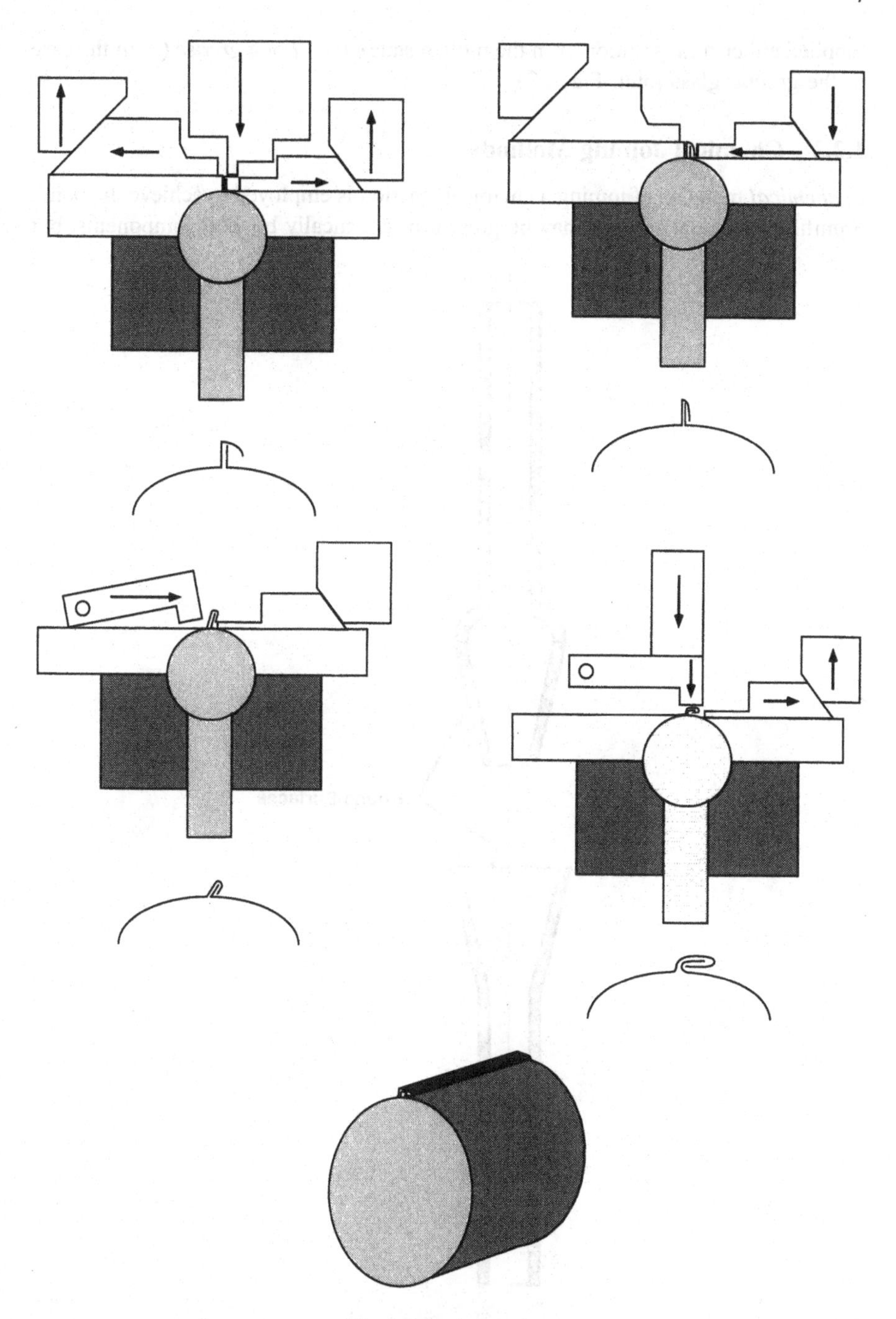

Fig. 1.2. (*continued*)

(applied either from solution or in the molten state), an *oil* or a *grease* (as in the case of the ground glass joint, Fig. 1.3).

1.2.3 Chemical Joining Methods

In *chemical* methods of joining a chemical reaction is employed to achieve the bond. Significant residual stresses may be present in chemically bonded components, but

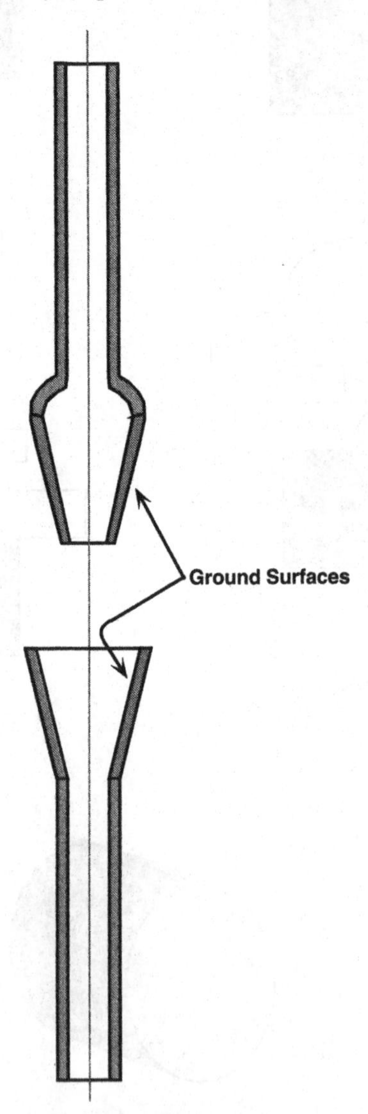

Fig. 1.3. A demountable 'cup and cone' ground glass joint is sealed with a layer of grease.

they are usually deleterious, acting to reduce the strength of the bonded joint. Strong *adhesives* or *glues* very often depend on the reaction between a liquid *precursor* and a *hardener*, which are mixed together before being applied to the joint. A common example is an epoxy resin, which hardens over a period of time after mixing the components. *Reaction-based* adhesives and glues can be very strong indeed when compared to solvent-based systems, but must be *cured* by a controlled heating cycle in order to attain their maximum strength. Chemical reactions may be a determining factor in non-polymer bonding systems, for example by modifying the ability to wet the components. The dissolution of copper oxide by a soldering flux is an example of chemical action, and such *fluxing reactions* are important in most soldering and brazing processes, as well as in welding operations. The successful brazing of *ceramic* components to metals also requires a *wetting reaction* between the brazing alloy and the ceramic. An example is the brazing of silicon nitride wear components to alloy steels, in which a copper/silver brazing alloy is used which contains titanium. The titanium in the molten braze reacts with the silicon nitride to form titanium nitride which is then precipitated. This process of *reactive brazing* results in a brazed contact free of residual porosity.

1.2.4 Physical Joining Methods

Physical methods of joining include all processes based on a phase transition from the liquid to the solid state. Examples would be the use of *solvent-based* adhesives and glues, as well as *welding*, *brazing* and *soldering*. In addition, we include *solid state* joining processes, most notably *diffusion bonding*, as well as *transient liquid state* processes, in which the transition from liquid to solid is the result of a *diffusion-controlled* process, rather than a heating and cooling cycle.

1.3 ENGINEERING REQUIREMENTS

As in the above classification of joining methods, it is convenient to discuss the *engineering requirements* for the joining and assembly of components under the same three headings: *mechanical*, *chemical* and *physical*.

1.3.1 Mechanical Requirements

Mechanical requirements include the *strength*, *toughness* and *stiffness* of the joint, and in bulk materials these are usually specified in terms of the mechanical properties: the uniaxial *yield strength*, the mode I *fracture toughness* and the elastic moduli (*tensile modulus*, *shear modulus* and *Poisson's ratio*). However, any joint in an assembled engineering system is a region of heterogeneity over which the material properties generally change dramatically, and sometimes discontinuously. It follows that the properties of the *assembly* cannot be described in terms of any simple average of the bulk properties of the constituent materials. That is, the bulk material parameters assume both homogeneity and continuity, which are absent at the join, and hence they lose much of their engineering significance in the assembly.

For tests on samples taken from an assembled system to be relevant in evaluating engineering design, or even for the control of quality, careful attention must be paid to the testing of the assembled module. For example, the fracture strength of a welded joint will depend on the weld *geometry* (the design of the joint) and the *welding cycle*, as well as on the composition of the *filler metal*, the dimensions of the test sample and the details of the fracture test procedure. The elastic compliance of an assembly and the load required to cause failure may be determined unambiguously, but it is much more difficult to specify an allowed *design stress* for an assembled system. If potential failure modes include *high temperature creep*, *stress-corrosion* or *mechanical fatigue*, then complete quality assurance can only be achieved through full-scale testing of the assembled system, although scaled-down test modules are also used.

1.3.2 Chemical Requirements

Chemical requirements for engineering systems include the effects of chemical attack by the *environment*, and degradation associated with *irradiation*. Ultra-violet radiation is a common cause of embrittlement and cracking in commercial plastics. High energy neutrons give rise to displacement damage in nuclear reactor pressure-vessel steels which raises their yield stress and reduces ductility, making the steel *notch-sensitive*, and hence susceptible to brittle fracture. Common forms of *corrosion* and *oxidation* are exacerbated by the chemical heterogeneities associated with the joining process, as well as by the effects of joint geometry on oxygen access to the corrosion site. Variations of *chemical potential* across the joint provide the driving force for *corrosion*. Thus, an insufficiently stabilized stainless steel may be susceptible to preferential corrosion at the *heat-affected zone*, in the vicinity of a weld line, a form of corrosion termed '*weld decay*'.

Riveted steel plates are frequently subject to *crevice corrosion* associated with the accumulation of H^+ ions in a reentrant *crevice* at the joint. The local decrease in pH (acidic conditions) is due to reduced access of oxygen in the crevice. This region then becomes *anodic* with respect to the rest of the surface, and at a sufficiently low pH ferrous ions are formed by direct dissolution of the metal, leading to intense, localized *pitting corrosion*. In this example, a local difference in chemical potential is generated by the difference in oxygen activity between the region of the crevice and the free surface of the assembled component. This difference in *chemical activity* in the solution provides the driving force for localized corrosion at the joint.

1.3.3 Physical Requirements

The *physical* requirements for a joint may be limited to the need to *seal* an enclosure from the surroundings, and thus prevent access or egress of gas or liquid. The physical requirements may include adequate *thermal* or *electrical conductivity* (or insulation), or control over the *optical* properties (reflection, transmission or absorption of *electromagnetic radiation*). It is important to define the engineering requirements, quantitatively as far as possible, but it may not be easy to isolate these requirements. For example, the loss of a plasticizer from a thermoplastic component by evaporation (a *physical* process) may lead to embrittlement of the component during welding of the plastic, so that the component no longer fulfils the *mechanical* requirements.

1.3.4 Joining Dissimilar Materials

It is important to distinguish between joints made between *similar* materials (whether they be metals, ceramics, composites or plastics) and joints which involve interfaces between *dissimilar* materials (steel bonded to copper, metal bonded to rubber or ceramic, or a metallic contact to a semiconductor). In the case of *dissimilar* (unlike) materials, the *engineering compatibility* of the two components must be considered. Mismatch of the elastic modulus is a common form of *mechanical incompatibility* which leads to *stress concentrations* and stress discontinuities at the bonded interface between the two materials. An example is illustrated in Fig. 1.4, in which a normal load is transferred across the interface between two materials with different elastic moduli. The stiffer (higher modulus) component restricts the lateral contraction of the more compliant (lower modulus) component, generating shear

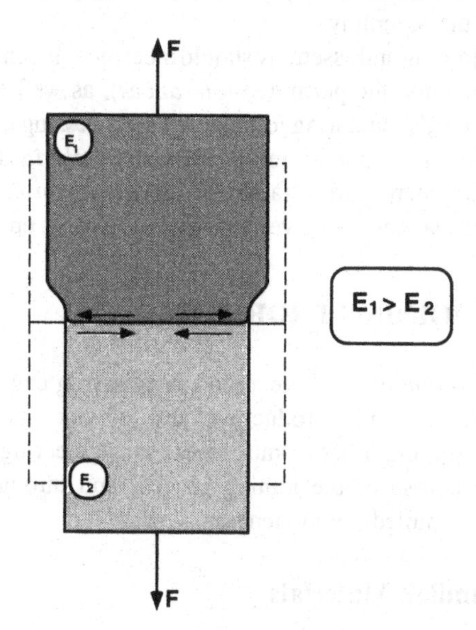

Fig. 1.4. Mismatch in the elastic constants of bonded components results in elastic constraint which generates shear stresses parallel to the interface under normal loading conditions.

stresses at the interface which may lead to debonding. *Thermal expansion* mismatch represents a lack of *physical compatibility* and is a common problem in metal/ceramic joints. Thermal expansion mismatch leads to the development of *thermal stresses* which tend to be localized at the joint and reduce its load-carrying capacity, ultimately leading to failure of the component. Poor *chemical compatibility* is commonly associated with undesirable chemical reactions in the neighbourhood of the joint. These reactions may occur between the components, for example the formation of brittle, *intermetallic compounds* during the joining process, or they may involve a reaction with the *environment*, as in the formation of an *electrochemical corrosion couple* due to a change in the *electrochemical potential* across the joint interface.

1.3.5 Defects and Tolerances

In addition to the problems of materials compatibility, associated with the joining of unlike materials, joints may be subject to a variety of problems related to either the *joining process* or the *joint structure*. We have mentioned chemical effects leading to microstructural changes, such as the precipitation of new phases during brazing or welding. Many *metastable* microstructures will be modified in the course of the joining process, with resultant modification of the properties. The mechanical strength of the joint usually differs from that of the parent components, as does the joint's resistance to environmental attack. Most joining processes give rise to *residual stresses* in the assembled components, which may either improve or degrade the performance of the assembly.

All processes for joining and assembly should meet recognized *standards* for both the *dimensional* requirements (the permitted *tolerances*), as well as for any deleterious *processing defects* introduced during joining. Such processing defects may be regions of *incomplete bonding* in a glued joint; or *porosity*, *inclusions* or *microcracks* in a welded structure. They may lead to failure to meet a *performance requirement*, for example an excessive leak rate at a given over-pressure on an engine gasket.

1.4 COMMON JOINING PROBLEMS

Just as *materials selection criteria* are aimed at satisfying engineering requirements by avoiding failure in both production and service, so the *selection* and *implementation* of a joining process must meet the engineering requirements for the *system*, both in the course of the joining process and throughout the subsequent *service life* of the assembled components.

1.4.1 Joining Similar Materials

The *parameters* of the joining process must be selected to ensure that the joint fulfils the engineering requirements. Successful engineering processes have a *working window* for the process parameters, within which acceptable performance can be assured for the system. In the assembly of engineering components, this working window may include a *heating* and *cooling cycle* required to cure an epoxy bond or braze a vacuum component. It may also involve the application of a critical *pressure*, as in diffusion bonding, or the maintenance of a *controlled atmosphere*, as in many welding processes. In every case, these process parameters must take account of *dimensional accuracy* in the location of the components prior to bonding and during the bonding process. If the processing parameters are allowed to fall outside this specified working window, then undesirable consequences may include dimensional distortions, imperfectly bonded components, excessive residual stresses and severe contamination of the bonded region.

Since the joint is a region of *heterogeneity*, many problems associated with the performance of the joint in service can be traced to the various sources of heterogeneity. Changes in *microstructure*—which occur in the *heat affected zone* (HAZ) that borders a weld—give rise to differences in chemical potential and corrosion susceptibility. They also change the local mechanical properties: either a reduction in the *yield strength*, and hence increased susceptibility to dynamic (mechanical) *fatigue*, or an increase in the *hardness*, and associated susceptibility to *brittle failure*.

Loading a bolted joint generates a complex system of local stresses and stress concentrations. These can be compensated by efficient *stress transfer* to spread the load and reduce the local *stress concentration* with the help of a compliant gasket. *Residual stresses* (for example, thermal shrinkage stresses or the stresses associated with solvent evaporation from an adhesive joint) may overload the joint to the point of failure, even in the absence of an applied load.

Dimensional mismatch may be accommodated by a *filler* whose performance in service depends on the constraints exerted by the assembled components. Most joints will be less than perfect, and will contain *some* defects in the form of *inclusions*, *microcracks*, *pores* and *imperfectly bonded regions*. The size, position and elastic compliance of these defects (compare a pore with a hard inclusion) frequently are major factors determining the final performance of the assembled components.

1.4.2. Joints Between Dissimilar Materials

A joint between dissimilar materials is commonly accompanied by mismatch in the mechanical, physical and chemical properties of the components which have been joined. A mismatch in the *elastic modulus* of the two materials will give rise to localized shear stresses when the joint is loaded in tension (as noted previously, Fig. 1.4), and may lead to mechanical failure.

Chemical reactivity between the components may lead to undesirable interface reactions and the products of these reactions are often brittle. Reactions accompanied by a volume change generate local stresses, and the mechanical integrity of the joint will be threatened. If the phases present in the components to be joined have significantly different *chemical potentials*, then differences in *electrochemical potential* may lead to localized *corrosion* in the presence of a suitable electrolyte.

Thermal expansion mismatch is a major concern in the bonding of brittle materials, especially those which are required to withstand *thermal shock* or repeated temperature cycles (*thermal fatigue*). Borosilicate glass tubing (Pyrex is a trade name) can be successfully joined to stainless steel, but only by bonding through a series of graded glass compositions. These provide a transition region over which the expansion coefficient is monotonically changed in controlled steps. The stainless steel must also be welded to a low-expansion nickel alloy tube (the common trade name is *Kovar*). The Kovar alloy is then bonded to a metal-wetting glass of matched expansion coefficient to minimize the elastic modulus mismatch at the interface. The result is a complex, but successful, *graded seal* (Fig. 1.5).

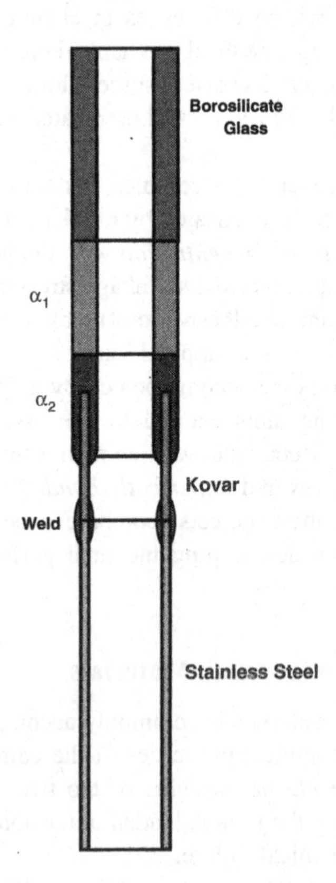

Fig. 1.5. A graded glass seal between stainless steel and borosilicate glass makes use of a low thermal expansion coefficient alloy (Kovar) and intermediate glass compositions in order to 'grade' the residual thermal stresses.

Summary

This text integrates an elementary description of the structure and properties of surfaces with an introductory treatment of the principle methods of joining components in an engineering system. The text assumes that the student is already familiar with basic concepts of materials science. Three major classes of joining process are distinguished:

1. Mechanical joining processes, which rely on the development of a beneficial residual stress pattern in the assembled components. This stress pattern ensures the integrity of the joint without endangering the mechanical performance of the system.
2. Chemical methods in which the integrity of the joint is dependent on a chemical reaction at the interface, typically involving an adhesive or glue. The adhesive strength of the bond between the components and the adhesive then determines the strength of the joint.
3. Physical methods, most commonly the metallurgical processes involved in the bonding of metals and alloys by welding, brazing or soldering, in which metal melted in the contact zone between two components subsequently solidifies to form a strong join.

The engineering requirements for all classes of joint must include not only the mechanical requirements of the system, but also the resistance of the joined components to environmental attack (typically corrosion in the presence of moisture), as well as any special requirements, such as those which involve thermal or electro-optical properties. When dissimilar materials are joined, mismatch in the mechanical, chemical or physical properties of the components across the interface may limit the performance of the joint in service.

In all cases, it is important to quantify the performance of the joint by using established standard test procedures in order to determine both the properties of importance for the specific application, and the allowable dimensional tolerances for the assembly. Non-destructive evaluation provides the tools for ensuring that defects associated with the joining processes are neither too large nor too numerous to limit performance in service of the assembled system.

Further Reading

1. W. D. Callister, Jr., *Materials Science and Engineering—An Introduction*, Third Edition, John Wiley & Sons, New York, 1994.
2. B. W. Niebel, A. B. Draper and R. A. Wysk, *Modern Manufacturing Process Engineering*, McGraw-Hill Book Company, New York, 1989.
3. R. E. Reed-Hill and R. Abbaschian, *Physical Metallurgy Principles*, Third Edition, PWS Publishing Company, Boston, 1994.

Problems

1.1 Give six examples of joints which you have encountered. *Define* the processes used to form these six joints in terms of mechanical, chemical or physical joining methods.
1.2 Every joint satisfies certain mechanical, chemical or physical *requirements*. For each of the examples you gave for problem 1.1, describe the mechanical, chemical or physical requirements of the joint needed to ensure its successful use in service.
1.3 An upright pole supports an antenna assembly and is welded to an inclined base-plate, which in turn is bolted to the rafters of a roof. List the design requirements for both the welded joint and the bolted plate. Consider the possible sources of joint failure, including environmental causes.
1.4 A flat copper sheet is to be bonded directly to a fine-ground plate of polycrystalline alumina. Why is mechanical joining not a viable option? Consider some other joining options. In each case describe the likely difficulties both during the joining process and in service.
1.5 Wrought iron rods are used to reinforce concrete by pouring the concrete slurry into a temporary mould which contains a framework assembled from the rods. To what extent does the interface between the iron and the concrete constitute a mechanical, a chemical or a physical joint.
1.6 The metal rear-view mirror assembly of a car has fallen off the glass windscreen to which it had been glued. To re-attach the assembly you have to select a commercial adhesive. Describe the design requirements for this joint, including surface preparation, the joining process and the performance of the assembly on the car (that is, the effect of mechanical vibrations and temperature variations).
1.7 Paint cans are to be produced by the interlock seam process illustrated in Fig. 1.2. What mechanical properties are required of the sheet steel for the can material? Describe in words the function of each manufacturing stage illustrated in the figure.
1.8 A vacuum flange assembly is shown schematically in Fig. 1.6. Describe in detail the mechanical, chemical and physical requirements for each of the components of this joint. Why not simply weld the two pipes together?

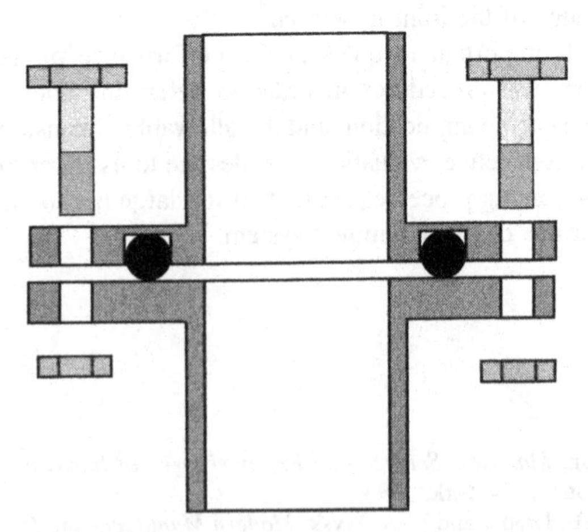

Fig. 1.6. A vacuum seal is made by uniformly compressing a rubber O-ring seated in a grooved flange with the help of equally spaced nuts and bolts sited peripherally around the circumference of the flange.

2

Surface Science

In *Materials Science*, much of the 'action' is at the surfaces and interfaces of the components. That is where loads are transferred from one component to another. It is where the system is exposed to the environment and where structural or microstructural instabilities are most likely to lead to subsequent engineering failure. Introductory courses in materials science all but ignore surface structure and properties, devoting most of the intellectual effort to understanding the bulk material. In this chapter we will try to redress the balance.

Scientists and engineers tend to think of surfaces in different ways. *Scientists* are for the most part concerned with *atomic* processes (catalytic mechanisms, electronic transitions, surface transport...), while *engineers* are more concerned with the *macroscopic* features and properties of surfaces (roughness, friction, heat transfer...). We attempt to bridge the gap between science and engineering by contrasting the *crystallographic structure* of the surface with the *topology* of engineering surfaces. We then discuss the *chemical* and *physical* characteristics of solid surfaces in atomic terms, and then contrast these with their *thermodynamic* properties. A discussion of the *mechanics* of surface contact will be deferred until Chapter 3.

2.1 SURFACE GEOMETRY

In this section we describe the free surfaces of *single crystals*, and then discuss modifications to the surface structure associated with the presence of grain and phase boundaries in *polycrystals*. We then describe possible interactions of the free surface with the environment and the formation of *surface films*, continuing with a geometric description of 'real', *engineering surfaces*.

2.1.1 Single Crystal Surfaces

The simplest possible model for a single crystal surface is derived by imagining the *crystal lattice* to be sectioned by a plane (Fig. 2.1). All the *lattice points* on one side of the plane of the section are discarded, breaking the atomic bonds which intersect the plane of the section, while all atoms associated with lattice points on the other side of the plane are assumed to retain their *equilibrium positions* in the bulk crystal. The *broken atomic bonds* between retained and discarded atoms are the source of the *surface energy*. This *unrelaxed surface*, defined by the plane of the section, thus consists of an array of *surface atoms* whose *atomic coordination* has been changed, but which remain in the atomic sites characteristic of the *bulk lattice*. Sections parallel to low index, close-packed crystal planes will result in dense arrays of surface atoms, while sections taken at a small angle to these close-packed planes contain *ledges* or steps at the edges of the close- packed planes. Surfaces which deviate slightly from these low-index planes are called *vicinal surfaces*. If the surface section is tilted away from the close-packed plane, but about an axis parallel

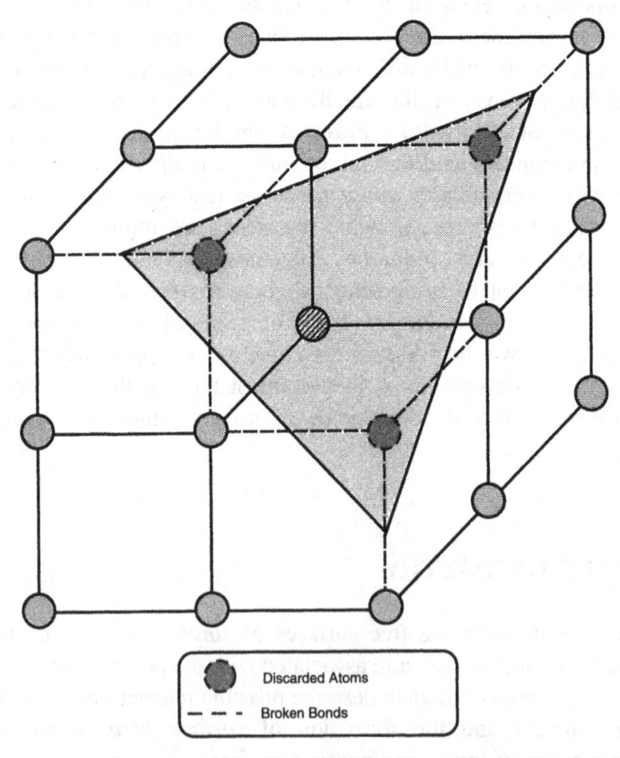

Fig. 2.1. A random planar section through the lattice defines a *surface plane*. All atoms to one side of the surface are *discarded* while all other atoms are assumed to retain their bulk equilibrium positions.

to an atomically close-packed direction, then the surface ledges will consist of *close-packed rows* of atoms, but if the axis of tilt deviates from an atomically close-packed direction, then the ledges will themselves be 'stepped', each ledge containing a regular array of *kink-site* atoms, whose spacing will depend on the deviation of the tilt axis from the close-packed direction. Figure 2.2 illustrates schematically the formation of *close- packed surfaces*, *ledges* and *kink-sites*.

It is revealing to consider the *bonding* of the surface atoms predicted by this geometrical model for the case of a pure metal. The close-packed planes in the *face-centred cubic* (FCC) lattice of a metal are the {111} planes (Fig. 2.3). A section taken parallel to a {111} plane removes three nearest neighbours from each close-packed surface atom, reducing the *bulk* nearest-neighbour coordination number from *12* to *9* for an atom A in the close-packed surface. At the close-packed edge of a {111} plane, a further two nearest neighbours are removed, so that the *ledge* atoms B, which sit on a vicinal surface tilted from the {111} plane about a close-packed ⟨110⟩ direction, have a coordination number of *7*. The *kink sites* C lose one more neighbour, leaving them with a nearest-neighbour coordination number of *6*, which

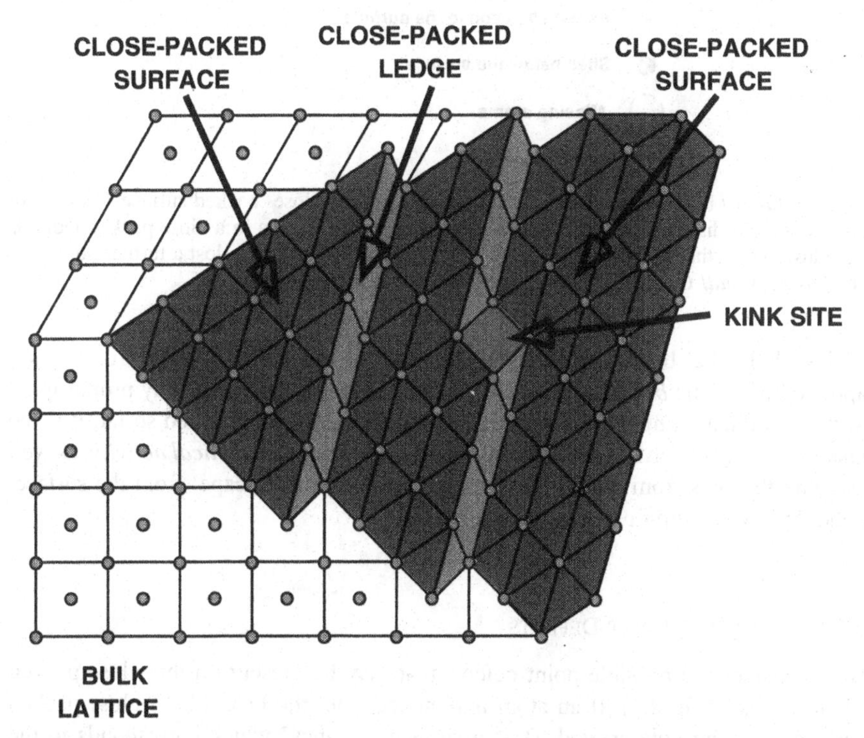

Fig. 2.2. Densely packed planes in the crystal lattice form *close-packed terraces*. At the edge of such a terrace a *close-packed step* or ledge is formed parallel to the close-packed rows of atoms. When ledges deviate from the close-packed direction *kink sites* result.

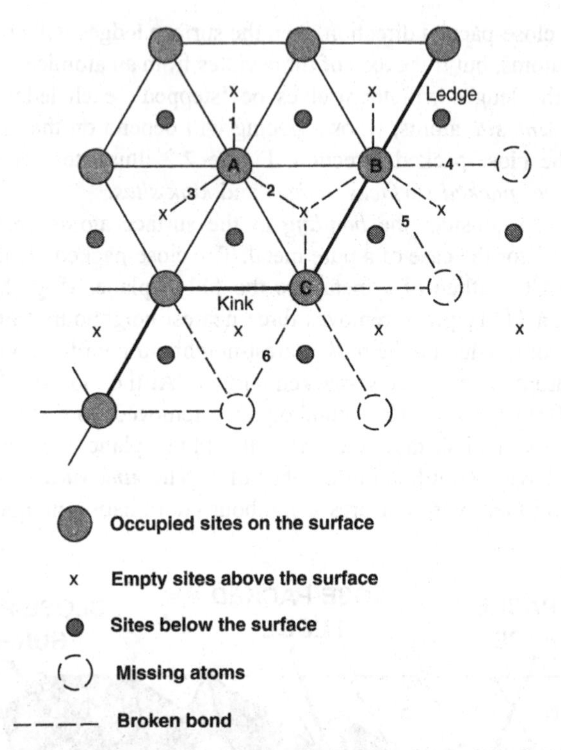

Fig. 2.3. On a {111} plane of an FCC metal atoms in the close- packed surface A have lost *three* bonds to discarded atoms above the surface (#1–3). Atoms in a close-packed ledge B have lost two further bonds (#4–5), while atoms at a kink site C have lost a further bond (#6) and have just *half* the coordination number of atoms in the bulk lattice.

is just half the bulk coordination number of the FCC crystal. To a good approximation, the *binding energy* of an atom to the surface is directly proportional to the coordination number. For the sites which have been discussed so far, it is the *kink site atoms* which may be expected to have the greatest *chemical activity*, as well as being the sites from which atoms may be expected to 'escape' from the surface, either by *evaporation* or by *surface diffusion.*

2.1.1.1 Surface Point Defects

Now consider the possible point defects that may be present on this close-packed {111} surface (Fig 2.4). If an atom is removed from the FCC {111} close-packed surface, then the hole created, D, is a surface 'vacancy' which corresponds to the destruction of *9* bonds. An atom sitting on top of a close-packed surface E has only *3* nearest neighbours and can be expected to be very mobile. If such an atom migrates to a neighbouring close-packed ledge F it can increase the number of nearest-neighbours from *3* to *5*, thus increasing its stability and reducing the overall energy of the surface. The atom will be trapped at the ledge with a binding energy corresponding to the extra two bonds. If this same atom can migrate along the edge to a kink site, there is a further reduction in energy (the atom now has *6* nearest neighbours, C in Fig. 2.3). The removal or addition of atoms at kink sites has no

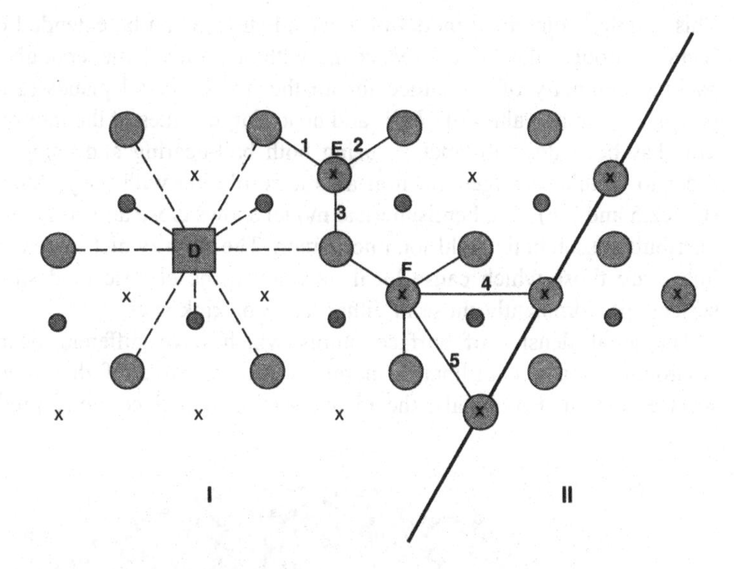

D Vacant surface site

Surface sites on terrace I

x Empty sites on upper terrace II

Subsurface sites

Occupied site on upper terrace

Fig. 2.4. A *missing* atom in a FCC close-packed surface D breaks *nine* bonds. An atom sitting on a close-packed terrace E is bonded to only *three* nearest neighbours and is highly mobile. At a ledge F, such an atom acquires two more neighbours (#4–5), and, if it migrates along the edge to a kink site (C in Fig. 2.3), it will have a coordination number of *six*.

effect on the surface geometry and should not change the surface energy. On the other hand, the nucleation of new kink sites at a ledge, the formation of surface vacancies (*missing atoms*), or the addition of isolated surface atoms to the close-packed surface, all imply *increases* in surface energy.

2.1.1.2 Crystallography of an Ideal Surface

This simple geometrical model of a crystal surface can be extended by making the section through the lattice *spherical*, with a radius large enough to reflect the packing geometry of the lattice for all the major crystal planes (those low index planes with small values of $\{hkl\}$, and hence large values of the interplanar spacing). This has been done for metals, using both ball-bearing and *computer models*, in order to simulate images taken at atomic resolution with the *field-ion microscope* (Figs 2.5 and 2.6). The hemispherical model shows close agreement with the atomic distribution seen in the field-ion micrograph. The atoms which appear in the field-ion image are those which cause local maxima in the electric field strength over the surface, predominantly those in either ledge or kink sites.

The areal density of surface atoms which have different nearest-neighbour environments and coordination numbers is a function of the orientation of the surface section. For metals, the planar section model correctly predicts the local

Fig. 2.5. A ball-bearing model of the surface defined by a *spherical section* taken through a BCC crystal lattice of a metal. A $\langle 110 \rangle$ axis of the crystal is normal to the plane of the image. Close-packed *facets*, surrounded by *ledges* and *kink sites*, are clearly visible (*Courtesy of G. Ehrlich.*)

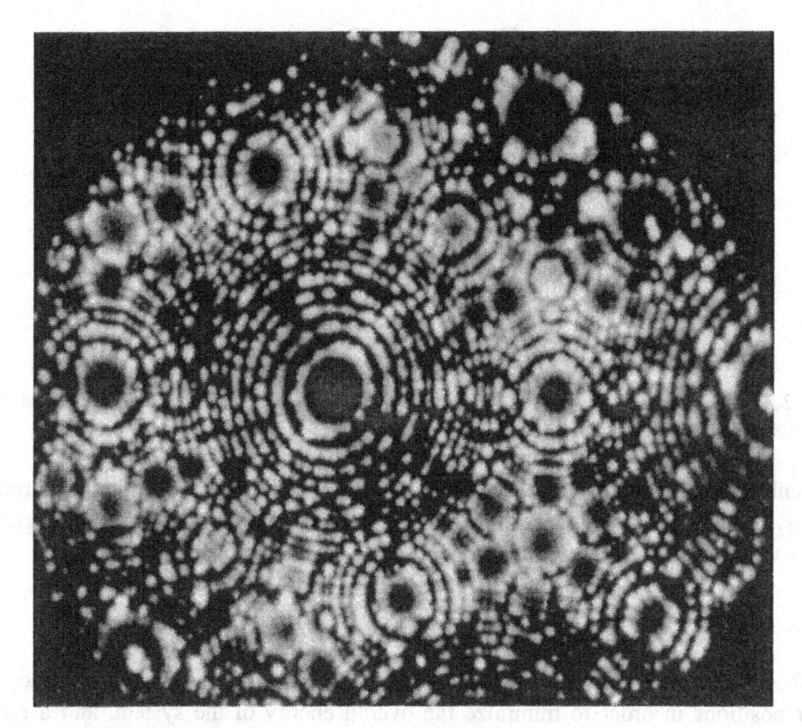

Fig. 2.6. The field-ion image of a tungsten crystal in the same orientation as that depicted in Fig. 2.5. The hemispherical surface section of Fig. 2.5 is an excellent first approximation to the atomic structure of the surface of this field-ion specimen. Reproduced by permission of John Wiley & Sons Inc

atomic environment and its dependence on the orientation of the surface with respect to the crystal lattice. Surfaces which are predicted to have a high density of surface ledges and kink sites are expected to have a higher energy than those surfaces which correspond to close-packed arrays of atoms.

The situation is somewhat more complicated in *ionic crystals*, for which the electrostatic energy has to be minimized. Retaining the FCC crystal lattice as our model, we note that the NaCl crystal lattice consists of two separate, interpenetrating arrays of *anions* (Cl^-) and *cations* (Na^+), displaced by a unit cell vector of $1/2 \cdot a\langle 100 \rangle$ (Fig. 2.7). Defining a crystal surface by sectioning on a $\{111\}$ plane yields two alternatives, with a surface array of either densely packed anions or cations (Fig. 2.8). In both cases the surface is highly charged (negative for the anions and positive for the cations), so that the *electrostatic* contribution to the surface energy will be very large and the surface is expected to be unstable. If we section on a $\{100\}$ cube plane, on the other hand, the surface consists of a densely packed, ordered array of both anions and cations with no net electrostatic charge. We would expect (and we would be correct) that the surface energy would be a minimum for

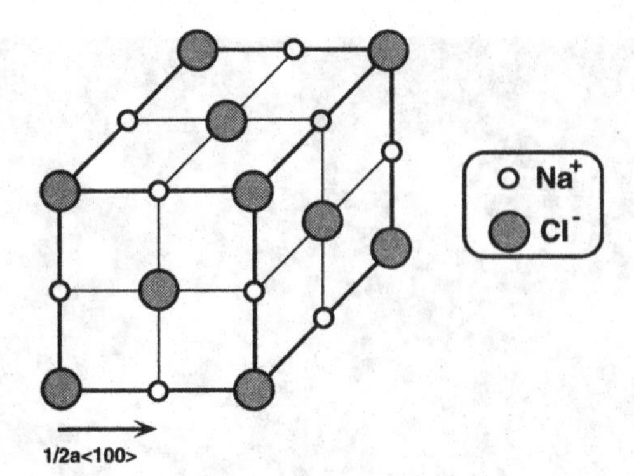

Fig. 2.7. An *ionically bonded* lattice—NaCl (FCC). The *anion* and *cation* lattices are displaced by a vector $1/2 \cdot a\langle 100 \rangle$.

the cube plane of the ionic crystal. The general rule is that the minimum surface energy of an ionic crystal corresponds to that surface orientation having the highest density of mixed cations and anions, with no nett surface charge.

2.1.2 Relaxation and Reconstruction

Surface atoms whose atomic coordination differs from that in the bulk will adjust their positions in order to minimize the overall energy of the system, and a rigid lattice model for the surface structure is unlikely to be a good predictor of the actual atomic positions at the surface. Moreover, the termination of the periodic crystal lattice at a free surface is expected to give rise to a redistribution of surface charge, regardless of the nature of the chemical bonding (metallic, ionic or covalent), a redistribution which is reflected in the *dipole moment* and the *polarizability* of the surface.

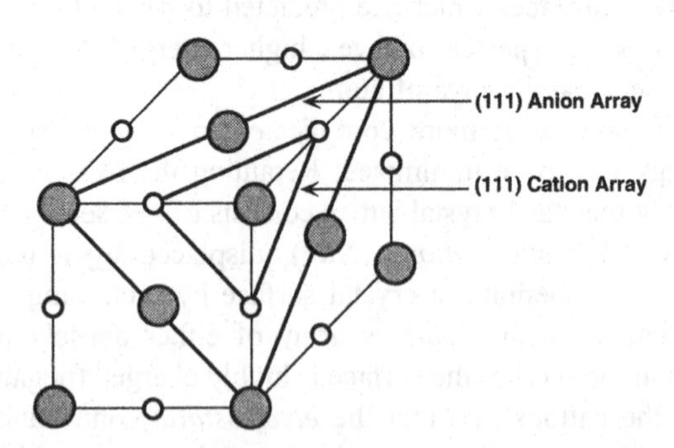

Fig. 2.8. If the NaCl crystal is sectioned on a $\{100\}$ plane, then the crystal terminates at a *neutral* surface containing equal numbers of anions and cations. If it is sectioned on a $\{111\}$ plane the crystal will terminate at a highly charged surface which may contain *either* anions (negatively charged) *or* cations (positively charged).

2.1.2.1 Surface Dipoles

In a metal, the balance of surface charge will be determined by redistribution of the free electrons between the ion cores. For a close-packed surface the metal ion cores are closely spaced and cannot protrude much beyond the free electron cloud, but for surfaces which contain a high density of kink sites, the positively charged kink-site ions can protrude appreciably. This leads to a positive *dipole moment* associated with the *double layer* of electric charge. Electron emission from solid surfaces can be excited by thermal activation (*thermal emission*), by photon-excitation (*photo-emission*) or by application of a large electric field (*field emission*). In all three cases energy must be supplied in order to remove an electron from just inside the surface to a point just outside this surface. The amount of energy which is required, the *electronic 'work-function'*, depends on the *dipole moment* and *polarizability* of the electrical double layer at the surface, and is reduced for those surfaces which have high densities of kink sites, and hence a large positive dipole moment.

2.1.2.2 Surface Relaxation

In addition to the redistribution of the electrons at the free surface of a metal, the position of the ion cores may also be expected to be modified at the free surface. Such *relaxation* effects can be observed by high resolution electron microscopy (Fig. 2.9). The sample is a thin film viewed in transmission, and the contrast observed is due to constructive interference of the electrons diffracted by the crystal lattice. The bright spots reflect the periodicity of the rows of atoms parallel to the axis of the microscope and perpendicular to the plane of the image. Towards the edge of the specimen (the free surface) the rows of spots (the projection of the close-packed planes in the crystal lattice) are displaced outwards, implying an increase in lattice spacing. Similar effects have been documented for semi-conductors and insulators using scanning tunnelling microscopy. The reduction in atomic constraint at a free surface also gives rise to surface modes of atomic vibration. Surface acoustic waves have important applications in electronic device technology.

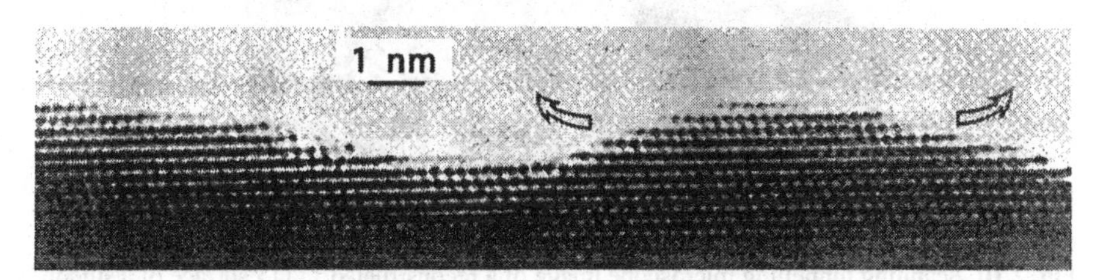

Fig. 2.9. Displacement of the lattice image at the edge of a thin gold film is interpreted as a change in the lattice spacing at the free surfaces. Zangwill, Physics at Surfaces, 1988, reproduced by permission of Cambridge University Press.

2.1.2.3 Surface Reconstruction

In covalently bonded solids more complex relaxation behaviour is sometimes observed. Figure 2.10 is a scanning tunnelling microscope (STM) image of a restructured {111} surface of silicon. In the STM the image is formed by scanning an electrically charged, sharp metal needle (usually tungsten) over the surface and recording variations in either the electric current or the electrical potential at constant current. If the system is mechanically stable and the sharp tip of the needle is close enough to the surface (typically of the order of 1 nm), it is possible to detect variations in the electrical potential which are associated with *individual surface atoms*.

The hexagonal array of silicon atoms predicted by a simple geometrical section parallel to the {111} planes (Fig. 2.3) would leave the surface atoms with 'dangling' covalent bonds and only three (covalently-bonded) neighbours, instead of the four required to satisfy the valency requirements. Figure 2.10 shows that the surface atoms rearrange to form a crystallographically distinct '7×7' surface array. In this new surface structure the dangling bonds at the surface are linked to form six- and twelve-membered rings of surface silicon atoms about a central site. This new, *reconstructed*, two-dimensional array has a *different unit cell* from the parent crystal, and is therefore a new (two-dimensional) phase unique to the surface.

In many crystal lattices specific surface sections may generate alternative surface structures. This is illustrated in Fig. 2.11 for the {100} cube surfaces of silicon. These surfaces consist of *close-packed rows* of atoms, with the rows in alternate layers rotated by 90°. A section inclined at a small angle about an axis parallel to one of the rows will generate a surface which can contain two *alternative* types of ledge, which are either *parallel* to the close-packed rows, or *perpendicular* to the close-

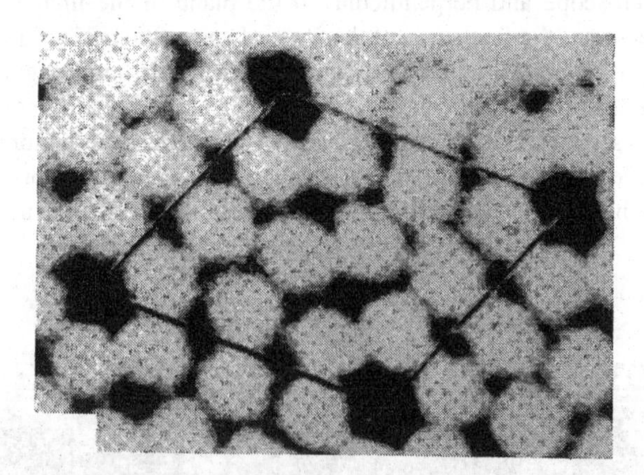

Fig. 2.10. Scanning tunnelling microscope image of a reconstructed {111} surface of a silicon crystal showing the 7×7 periodicity. Reprinted from Walls and Smith, Surface Science Techniques 1994, with kind permission from Elsevier Science-NL, Sara Burgerhartstraat 25, 1055 KV Amsterdam, The Netherlands.

packed rows. In practice, these ledges on the cube surface of silicon can associate to form pairs, and at low temperatures *double ledges* are formed with only the {100} planes having atom rows parallel to the ledges exposed. At high temperatures the ledge doublets dissociate into single steps, and both variants of the {100} planes are then observed. The difference in stability of the two ledge configurations results in different ledge *morphologies* and the density of kink sites is observed to be very much less along the more stable ledges (Fig. 2.11).

2.1.2.4 SURFACE FACETTING

Finally, many crystals can lower their surface energy by forming surface *facets*. Ionic crystals are typical of this group, since most geometrically defined surfaces contain only ions of one charge, and hence are inherently unstable. In the FCC lattice, typified by NaCl and MgO, the {100} planes are the most densely packed planes which contain both cations and anions, and hence have by far the lowest surface energies. It is therefore not surprising that the stable morphology for such crystals is a well-formed cube with {100} surface facets.

The atomic ledges present on a vicinal surface become mobile when the sample is heated. Ledge migration occurs by the *emission* and *adsorption* of atoms from kink

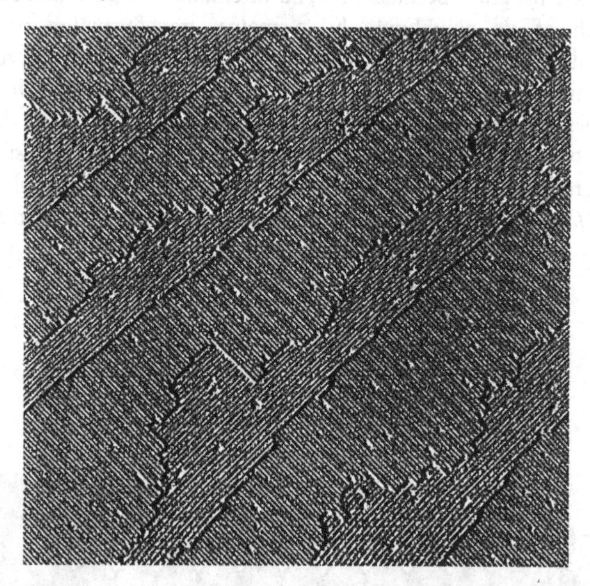

Fig. 2.11. Reconstruction of a {100} silicon surface showing 0.14 nm steps separating terraces with *alternate* directions of dimerization (rows with paired bonding separated by 0.77 nm). The kink formation energy along the ledges on *alternate* terraces is clearly different. Reprinted from Walls and Smith, Surface Science Techniques 1994, with kind permission from Elsevier Science-NL, Sara Burgerhartstraat 25, 1055 KV Amsterdam, The Netherlands.

sites. Pile-up of these mobile steps then gives rise to facets that accommodate the misorientation between the plane of the section and the low energy surface. Facets formed by the pile-up of atomic ledges may be either *rational* or *irrational.* A rational facet will be planar and will have Miller indices which correspond to another surface of low energy, while irrational facets are non-planar.

Frank was the first to demonstrate that facetting of a crystal growing from a supersaturated solution is to be expected on *kinetic* rather than *thermodynamic* grounds. The rate of arrival of atoms at a step depends on the catchment area of the low energy terrace between the steps (Fig. 2.12), so that the *closer* a step is to its neighbour, the *slower* the neighbouring step migrates. While crystal growth from solution usually leads to strongly *facetted* surfaces, the reverse is the case for a dissolving crystal. Dissolution is promoted by the high density of kink sites at the *edge* of a facet, leading to the *dispersion* of atomic ledges, as opposed to their pile-up, with the resultant rounding of sharp edges and corners.

2.1.3 Polycrystals

Most engineering materials are polycrystalline, and consist of an aggregate of single crystals separated by grain boundaries. The orientation of an individual crystal is related to each of its neighbours by a matrix transformation of the crystal coordinates. This transformation corresponds to a rotation about a given axis by a specific angle, and corresponds to three degrees of angular freedom (the three *Euler angles* relating the two sets of crystal coordinates). The normal to the boundary separating the two crystals can be specified by a *unit vector*, corresponding to a further two degrees of freedom, so that every boundary in a polycrystal is associated with a total of *five degrees of freedom* and requires five parameters to specify both the orientation relationship between neighbouring grains and the plane of the boundary.

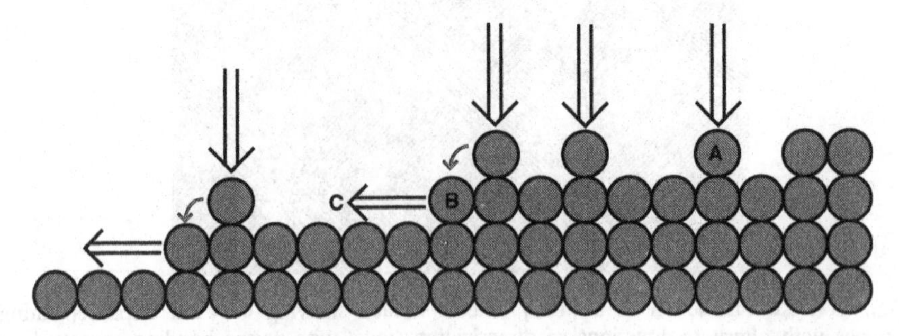

Fig. 2.12. Atoms A, arriving from solution (or from a gas phase) *migrate* on the terrace until they are *trapped* at a kink site B, leading to migration of the ledge C. *More* atoms arrive at a large terrace than at a smaller one, leading to *faster* migration of the ledges which bound the *larger* terraces. Eventually, the ledges pile up to form a secondary facet.

In the bulk polycrystal three neighbouring grains will intersect at a line, and on the free surface this *triple grain junction* will intersect the surface at a point (Fig. 2.13). If we assume that a random planar section is taken through the polycrystal, and that the material to one side of the plane is discarded, then the surface of each exposed grain reflects the orientation of the crystal lattice with respect to the plane of the surface section. The *intercept area* of each grain intersected by the section will depend on the position of the plane of the section with respect to the centre of gravity of the grain. Grains sectioned close to their centre of gravity will have an intercept area close to the maximum possible cross- sectional area of the grain, while grains sectioned far from their centre of gravity will have only a small intercept area in the plane of the section (Fig. 2.14).

A good approximation to such a random section is obtained during the *mechanical polishing* of a specimen in preparation for optical microscopy, which requires that the surface be *optically flat*. If the material is *optically anisotropic*, then the variation in the orientation of the exposed surface from grain to grain is often visible in *polarized light* (Fig. 2.15).

The grain boundaries are paths of easy diffusion, and mass transport to and from a free surface takes place readily at the grain boundaries, so the grain boundary area per unit volume and the boundary intercept length per unit area of surface are important microstructural parameters. Both are related to the *grain size*.

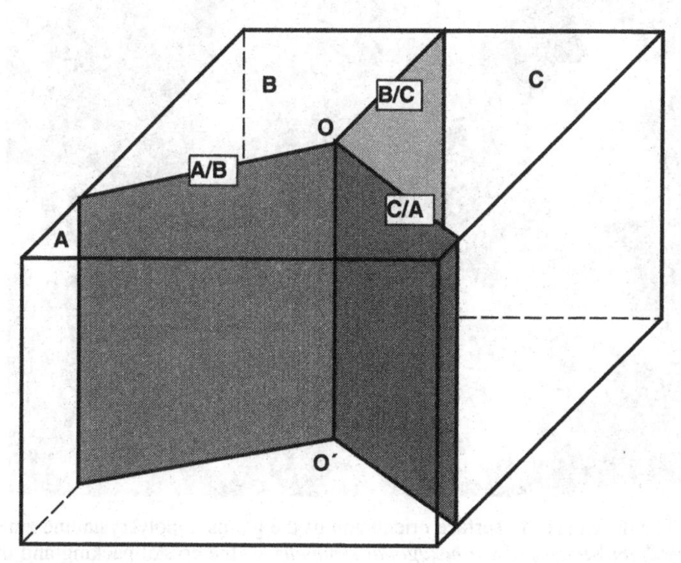

Fig. 2.13. Three *neighbouring* grains, A, B, C, meet along a *line* (triple junction) OO′ which intersects the free surface at a point O.

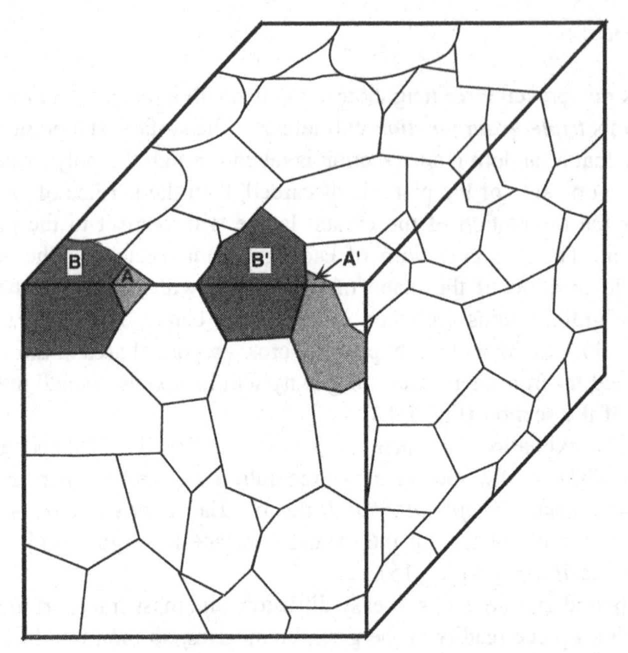

Fig. 2.14. The surface area of a grain seen on a section is a poor indication of the *size* of the grain: grains intercepted far from their centre of gravity appear small (A, A′) while those intercepted close to their centre of gravity appear large (B, B′).

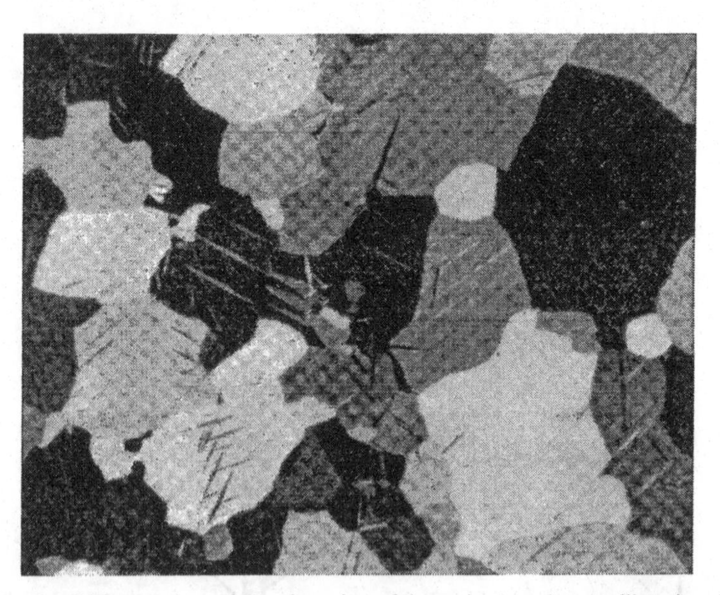

Fig. 2.15. The differences in surface orientation of the grains in polycrystalline zinc show up in *polarized light* because of the *hexagonal symmetry* of the crystal packing and the *optical anisotropy* of the reflection coefficient. From E. C. W. Perryman: *Polarized Light in Metallography* (Ed. Conn, G. K. T. and Bradshaw, F. J.), 1952, Butterworths, photo from British Non-Ferrous Metals Research Association.

2.2 SURFACE EQUILIBRIA

The polished surface of a polycrystal is not in equilibrium, since the grain boundary energy exerts a *surface tension* which is not balanced at the free surface. Surface equilibrium is only possible when the surface tension exerted by the two free surfaces of neighbouring grains at their common line of intersection with the surface are in equilibrium with the surface tension of the grain boundary (Fig. 2.16). That is:

$$\gamma_{A_s} \cdot \cos\theta + \gamma_{B_s} \cdot \cos\phi = \gamma_{GB} \qquad (2.1)$$

In fact, the polished surface is in *metastable equilibrium* and the surface tensions are balanced by elastic forces in the neighbouring bulk crystals.

2.2.1 Thermal Etching

If the polished surface of a polycrystal is heated in an inert atmosphere to a temperature at which *surface diffusion* can occur, then the surface forces can *relax* by thermal grooving at the boundaries, also shown in Fig. 2.16. An example of a groove formed at the interface between a polycrystalline solid and a free surface is shown in Fig. 2.17. The angles of the boundary grooves can be measured by the shifts of the *optical interference fringes* and used to calculate the ratio of surface to grain-boundary energy. For *metals* this ratio is typically of the order of *three*, consistent with the lack of directionality of the metallic bond and the consequent ability of the two lattices on either side of the boundary to adapt to the mismatch. In *ceramics* (with mixed covalent and ionic bonding) the ratio is closer to *one*,

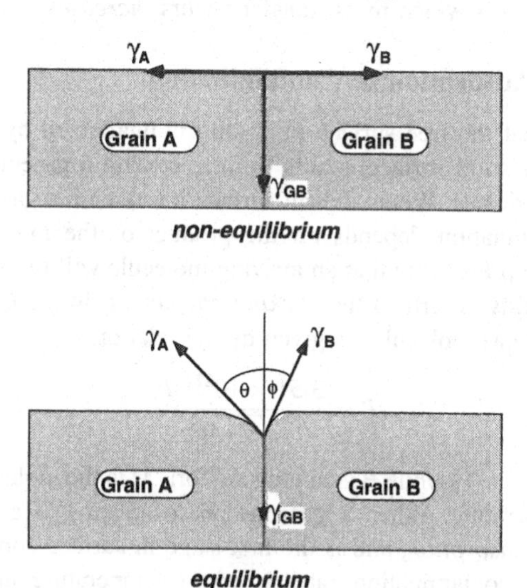

Fig. 2.16. The balance of surface forces along the line of intersection of a grain boundary with a free surface is achieved by *thermal grooving.*

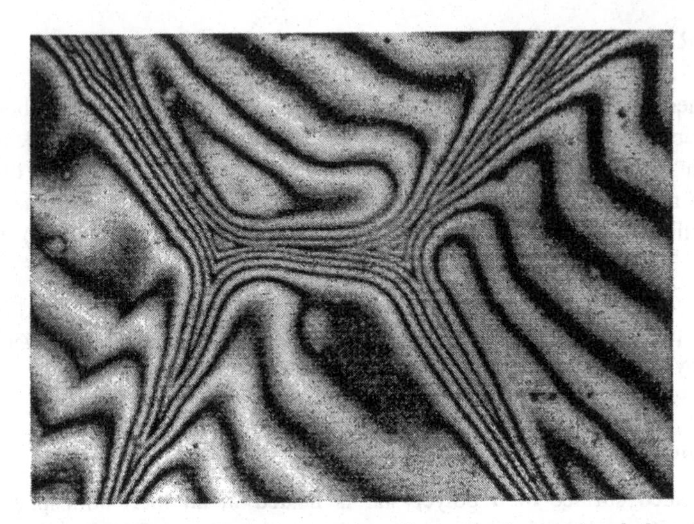

Fig. 2.17. The grooves formed at the intersection of copper polycrystal grain boundaries with a free surface (interference microscopy). Reproduced from McLean: *Grain Boundaries in Metals*, Oxford University Press.

reflecting the difficulties of achieving electrical neutrality (ionic packing) and satisfying the valency requirements (covalent bonding).

In the approach to equilibrium the surfaces of the two grains on either side of the boundary develop *curvature* by the transfer of material away from the boundary groove (Figs 2.16 and 2.17). This curvature *decreases* as the groove depth *increases*, while the distance over which mass transfer occurs *increases*.

2.2.2 Surface Adsorption & Contamination

It is only rarely that the free surface of a solid is unaffected by the *environment*. More typically, the solid surface is able to interact with a gaseous or liquid phase (air, water or a lubricant). When a clean surface reacts with a gaseous *contaminant* the rate of contamination depends on the product of the rate of arrival of gas molecules with the probability that an arriving molecule will remain on the surface. This latter probability is termed the *sticking coefficient*. In a gaseous environment the arrival rate for gas molecules is given by the relation:

$$R = \frac{3.51 \times 10^{22} \cdot P}{\sqrt{T \cdot M}} \tag{2.2}$$

R is in units of cm^{-2} s^{-1}, while P is in units of Torr, M is the molecular weight and T the absolute temperature. When a gaseous phase adheres to a clean surface the process is termed *adsorption*, and is the first stage in surface contamination by the environment. The contamination rate, at room temperature and a pressure of 10^{-6} Torr, by an active gas with unit sticking coefficient is about *one monolayer per second.* This is in the range of *high vacuum. Ultra-high vacuum* equipment, for the manufacture of solid state devices, is typically required to operate at pressures of 10^{-9} to 10^{-10} Torr, which ensures that a surface can be preserved for several hours without appreciable surface contamination.

Adsorption may occur without any chemical reaction to form an additional phase, resulting only in the reduction of the surface energy. The adsorbate is *surface active.*

2.2.2.1 SURFACE REACTIONS

If the bulk solid does react chemically with the environment, then new phases are formed at the surface. The commonest example is the *oxidation* of a metal. Gold is the only metal which is stable in air at room temperature. All other metals are oxidized to a greater or lesser extent. Growth of an oxide requires that both *cathodic* and *anodic* reactions occur (Fig. 2.18). In dry air the *cathodic reaction* at the surface of the oxide requires free electrons to create the negatively charged oxygen *anions*:

$$\frac{1}{2}O_2 + 2\varepsilon = O^{2-} \tag{2.3}$$

The corresponding *anodic reaction* occurs at the interface between the oxide and the metal and releases the free electrons to create positively charged cations:

$$M = M^{2+} + 2\varepsilon \tag{2.4}$$

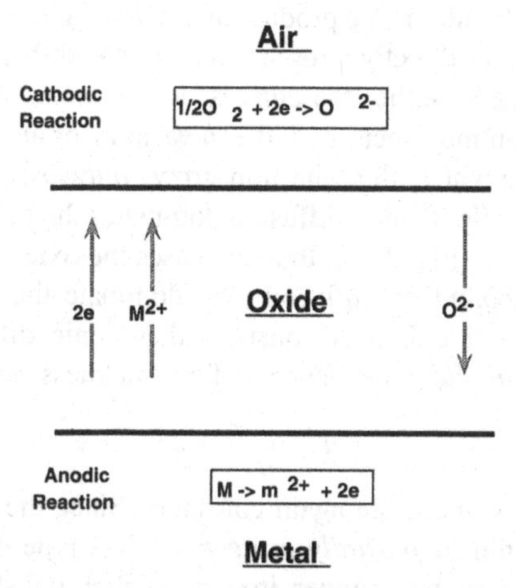

Fig. 2.18. During oxidation, *cathodic* and *anodic* reactions take place at the free surface and at the oxide–metal interface respectively. Oxidation requires both *electron transfer* and *diffusion* of either anions or cations.

2.2.2.2 Growth Laws for Surface Films

If the oxide is a *stoichiometric insulator*, then the electrons can only be transported across the film by *quantum tunnelling* across the potential barrier created by the non-conducting film. The tunnelling probability for electrons decreases *exponentially* with increasing thickness of the oxide, so that the rate of growth obeys a logarithmic law (and becomes negligible at room temperature when the film thickness is of the order of 2 nm). This is the basis for the *protective* (passivating) oxide films on metals and alloys which form *stoichiometric* oxides: *alumina* on aluminium alloys and nickel-based superalloys, *titania* on titanium alloys and *chromia* on chromium-plated components and chromium-containing stainless steels. These films are much thinner than the wavelength of visible light, so that the polished metal surface retains its metallic reflectivity despite the presence of the thin oxide overlayer.

Logarithmic growth is described by the equation:

$$d = k \cdot \log(at + b) \tag{2.5}$$

where d is the thickness of the film, t is the time and the parameters k, a and b are constants.

Non-stoichiometric oxides are usually semiconductors whose electrical conductivity ensures adequate electron transport through the growing oxide. The growth rate of the oxide film is then controlled by the rate of *ionic diffusion* in the oxide. The *driving force* for ionic diffusion is usually provided by the difference in *chemical potential* associated with the ionic concentration gradients near the sites of the anodic and cathodic reactions. Alternatively, the driving force for diffusion may derive from a *surface charge* developed by the anodic and cathodic reactions, which creates an electric field across the oxide film.

The growth rate of the film is the product of the *driving force* for diffusion and the *ionic mobility*, which is directly proportional to the diffusion coefficient of the migrating ions. In general, either the *cations* or the *anions* dominate the diffusion process, while diffusion may occur by either a vacancy or an interstitial mechanism. If cation diffusion dominates, then the film grows *outwards* from the original free surface of the metal, while if anion diffusion dominates the film grows *into* the metal from the original surface (Fig. 2.19). In many cases the oxide film is polycrystalline, in which case *grain boundary diffusion* may dominate the rate of growth. If the diffusion rate in the oxide film is constant, then ionic diffusion-controlled film growth leads to a *parabolic* dependence of film thickness on time:

$$d = k\sqrt{t} + a \tag{2.6}$$

where the parameters k and a are again constants. Since the growth rate decreases with time, the oxide film is *partially protective*. This type of growth is typical of many transition metals, such as copper, iron and nickel. Polished specimens of such metals eventually lose their reflectivity after exposure to dry air, since the film growth can continue to thicknesses well in excess of the wavelength of visible light

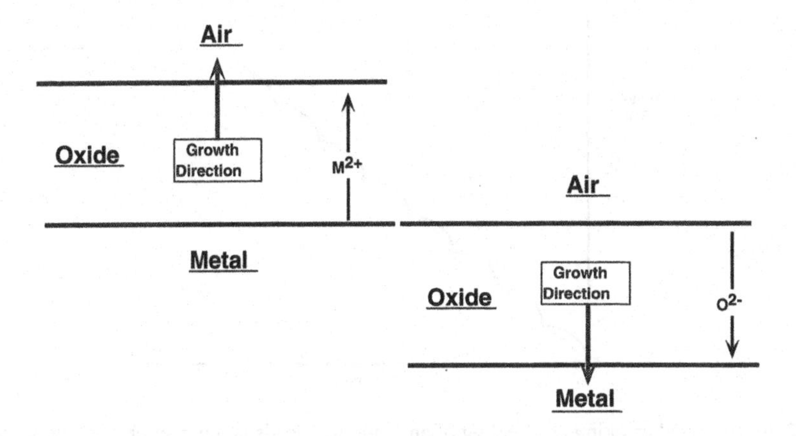

Fig. 2.19. Oxide films grow *outwards* from the original metal surface, by diffusion of *cations* to the free surface, or *into* the metal by the diffusion of *anions* to the metal–oxide interface.

($\sim$0.5 μm). The ionic *diffusion coefficient* D is an exponential function of the temperature, so that the rate of film growth for non-stoichiometric oxides increases rapidly with increasing temperature:

$$D = D_0 \exp\left(\frac{-Q}{RT}\right) \tag{2.7}$$

where D_0 is a pre-exponential factor, Q is the activation energy, T is the absolute temperature and R is the gas constant.

Finally, under some conditions the rate of film growth can be a constant, so that the thickness increases *linearly* with time:

$$d = kt + a \tag{2.8}$$

The rate of oxidation of steel heat-treated in air at elevated temperatures is often linear, because stresses developed during film growth lead to cracking and peeling of the oxide and the consequent exposure of fresh metal surface. In the case of copper, a transition is observed from parabolic growth at low temperatures, to a linear growth law. This is again associated with the repeated cracking and peeling of the thicker oxide films which are formed at higher temperatures (Fig. 2.20). Cracking and peeling of a surface film is a complex process which depends on the nature of the *internal stresses* in the film, the strength of *adhesion* between the film and the substrate, and the mechanical strength of the film. Thick films of brittle materials will *always* crack and peel above some critical thickness, leading to direct access of the environment to the exposed surface.

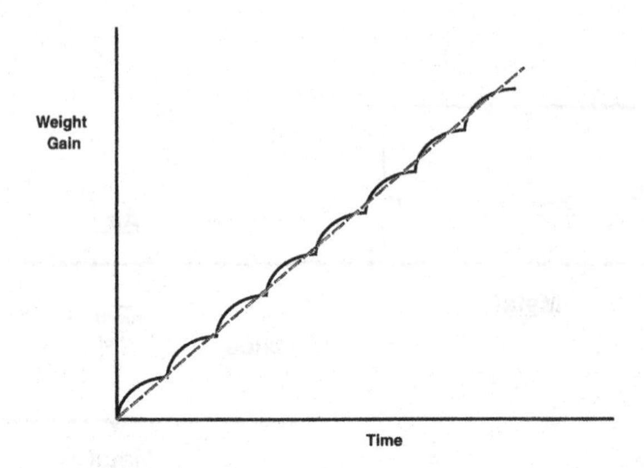

Fig. 2.20. Repeated cracking and peeling of an oxide film leads to a transition from *parabolic* growth to a *linear* growth law.

Linear growth is also observed under conditions where the oxidation product is *volatile*. An example is the oxidation of molybdenum above 900°C, when the evaporation of molybdenum oxide from the surface leads to a weight loss which is linear in time. Other refractory metals (niobium, tantalum and tungsten) behave similarly, so that the otherwise excellent properties of these metals at elevated temperatures are only available to the engineer in a non-oxidizing, protective environment. Oxidation may sometimes occur at very low oxygen partial pressures, thus the tungsten filament in an electric light bulb is evaporated as *oxide*, but the oxide vapour can break down on the surface of the cold glass envelope of the bulb. Vapour phase transport of tungsten is being driven by the *temperature gradient* between the filament and the glass envelope.

2.3 SURFACE ROUGHNESS

It is convenient to describe many surface characteristics as though the surface were either planar or bounded by a smoothly curved surface, but in practice most engineering surfaces have a measurable *roughness* which depends on the finishing process for the component. Table 2.1 gives some examples for the roughness which results from surface finishing. Roughness is usually defined in terms of the deviations from an average position of the surface as measured by a test probe or *stylus* (Fig. 2.21). If n measurements are made of the position x of a flat surface with respect to a parallel test plane, then the values x_i can be averaged:

$$x_{av} = \sum \frac{x_i}{n} \tag{2.9}$$

The square root of the mean of the squares of the deviation from this average position (*RMS—the root mean square roughness*) is then calculated:

$$RMS = \sqrt{\sum(x_i - x_{av})^2} \qquad (2.10)$$

Although RMS values provide a simple and reproducible method of evaluating surface roughness, they do not include any measure of *wavelength* or *directionality* (for example, due to machining). The *compliance* of the solid is also important in determining the mechanical response of the surface at points of contact. One possible roughness parameter which takes all three of these factors into account, and which is particularly useful in comparative studies of surface friction, is given by $(RMS)^2/(\lambda \cdot E)$, where λ is the wavelength of the surface roughness and E is the tensile modulus. The parameter $\lambda \cdot E$ has the dimensions of energy per unit area and is proportional to the elastic energy density which would be stored in an area of contact.

From Table 2.1 it is clear that engineering surfaces only approach atomic perfection in exceptional circumstances. *Cleavage* faces of some crystals are one example. Cleaved sheets of *mica* (a silicate mineral in which the bonding between the planes of silica tetrahedra is very weak) have been used for experiments on 'ideal' surfaces. Cleaved sodium chloride (table salt) crystals are often used as substrates for *epitaxial growth* (the growth of a crystal in an orientation determined by the periodicity of the substrate). Fire-polished glass approaches atomic perfection due to the *surface tension* forces acting on the viscous surface layer of semi-molten glass. At the other end of the scale of roughness, sand-cast metal components may have RMS surface roughness approaching 1 mm, and will require machining of the contact surfaces for the assembled system to meet the *tolerance* requirements. Finely *machined* surfaces may have an RMS roughness of between 1 and 10 μm, while *polished* reflecting surfaces will have an RMS roughness of 0.1 to 0.01 μm (well below the wavelength of visible light: $\sim$0.5 μm).

Table 2.1. Amplitude of surface roughness as a function of surface finish

Surface finish	*Average amplitude* (μm)
Polished	0.1–0.02
Lapped	0.5–0.1
Ground	2.5–0.5
Machined	10–2.5
Die cast	5–0.5
Sand cast	500–50

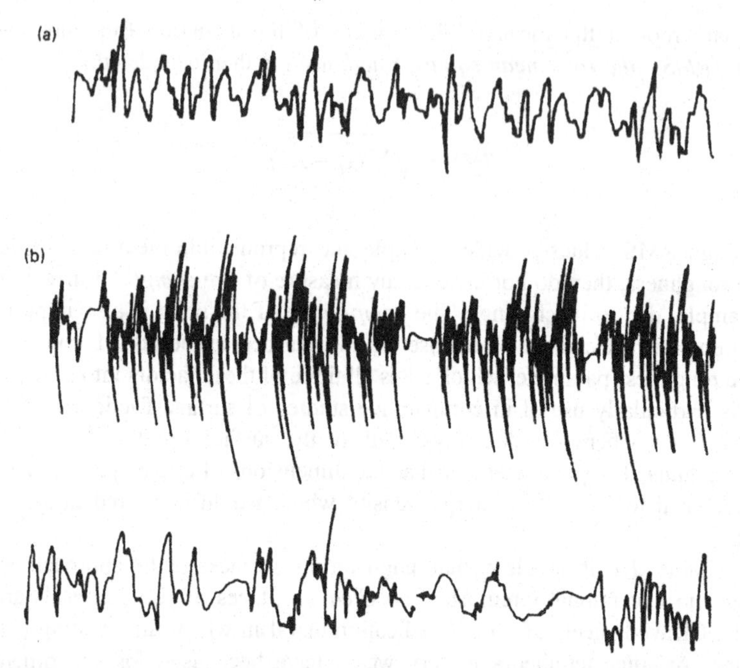

Fig. 2.21. Some examples of the surface *roughness* as measured by a travelling stylus. The *vertical* magnification is 40 000 and the *horizontal* 3.3. (a) Stainless steel abraded to a 2 μm finish. (b) Fine-ground nickel *perpendicular* (upper curve) and *parallel* (lower curve) to the grinding direction. Reproduced with permission of Academic Press Ltd, from *The Science of Adhesive Joints* (1968) J. J. Bikerman

2.4 SURFACE CHEMISTRY

The chemical behaviour of solid surfaces is determined by the nature of the inter-atomic bonding, classified as *metallic*, *ionic*, *covalent* or *polar* (van der Waals forces). *Hydrogen bonding* is a particularly strong variant of polar bonding which results from the very small ionic radius of the singly charged hydrogen *proton.*

2.4.1 Metallic Surfaces

In *metals* the mobility of the free electron cloud leads to the formation of an *electrical double layer* at the surface, discussed in Section 2.2.2.2. On the close-packed planes of the free surface there is very little protrusion of the ion cores, and the double layer is restricted, but on the loosely packed, high-index planes, which contain a large proportion of kink sites, appreciable ion core protrusion can occur.

The surface density of kink sites can have a large effect on the rate at which chemical reactions occur, and the thickness of an oxide film frequently depends on the orientation of the underlying metal surface. These effects of the surface crystallography on the rate of growth of a surface film can sometimes be used to develop striking microstructural images, associated with optical interference which depends sensitively on the film thickness (Fig. 2.22).

2.4.2 Ionic Compounds

Ionic crystals are precipitated from solution with complex, facetted shapes and it is the low energy, electrostatically *neutral* surfaces which are exposed. During growth from a dilute solution ions arriving at the free surface will migrate to the nearest surface step (Section 2.1.2.4) and those steps which are attached to the largest trapping area for arriving atoms will migrate the fastest (Fig. 2.12), leading to the formation of a *facet*. During the *dissolution* of an ionic crystal, on the other hand, ions are only removed from the kink sites, leading to *dispersion* of the steps and rounding of the crystal at the edges of the facets.

The observed shapes of ionic crystals which have been grown from the melt differ from those grown from dilute solution, since the rate-limiting step is no longer the rate of arrival of ions deposited on a terrace from (dilute) solution, but rather the nucleation of new surface steps by the aggregation of the high concentration of surface ions deposited on the terraces with low coordination numbers (Fig. 2.4). Highly facetted crystals are only to be expected during growth from *dilute* solution. In most processes of *precipitation*, *dissolution*, *solidification* and *melting*, it is the

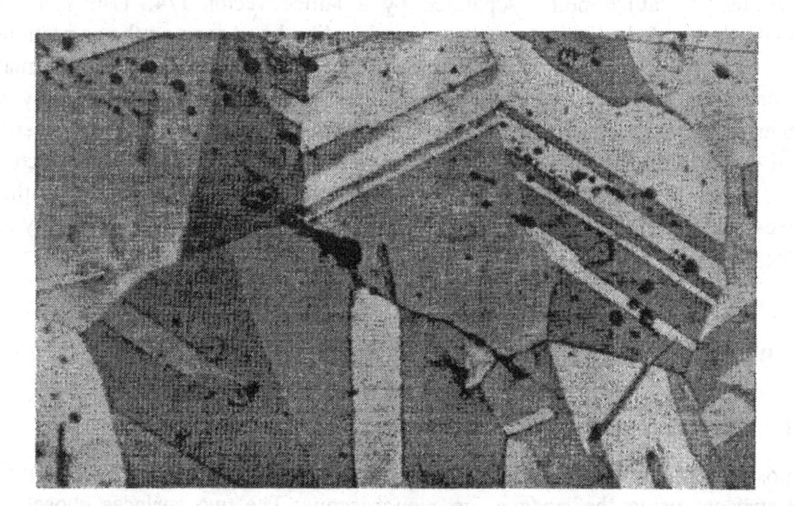

Fig. 2.22. Polycrystalline copper, chemically stained. The thickness of the surface film responsible for staining is sensitively dependent on the orientation of the crystal surface, and reflects the underlying crystal symmetry. Reproduced from Tomer: *Metals through the Microscope* (1988).

anisotropy of the surface *mobility* which dominates the shape of the crystal, rather than the anisotropy of surface energy.

Some of the most interesting *morphological* effects are observed in *biomineralization* processes, in which the crystal growth takes place from dilute solution close to equilibrium conditions, while the growth morphology is closely controlled by surface active organic molecules and membranes. *Aragonite* (hexagonal calcium carbonate) crystals may have an *acicular* morphology (preferred growth parallel to the *c*-axis of the hexagonal unit cell) in the spines of a sea urchin, but a *plate-like* morphology, parallel to the basal plane, in an abalone shell. Both morphologies are observed for growth in sea water at the same ambient temperature.

2.4.3 Covalent Bonding

At the free surface of a covalently bonded crystal it is difficult to satisfy the covalent bonding requirements, which are dictated by the *valencies* of the elements. The simple geometrical model, derived from sectioning a crystal with a random surface and discarding the atoms lying to one side of the plane of the section, leads to a surface density of unsatisfied '*dangling*' *bonds* which depends on the plane of the section. In general, the high energy of the covalent bond will then lead to appreciable surface *relaxation* and *reconstruction*, typically with a loss in symmetry of the surface structure. Perhaps the most extensively studied example of such surface *restructuring* has been for the $\{111\}$ planes of silicon (Fig. 2.10).

Silicon has the diamond structure, that is, it is a face-centred cubic crystal with two atoms per lattice point, separated by a lattice vector 1/4, 1/4, 1/4. The tetrahedrally-bonded silicon atoms have one dangling bond each on the as-sectioned $\{111\}$ surface which reconstructs (Section 2.1.2.3, Fig. 2.12). In many semiconductors (III–V and II–VI compounds) the surfaces may be terminated by one or other of the atomic species. For example either Ga or As in a gallium arsenide crystal could terminate a $\{111\}$ plane, leading to two alternative surface structures for the same crystal plane, each having a different surface energy. Either of these surfaces could then reconstruct to form two-dimensional arrays of lower energy and symmetry.

2.5 SURFACE PHYSICS

2.5.1 Surface Forces

It is possible to monitor the force–distance curve for controlled contact between two solid surfaces using the *surface force microscope*. The two surfaces chosen are typically formed from curved sheets of freshly cleaved mica. In classical mechanics, the loaded contact between two crossed cylinders (Fig. 2.23) should give a circle of elastic *Hertzian contact* whose radius r depends on the cylinder radius R, the elastic

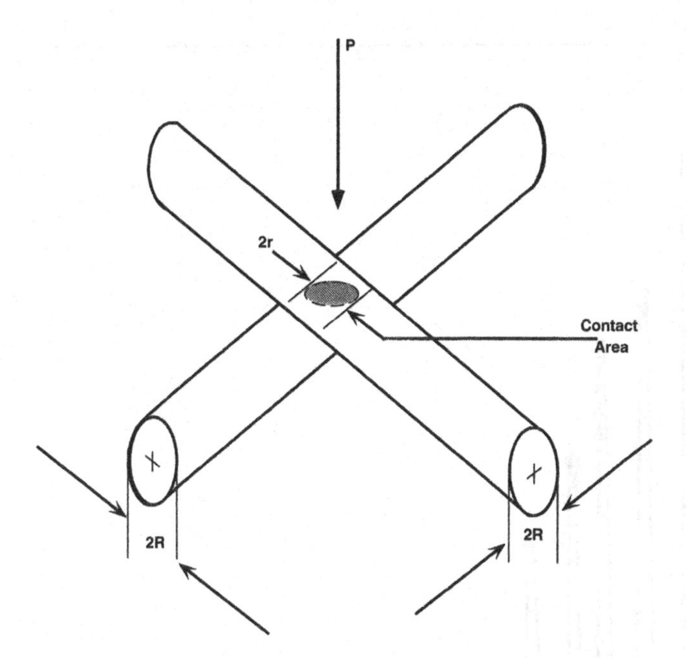

Fig. 2.23. The loading geometry for identical crossed cylinders.

modulus of the cylinders E and the force across the contact area P. The observed force–distance curve deviates from the classical *monotonic* law, exhibiting *oscillations* which reflect the presence of *adsorbed layers* of atomic and molecular species at the surface (Fig. 2.24). The *partial coverage* of the surface by an adsorbed species can be described by an *adsorption isotherm* (that is, at constant temperature) associated with the surface trapping of molecular species. The surface force microscope is a powerful tool for investigating the nature and structure of *boundary lubricants* and other surface active molecular species.

2.5.2 Surface Films

Once the surface of the solid has been completely covered by a film of adsorbate (the *coverage* θ of the surface by the adsorbate is then equal to unity), the surface properties will be dominated by the *adsorbate* rather than the *substrate*, and it is logical to speak of a *surface film*, whether or not this film involves a true chemical reaction with the substrate (as in the oxidation of a metal). In the absence of a chemical reaction, the adsorbate is said to be *physically adsorbed*, but if a chemical reaction has taken place, then the adsorbate is said to be *chemically adsorbed*.

Any surface film will be stable if its formation reduces the energy of the system. It is the *polarization forces* associated with the difference in dielectric constant of the

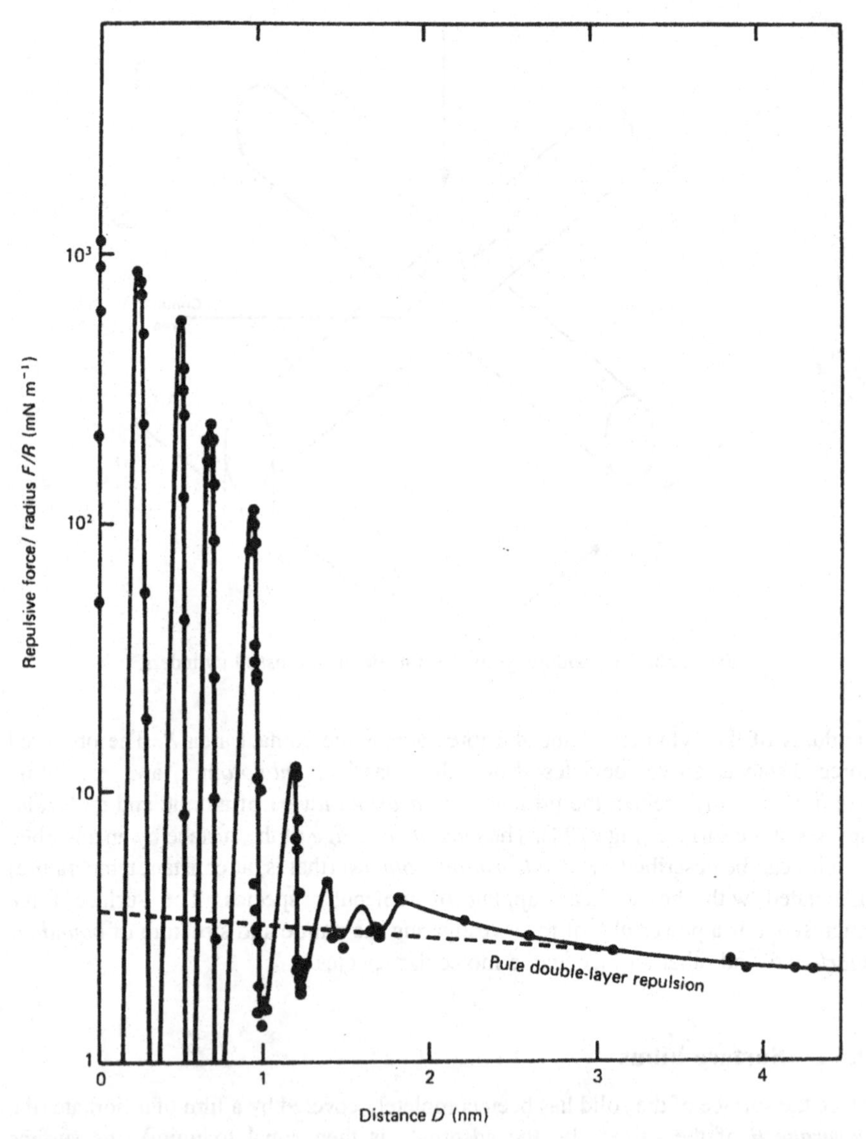

Fig. 2.24. A force–distance curve for crossed mica cylinders in the presence of a surface-adsorbed layer. Reproduced with permission of Academic Press Ltd, from *Intermolecular and Surface Tones*, (1985).

substrate and the surface film which determine whether or not a surface film is stable. A simple physical model would predict a monotonic decrease in surface energy γ with increasing film thickness h (Fig. 2.25), but at thicknesses of the order of molecular dimensions an inflection point is expected. In regions of film thickness below a critical value h_0 the system can then reduce its energy by *partial coverage* of the clean surface. The surface then consists of two regions, one covered by a surface film of the critical thickness h_0, and the other region free of any surface film. The *coverage* θ now defines the fraction of the surface covered by the film.

Many *boundary lubricants* consist of large organic molecules which attach themselves to the surface at a polar group. Several layers of the adsorbate may be present, with each additional nth layer of adsorbate being stable over a limited range of nominal thickness, in contact with regions of the surface covered by either $(n-1)$ or $(n+1)$ layers.

Wetting of the surface by a liquid phase will depend on the relative surface and interface energies, and the surface will be totally wetted if the surface energy of the liquid exceeds the sum of the surface energy of the solid and that of the solid/liquid interface:

$$\gamma_s > \gamma_l + \gamma_{ls} \tag{2.11}$$

The difference $\gamma_s - (\gamma_l + \gamma_{ls})$ is termed the *spreading force* and must be positive if complete wetting is to occur. One way to ensure wetting is to add a chemically active

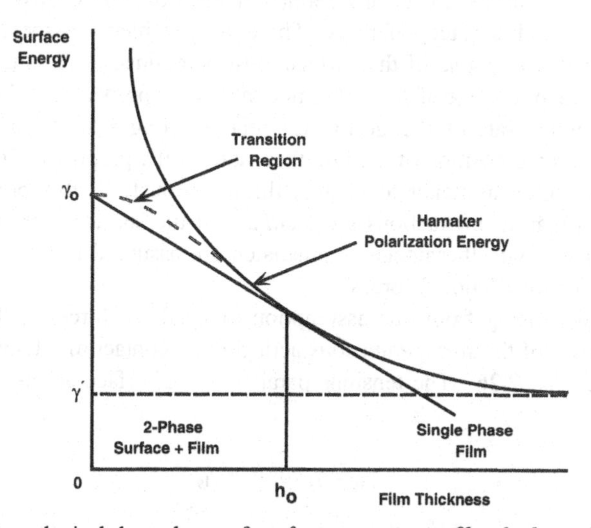

Fig. 2.25. The hypothetical dependence of *surface energy* γ on *film thickness* h for a stable adsorbate. The continuum (Hamaker) model for the polarization energy breaks down at molecular dimensions, as the energy approaches the clean surface value γ_0. The *tangent line* from the clean surface value to the polarization curve defines the *equilibrium* value of the film thickness h_0 in contact with the clean surface, and hence delimits the stable 'two-phase' region of *partial coverage*.

component to the liquid phase. *Surfactants* can fulfil this role in ceramic powder technology and mineral separation processes. Surfactants have polar groups which can bond ionically to the surface of solid particles. *Active brazes* can be used to bond otherwise inert ceramic components. An example is the use of a titanium-containing copper–silver alloy to braze silicon nitride wear components onto steel. The titanium reacts with the silicon nitride to release silicon, which is then dissolved in the molten braze alloy:

$$\langle Si_3N_4\rangle + 4[Ti]_{Cu} = 4\langle TiN\rangle + 3[Si]_{Cu} \tag{2.12}$$

where the square brackets [ ] indicate *solution* in the melt and the angle brackets $\langle\ \rangle$ indicate the *solid* phases. The equilibrium constant for this reaction is given by $K = a_{Si}^3/a_{Ti}^4$, so that the chemical activity of the braze can be controlled through its composition (see Chapters 7 and 8).

2.6 SURFACE THERMODYNAMICS

Thermodynamics is concerned with the temperature dependence of the time-independent, equilibrium state of the material. Thermodynamic equilibria at an interface or a free surface are usually defined in terms of *two-dimensional* balance of surface forces and surface energies. There are problems associated with this approach. Firstly, the *plane* of the surface must be defined. This is usually done in terms of the rate of change of some thermodynamic property across the surface. The point of maximum rate of change of the property along a direction normal to the interface defines the position of the interface, and for the purposes of calculation the properties are often assumed to change discontinuously at this point. A similar assumption is that all interactions are *localized* at the surface. For example, shear forces associated with thermoelastic mismatch are assumed to be localized at the interface, as are all frictional forces.

An anomaly arising from this assumption of localized forces is that associated with the balance of the *surface tensions* acting at the contact of a liquid drop with a solid surface (Fig. 2.26). The tensions parallel to the surface are assumed to be in balance:

$$\gamma_s = \gamma_l \cos\theta + \gamma_{ls} \tag{2.13}$$

where θ is the contact angle, but the tension perpendicular to the surface, $\sigma_l \sin\theta$, must be balanced by body forces, notably elastic strains in the solid beneath the line of contact. If the sample is now heated to a temperature at which diffusion and stress relaxation is possible, then mass transport will occur until the surface tensions are in balance, as indicated by the vector diagram in Fig. 2.26.

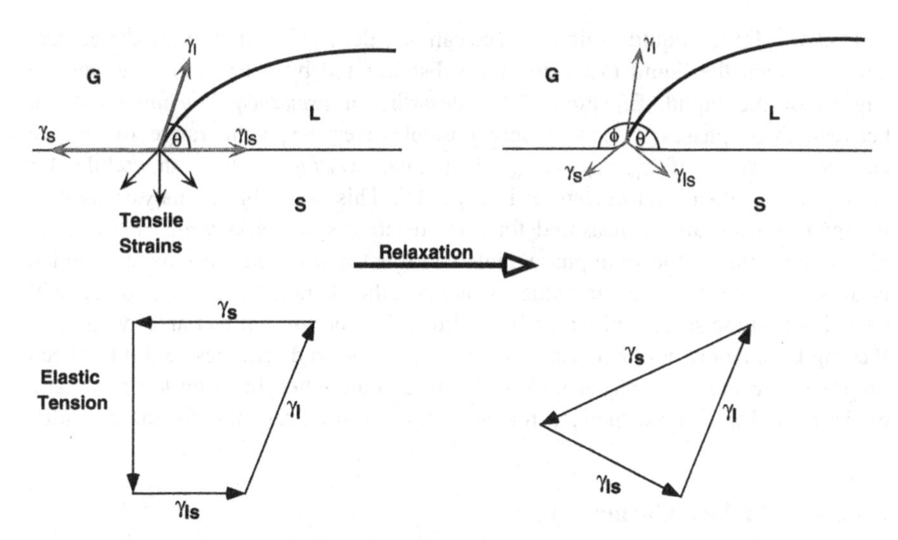

Fig. 2.26. In the absence of solid-state diffusion, the vertical component of the liquid surface tension is balanced by *tensile strains* in the solid. These strains can relax at high temperatures to balance the surface tension forces by *thermal grooving*.

2.6.1 Surface Equilibria

The *surface tension* of a solid is defined as virtual work:

$$\sigma = \gamma + A\frac{\partial\gamma}{\partial A} \tag{2.14}$$

where σ is the surface tension, γ is the surface energy and A is the surface area. In the absence of viscosity, no virtual work is done on a liquid, since the molecules are free to flow, that is $\sigma_{liq} = \gamma_{liq}$, and the force balance illustrated in Fig. 2.27 leads to the Young equation defining an *equilibrium wetting angle* θ:

$$\gamma_{sv} = \gamma_{ls}\cos\theta + \gamma_{ls} \tag{2.15}$$

The thermodynamic work of adhesion W_{ad}, defined by Dupré:

$$W_{ad} = \gamma_{sv} + \gamma_{lv} - \gamma_{sl} \tag{2.16}$$

is the reversible work per unit area that must be performed to separate a solid/liquid interface and create the solid/vapour and liquid/vapour interfaces. Combining equations (2.15) and (2.16) we obtain the Young–Dupré equation:

$$W_{ad} = \gamma_{lv}(1 + \cos\theta) \tag{2.17}$$

Thus W_{ad} for the liquid–solid interface can be calculated by measuring the contact angle between the liquid metal and the substrate, and by determining the surface tension of the liquid. Equation (2.15) describes a *pseudoequilibrium* condition between three phases, which is only possible over a specific range of surface energies. Clearly, if $\gamma_{lv} + \gamma_{sl} < \gamma_{sv}$ then total *wetting* will occur, while for $\gamma_{lv} + \gamma_{sv} < \gamma_{sl}$ then total *dewetting* is expected. This is the basic analysis used to interpret contact angles measured for non- reactive systems in which there is no plastic relaxation in the solid phase. Note that it is the surface energy balance which is important, not the specific value of one or other parameter. A drop of oil will spread out on the surface of water, but a drop of water will not wet an oily surface! Wetting is generally poor between non-reacting metals and ceramics, and it has been proposed that the weak physical (Van der Waals) attraction between the metal and oxide, related to either dispersion forces or image forces, accounts for these results.

2.6.2 A Surface Carnot Cycle

As in the case of classical thermodynamics, it is possible to define a *Carnot cycle* for the surface in order to derive the temperature dependence of the surface tension. Four steps are required to complete the cycle:

1. The surface area is first *increased* at *constant* temperature, requiring an amount of work equal to $\sigma \cdot \delta A$ and resulting in an *increase* in the internal energy of the system of $E_s \cdot \delta A$. The heat absorbed from the surroundings during this step is:

$$(E_s - \sigma) \cdot \delta A \tag{2.18}$$

2. On *cooling* by δT at *constant* area no work is done, but heat is transferred to the surroundings.

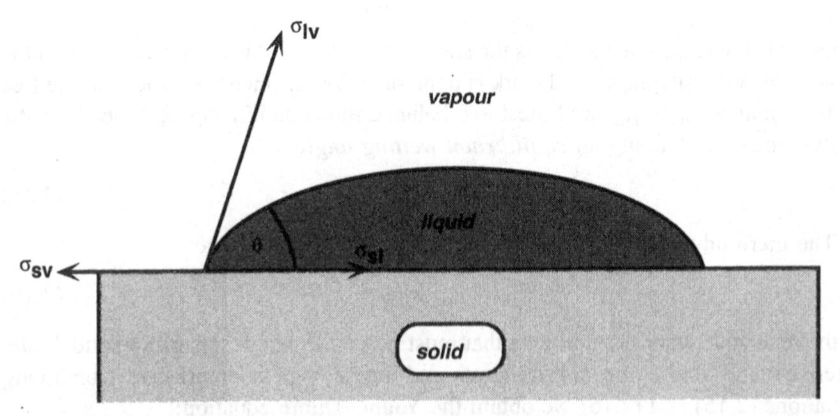

Fig. 2.27. The force balance for the line of contact of a *liquid* on a *solid surface*, ignoring elastic strains.

3. Now *reduce* the surface area by δA at the lower temperature, $T - \delta T$. The work done at this stage is:

$$[\sigma - \delta T \cdot (\mathrm{d}\sigma/\mathrm{d}T)] \cdot \delta A \tag{2.19}$$

4. On *reheating* to the original temperature T at *constant* area no work is done.

The ratio of the *total work done* in the cycle to the *heat absorbed at the higher temperature* is then equal to $\delta T/T$, so that:

$$\delta T/T = \{[\sigma - \delta T \cdot (\mathrm{d}\sigma/\mathrm{d}T)] \cdot \delta A - \sigma \cdot \delta A\}/[(E_s - \sigma) \cdot \delta A] \tag{2.20}$$

or $\delta T/T = -\delta T(E_s - \sigma) \cdot (\mathrm{d}\sigma/\mathrm{d}T)$, that is: $\sigma = E_s + T\,\mathrm{d}\sigma/\mathrm{d}T$, where the term $\mathrm{d}\sigma/\mathrm{d}T$ is negative.

The surface tension is therefore expected to *decrease* with *increasing* temperature, and should go to zero at a high enough temperature. Indeed, this is what happens at the critical point for the disappearance of a phase transition. Well known examples are carbon dioxide and superheated steam—no phase boundary exists above the critical point, and the distinction between the gas and liquid phase is lost.

2.6.3 Equilibrium Reactive Wetting

Chemical reactions at the interface strongly affect the process of wetting, and the resultant interface phases will then influence the *contact angle*. An interfacial chemical reaction involves two *additional* energy terms which influence the equilibrium state and *reduce* the wetting angle:

$$\cos\theta_{\min} = \cos\theta_0 - \frac{\Delta\sigma_r}{\sigma_{lv}} - \frac{\Delta G_r}{\sigma_{lv}} \tag{2.21}$$

where $\theta_{\min}$ is the new contact angle due to the chemical reaction, θ_0 is the contact angle in the absence of a reaction, $\Delta\sigma_r$ is the total change in the interfacial energies due to the reaction and ΔG_r is the change in the free energy per unit area which accompanies the reaction. Although ΔG_r determines which transformations can occur, $\Delta\sigma_r$ is the dominant parameter which determines the wetting angle during the interfacial reaction. The possible reactions also depend on the compositions of the vapour and liquid phases: the oxidation of a liquid phase will be a function of the oxygen *partial pressure* in the gas phase, while the *activity* of the constituents in the liquid phase will determine the extent of reaction with the solid.

2.6.4 Kinetic Effects in Wetting

Most wetting experiments measure a *pseudoequilibrium* contact angle but there are a number of dynamic effects which are also observed during wetting. Hysteresis results in a different contact angle at an *advancing* line of contact between a liquid and a solid substrate, as opposed to a *receding* contact line on the same substrate.

Hysteresis effects may be associated with *surface roughness* or they may be due to hydrodynamic effects (viscous flow of the liquid). A frequently observed dynamic effect is the presence of a *precursor film* ahead of an advancing liquid/solid interface. Such films may also exist when the liquid is stationary. A precursor film followed by an interfacial reaction may lead to time-dependent changes in contact angle.

2.6.5 Surface Energy Anisotropy

The crystallographic orientation of the substrate surface during wetting can have a significant effect, especially in ceramic systems, for which ionic or covalent bonding can lead to significant *anisotropy* in the *surface energy*. This is illustrated by the *Wulff plot*, in which surface energy is plotted as a vector normal to the plane of the surface. The low energy planes appear as *cusps* in the Wulff construction (Fig. 2.28) and the crystal can reduce its surface energy by *facetting* on these low energy planes. In the limit of deep cusps the equilibrium shape of the crystal may consist *only* of low energy facets.

2.6.6 Segregation

Most materials contain solute additives or impurities, and in general these *solutes* are partitioned between the surface and the bulk. The excess concentration of a minor solute constituent at a free surface, an internal boundary or an interface between two phases is termed *segregation*, while the reverse process, in which the concentration

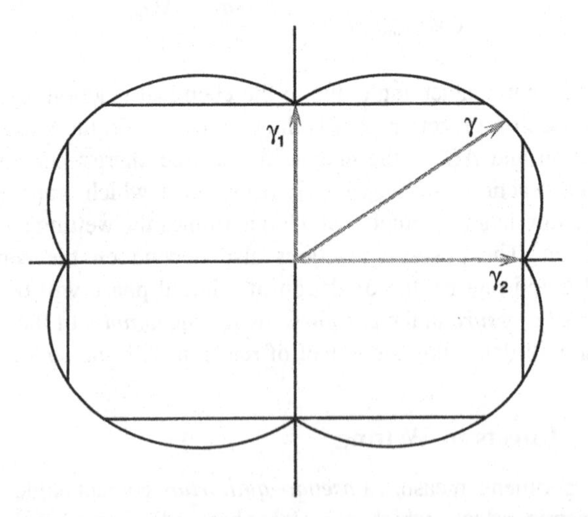

Fig. 2.28. Hypothetical Wulff plot in which two low surface energy orientations generate *minima* in the *surface energy* plotted as a *vector* normal to the surface considered. The minima predict *facetting* over specific ranges of the surface orientation.

of the constituent is reduced at the surface, is termed *desegregation*. The segregation process can be approximated as a chemical reaction with an *equilibrium constant* for surface segregation, K_s. For a binary AB alloy:

$$A_s + B_b = A_b + B_s \quad \text{and} \quad K_s = a_b b_s / a_s b_b \tag{2.22}$$

where the subscripts refer to the bulk and the surface, respectively, and a and b refer to the activities of the two components A and B at the surface and in the bulk. When the equilibrium constant K_s exceeds unity, *segregation* results; where K_s is less than unity, *desegregation* is expected.

The *surface chemical potential* of a segregated species A is:

$$\mu^{A_s} = -RT \ln a_s \tag{2.23}$$

and the free energy of segregation is:

$$\Delta G_s = -RT \ln K_s \tag{2.24}$$

Segregation *lowers* the free energy of the surface and may inhibit wetting. For a free surface, it is the surface energy per atom that determines the extent of segregation, and it is those solute constituents with the lowest *sublimation energy* which are expected to segregate strongly, so that the tendency to segregate correlates with the *volatility* of the solute. In an aluminium-magnesium alloy, the magnesium is very much more volatile than the aluminium. Magnesium reduces the free energy of the bulk alloy, but nevertheless segregates preferentially to the surface of an aluminium alloy melt, lowering the surface energy (by an amount ΔG_s) below that expected for a simple rule of mixtures (Fig. 2.29). By contrast, aluminium dissolved in a magnesium melt has little effect on the surface energy of a magnesium alloy and is expected to *desegregate* from the magnesium free surface.

Summary

The molecular geometry of crystal surfaces can often be derived from a knowledge of the atomic sites in the bulk lattice, irrespective of the chemical nature of the solid. In many cases the surface crystallography is modified by the presence of surface defects, by relaxation of the exposed atoms from their normal bulk positions, or by reconstruction of the surface and consequent modification of the symmetry, leading to the formation of a new, two-dimensional surface phase.

Many crystals are facetted, and this may be due to anisotropy of either the surface energy or the rate of crystal growth. Most engineering solids are polycrystalline aggregates of single crystal grains, and the balance of surface and boundary energies dictates the equilibrium configuration at the triple junction where the grain boundaries intercept a free surface. The surface energy of a free surface is further modified by the presence of contamination. The adsorption and reaction of gaseous species are common occurrences at the surface of a solid.

Oxidation occurring at the surface requires both cathodic and anodic reactions in order to form the ionic oxidation product. Growth laws for the formation of oxide

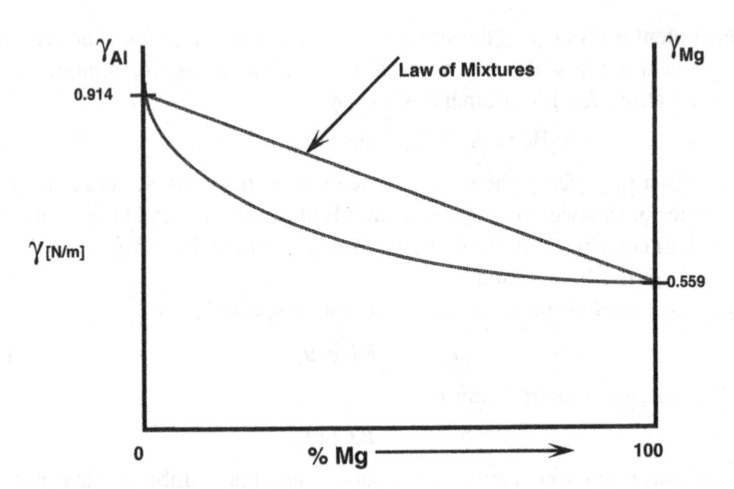

Fig. 2.29. The surface energy of molten aluminium–magnesium alloys. *Volatile* magnesium *reduces* the surface energy of aluminium and hence segregates to the surface, but aluminium has little effect on the surface energy of magnesium (and *desegregates* from the surface).

films on clean metal surfaces depend on the particular growth mechanism, which is in turn a function of the electronic and ionic transport processes, and of the integrity of the oxide film.

In addition to the presence of oxide or other surface films, very few engineering surfaces are atomically smooth. The surface roughness is dependent on the finishing process, varying in both amplitude and wavelength over many orders of magnitude.

The surface properties of a solid reflect its chemical characteristics and crystal structure. Thus covalently bonded crystals often have restructured surfaces, while the facetted surfaces of ionic crystals reflect the densely packed, electrostatically neutral planes in the bulk crystal. The surface forces which are experienced by an atom or molecule at the surface dominate both the adsorption of contaminant species and the wetting of a solid surface by a liquid phase. Suitable additives can be used to promote or prevent the wetting of a solid surface by molten flux, liquid metal or viscous adhesive. The relative values of the surface energy for the interfaces separating the solid, liquid and gas phases determine whether the solid surface will be wetted, dewetted or partially wetted. Surface energy decreases with increasing temperature and may be strongly affected by the segregation of impurity or dopant species to the interface. Segregation kinetics can also play a decisive role in controlling surface wetting.

Further Reading

1. J. M. Blakely (ed.), *Surface Physics of Materials*, Volumes I and II, Academic Press, London, 1975.
2. L. J. Clarke, *Surface Crystallography—An Introduction To Low Energy Electron Diffraction*, John Wiley & Sons, New York, 1985.

3. R. A. Laudise, *The Growth of Single Crystals*, Prentice-Hall, Englewood Cliffs, NJ, 1970.
4. A. P. Sutton and R.W. Balluffi, *Interfaces in Crystalline Materials*, Clarendon Press, Oxford, 1995.
5. D. Wolf and S. Yip (eds.), *Materials Interfaces—Atomic-Level Structure and Properties*, Chapman & Hall, New York, 1992.
6. N. Birks and G. H. Meier, *High Temperature Oxidation of Metals*, Edward Arnold, London, 1983.
7. G. Wrangler, *An Introduction To Corrosion and Protection of Metals*, Chapman & Hall, New York, 1985.
8. C. B. Duke (ed.), *Surface Science–The First Thirty Years*, North Holland, Amsterdam, 1994.
9. J. M. Walls and R. Smith (eds.), *Surface Science Techniques*, Pergamon Press, Oxford, 1994.
10. D. P. Woodruff and T. A. Delchar, *Modern Techniques of Surface Science*, Cambridge University Press, Cambridge, 1986.

Problems

2.1 Define an *ideal* surface. Explain briefly why such a surface *cannot* exist under equilibrium conditions.

2.2 Explain the difference between relaxation and reconstruction of a free surface.

2.3 Given an *ideal* low-index surface of gold and silicon, which would you expect to *relax* and which would you expect to *reconstruct*? Why?

2.4 Chromium has a BCC lattice with one atom per lattice point.

a. Sketch the ideal surface structure for the {100} and {110} planes and calculate the coordination numbers for atoms on these two surface planes. What are the nearest neighbour separations for atoms on each of these surface planes?

b. Based on the assumption that the bonds between atoms are pure metallic (non-directional), which surface plane should have the lower energy? Why?

c. Chromium oxidizes to form Cr_2O_3. Can you use the same model as in part (b) to predict the relative surface energies?

2.5 Compare the bonding on the free surface of an ionic crystal with that on a covalent crystal surface. What is the main driving force for reconstruction for each of these two different types of surface?

2.6 Why do contaminants alter (reduce) the bonding between two surfaces?

2.7 The figure shows the approximate dependence of the surface tension of liquid copper on the concentration of dissolved antimony. Explain this effect.

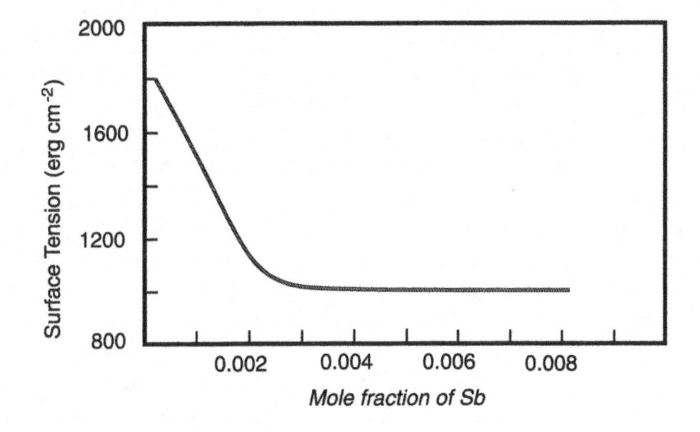

2.8 Calculate the rate of arrival of nitrogen molecules on a flat surface at 300 K and a pressure of 10^{-8} Torr. How much time is required to accumulate a monolayer on the surface under these conditions? Assume a sticking probability of 0.1 and a molecular surface density of 10^{15} cm^{-2}.

2.9 A knowledge of the surface energies of liquid metals and their contact angles on solid surfaces is critical in the design of solder and braze joints. Describe *one* experimental system to measure γ_{ls}. Try to locate the main sources of error for the method, and consider possible segregation effects, similar to that shown in problem 2.7, both for dissolved impurities and for the partial pressures of impurities in the gas phase.

3

The Mechanics of Joining

In this chapter we explore the mechanical implications of the presence of joints in engineering systems. The concept of *contact area* will be defined, followed by a review of the mechanical properties used to characterize materials behaviour, together with a discussion of the limitations associated with applying concepts such as *stress* and *strain* to the area of a joint. The concept of a *stress concentration* will be introduced, and the *stress intensity factor* will be defined. *Fracture toughness* will be discussed, and the *failure modes* at a joint will be described. Finally, some common methods for determining the mechanical *performance* and *integrity* of joints will be described.

3.1 CONTACT AREA

If two identical cylinders are rotated by 90° with respect to one another and then brought into contact, the *area of contact* will be a circle whose radius depends on the applied load normal to the cylinder axis, the elastic moduli of the material and the radius of curvature of the cylinders (Fig. 2.23). The processes that occur at the contact surface when two solid bodies are brought into contact are determined by the mechanical properties of the two solids and the rate at which the contact is established. *Equilibrium* contacts are established slowly, to maintain isothermal conditions in the system, while the impact of meteorites and high velocity projectiles converts *kinetic energy* into both ballistic damage and *heat*. Usually the contact between two engineering surfaces is established under conditions which are close to equilibrium, but there are some important exceptions: technological processes which depend on a high rate of loading to establish a strong interfacial bond. *Explosive bonding* generates a *shock wave* at the line of contact between a stationary substrate and a second component travelling at high velocity. The kinetic energy of the second body generates high pressures in the impact zone, which do plastic work and result in *adiabatic heating* (Fig. 3.1). A jet of contamination is forced out of the contact zone by the high pressure, allowing a clean, strong bond to form. *Friction welding* depends for its success on the heat generated by high rates of *plastic shear* at the contact surface between two bodies held in sliding contact. In this case, it is the *hysteresis loss*, associated with the repeated making and breaking of contacts between the two sliding bodies, which is responsible for the *frictional heating* and eventual bonding at the interface.

Under conditions of reversible equilibrium, when two bodies first make contact they start to *rotate* about an axis passing through the *point of contact* and the normal to the plane defined by the point of contact and their *centres of gravity* (Fig. 3.2a). When a second point of contact is established, the centres of gravity of the two bodies continue to approach each other by further rotation about an axis which joins the two points of contact (Fig. 3.2b). Once a third contact point is established (Fig. 3.2c), the centres of gravity of the two bodies can continue to reduce the distance between them, increasing the *areas of contact* by applying a force normal to the plane defined by the three contact points. This applied load then establishes *finite* areas of *interfacial contact* between the bodies.

It is extremely difficult to prepare polished surfaces to an accuracy that allows a uniform distribution of the *contact pressure* over the interface. That would require a uniform separation on the atomic scale of distance (less than 1 nm). In practice, all geometrical contacts between two bodies are established at *isolated* contact points, as illustrated in Fig. 3.2, so that *engineering surfaces* are only brought into physical contact at a limited number of points whose position and spacing are dictated by the *surface topology* and *surface finish* (roughness of the surfaces). We will now describe the nature of these contact points and their growth under an applied compressive load.

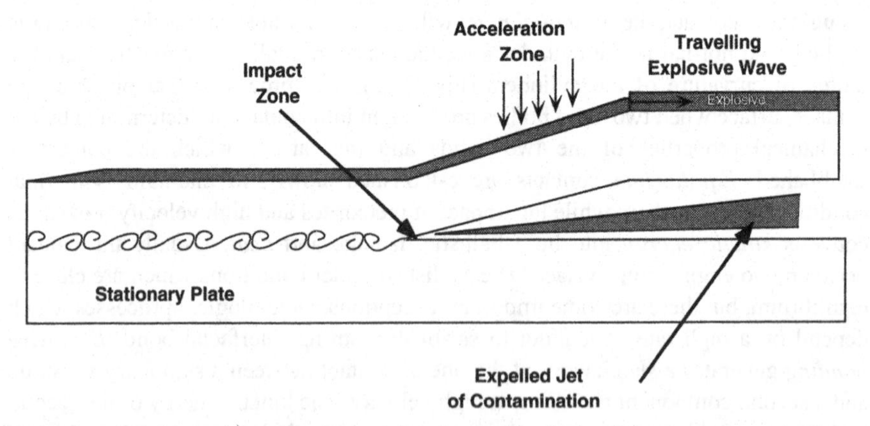

Fig. 3.1. An *explosive bond* is formed by the high velocity, low angle impact of one component on another. Surface contamination is forced out of the impact line by the high pressures which are generated.

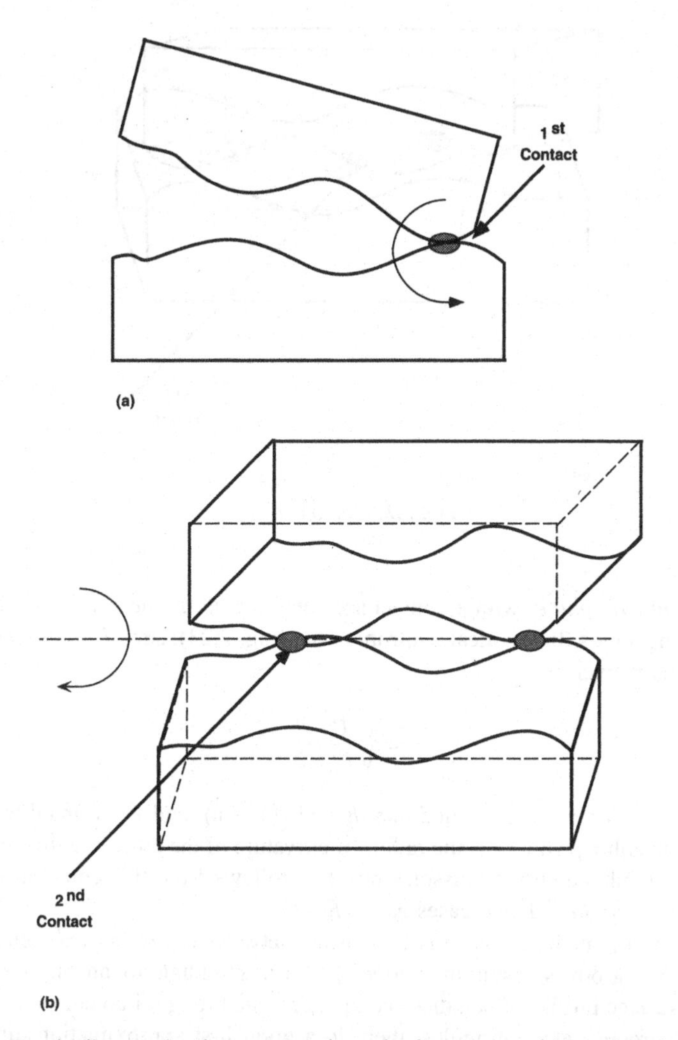

Fig. 3.2. Two bodies coming into contact do so in three stages: (a) rotation about the axis passing through the first contact point and normal to the plane containing the first point of contact and the two centres of gravity; (b) rotation about an axis containing the first and second contact points until a third contact is made; (c) displacement parallel to the loading vector with concomitant growth in the total number of contacts and the areas of contact.

3.1.1 Hertzian Elastic Contact

Consider first a single, circular area of contact established between two crossed cylinders by a load applied normal to the plane of the contact area (Fig. 2.23). As the load P increases, the response of the two surfaces becomes sensibly independent of any surface films associated with the environment. The two solid bodies are said to

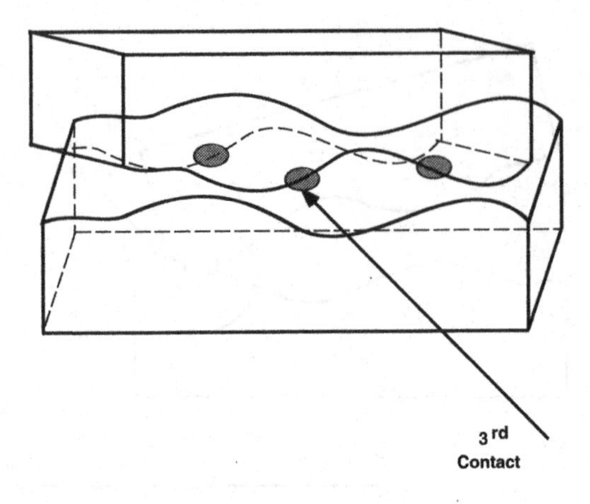

Fig. 3.2. *(continued)*

be in *mechanical contact* with a contact area determined by the *elastic* response of the contacting solids. In the ideal, isotropic case (Fig. 2.23), such *Hertzian contacts* have a radius r given by:

$$r = \sqrt[3]{\frac{P}{E_b} \cdot \frac{R}{2}} \tag{3.1}$$

where E_b is the *biaxial* elastic modulus, $E_b = E/(1 - \upsilon)$, and is assumed to be the same for both solids, while R is the radius of curvature of the crossed cylinders. (E is the tensile modulus and υ is Poisson's ratio). It follows from this equation that the *area* of *elastic contact* A_e increases as $(P/E_b)^{2/3}$.

Let us now assume that contact is established between two *machined* components whose surface grooves result in a topology corresponding to an approximately sinusoidal surface profile. When these components are brought into contact, with the two sets of grooves at right angles, then, to a good first approximation, the *same* formula for Hertzian contact will apply, but with each contact point supporting a *local load* p_i, which gives rise to a *local area of contact* a_i. The *total applied* load will still be the sum of the locally supported loads, and the *total* area of contact will still be that due to an elastic Hertzian response:

$$P = \sum p_i \quad \text{and} \quad A_e = \sum a_i \tag{3.2}$$

The true area of elastic contact for any two bodies which have been loaded in *elastic* compression normal to their contact surface is then given by:

$$A_e = \alpha \left(\frac{P}{E_b} \right)^{2/3} \tag{3.3}$$

where α is a constant of proportionality of the order of unity. Substituting $P = \sigma A_0$, where σ is the average applied stress normal to the apparent area:

$$\frac{A_e}{A_0} = \alpha A_0^{-1/3}\left(\frac{\sigma}{E_b}\right)^{2/3} \tag{3.4}$$

from which it is apparent that the *true* area of contact depends primarily on the ratio of the average applied stress to the elastic modulus. Since engineering materials normally fail at applied stress levels which are only a small fraction of the elastic modulus, it follows that the *true* area of elastic interfacial contact is normally only a small fraction of the nominal area of contact.

3.1.2 Plastic Contact

Plastic deformation will initiate in a region of contact when the *yield stress* σ_y is exceeded in this contact region. Assuming that no significant work-hardening occurs and that the stresses are uniaxial, then the expected *area of plastic contact* is just: $A_p = P/\sigma_y$, and the plastic contact area should increase linearly with the applied compressive load. It follows that the ratio of the *true* area of plastic contact to the apparent (nominal) area of contact should be given by $A_p/A_0 = \sigma/\sigma_y$. The material in the contact area is always constrained to some extent by the surrounding matrix, so that the stress cannot be uniaxial. *Slip line theory* predicts a *maximum* triaxial *constraint factor* of: $(2 + \pi)/\sqrt{3} \sim 3$, so that $A_p \sim P/3\sigma_y$. This constraint factor is equal to the commonly observed ratio of the *hardness* of a material (determined under conditions of maximum plastic constraint) to its *yield stress*, that is:

$$\frac{H_v}{\sigma_y} \approx 3 \tag{3.5}$$

If appreciable work hardening does occur, then this will increase the load required to achieve a given area of contact. However, work hardening under *uniaxial stress* is difficult to relate to work hardening in *triaxial constraint*, and the approximate plastic constraint factor of 3 can still be assumed to apply.

Now consider the case of two solid bodies placed in contact. While only three contact points are necessary for *mechanical stability*, the actual number and distribution of the contact points will depend on the *surface topology*, including the roughness, and on the applied load. The *total contact area*, obtained by summing the individual contact areas over all the points of contact, reflects the response to the applied load over the *nominal area of contact* A_0. Under constrained compression most engineering solids exhibit some ductility, with the notable exception of concrete. Although cast iron may be brittle in tension, it is ductile in compression and plastic yielding precedes brittle crushing. Localized ductile flow is initiated at the areas of primary contact, and in this ductile regime the *total contact area* will be given by $A_p = \Sigma A_i = P/\sigma_y$ for unconstrained contacts. The *nominal* applied compressive stress across the region of contact is given by $\sigma_a = P/A_0$, so that $A_p/A_0 = \sigma_a/\sigma_y$, and complete contact is established when the nominal applied stress reaches the bulk yield stress.

3.1.3 The Elastic–Plastic Transition

The transition from elastic to plastic contact can be approximated by considering the case of a *rigid punch* penetrating an *ideally plastic* matrix (Fig. 3.3). Yielding initiates at the corners of the punch and, if the *Von Mises criterion* for plastic yielding is satisfied, will begin when $\sigma_a = 1.15\sigma_y$. The plastic zone will spread to the centre of the area of elastic contact as the applied stress is increased, reaching a maximum value of approximately $3\sigma_y$, as noted above, at which point, in the absence of work-hardening, the punch will start to sink into the matrix. These simple models for elastic and plastic contact are not easily combined, but whereas elastic contact is dominated by the ratio of the applied stress to the *elastic modulus*, plastic contact is dominated by the ratio of the applied stress to the *hardness* of the material. In Table 3.1 some typical values of the yield strength, hardness and elastic moduli of engineering solids are given. The ratio of the *hardness* to the *biaxial elastic modulus* is generally small. Only in the case of *elastomers* (rubber-like solids) and *ceramics* is *elastic* contact at all likely to dominate the transfer of load across a contact surface.

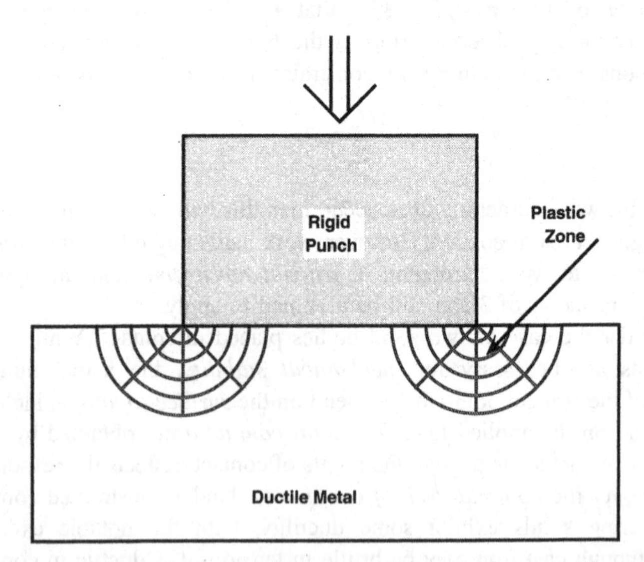

Fig. 3.3. When a rigid punch is pressed into a ductile metal *plastic yielding* initiates at the edge of the contact zone at a stress of 1.15 σ_y and spreads towards the centre of the zone. Complete plasticity (*general yielding*) beneath the punch corresponds to an applied stress of about 3 σ_y.

Table 3.1. Mechanical property ranges for engineering materials

Material	Hardness (GPa)	Yield strength (MPa)	Tensile modulus (GPa)
0.1% C Steel	0.9	310	220
Aged 2024 Al alloy	1.2	320	70
4340 Structural steel	2.7	1280	220
Diamond	84	5310	1030
Alumina	26	1100	390
Polyethylene	–	1.2	0.12
Natural rubber	–	3.2	0.0018

There are important implications for both *seals* and *bearing surfaces*. Materials which have been selected for these applications have unique mechanical characteristics: *rubber elasticity, high hardness* or exceptional *plasticity*.

3.1.4 Rolling & Sliding Contacts

In many applications *movement* is expected to occur between two contacting surfaces, and this movement may be of two types: *rolling* contacts, in which one component rolls over the other, making and breaking contacts by displacements which are *vertical* to the contact surface and dominated by *compressive* and *tensile stresses* at the contact points and *sliding* contacts, in which *shear displacement*, parallel to the plane of contact, controls contact formation and failure in a *shear stress* regime.

In a *rolling contact* the tensile response of the material determines the magnitude of the energy losses. If the making and breaking of a contact only involves elastic deformation, then these cyclic, hysteresis energy losses are small, although *some* energy is always lost in rolling friction. (For example, the heat generated in a rubber tyre supporting a travelling vehicle.)

In *sliding contacts* the *surface roughness* of the components results in geometrical interlocking at contact points (Fig. 3.4). *Polished* surfaces reduce the

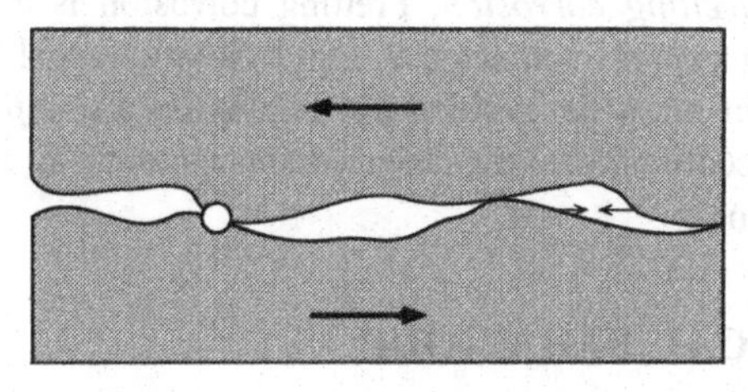

Fig. 3.4. *Surface roughness* leads to mechanical interlocking of sliding surfaces, the volume of material affected being a function of the amplitude and wavelength of the surface roughness.

thickness of the affected layer, while *hard coatings* of high elastic modulus also reduce the total area of true contact. *Lubricants*, with low shear strength, are used in hydrostatic bearings to minimize frictional losses in shear. *Boundary lubricants* with strong polar groups bond to the surfaces and inhibit the formation of *plastic contacts* by forming a compliant surface layer.

3.1.5 Contact Damage & Wear

Plastic deformation and fracture at contact points is the primary damage mechanism responsible for *wear*. The *wear rate*, the rate of weight loss per unit area of sliding surface, is dependent on the normal applied load, the working environment and the material properties of the *wear couple*. Two wear regimes are commonly observed. At low loads, and for hard, polished, bearing surfaces, *elastic contact* dominates. The rate of wear is controlled by microcracking, fracture and abrasion of a *surface film*, typically a coherent oxide which is replaced by further oxidation as wear proceeds. Fine particles of *wear debris* are formed at the sliding interface with the composition of the surface oxide film. These *wear particles* act as an *abrasive* and contribute to the wear rate.

At higher loads, and for more ductile materials, *plastic contact* dominates the transfer of load across the interface. The contact points are regions of *microwelding* which form and fracture as sliding proceeds. Wear particles are 'torn' from the substrate by a process of *ductile shear*. The wear particles in this regime of *severe wear* are much coarser than those in the regime of *mild wear*, and their composition approximates that of the substrate from which they are formed. The transition from *mild wear* to *severe wear*, commonly observed at a critical normal load, is marked by a rapid increase in the *wear rate* and correlates with an increase of surface roughness and sub-surface damage along the *wear track*.

When a mechanical joint is subjected to limited cyclic shear *mild wear* may cause cumulative damage. The shear displacements are now on the scale of the contact areas and the surface is gradually eroded at these contact points. Erosion relieves the load at the contact point, and the load is then transferred to other areas, so that the *cumulative wear* is uniformly distributed over the nominal contact area. The wear particles (of finely divided oxide) act as an abrasive medium in a form of degradation commonly termed *fretting corrosion*. Fretting corrosion is a common feature of mechanical joints in systems subjected to rapid vibration, such as heat engines and road vehicles. It can often be avoided by employing a *compliant seal* or *gasket* which serves to accommodate the shear displacements at the joint and damp mechanical vibrations.

3.2 MECHANICAL BEHAVIOUR

The selection of structural engineering materials is dominated by their mechanical properties: *stiffness* and *tensile strength*, *creep* and *fatigue* resistance, and susceptibility to *oxidation* or *corrosion-assisted failure* processes. It is the *mechanical properties* of a material which are the first concern of the design engineer. To optimize the design, he needs to know the *elastic moduli*, the *yield stress*, the *fracture toughness* and the *fatigue limit*. He needs to know how these properties are affected by *heat treatment* of the components, by changes in the operating *temperature* and by the operating *environment*.

When components are joined together in an engineering system additional questions present themselves. How is load transmitted across the interface between the components? What mechanical parameters best describe the performance of the joint? Can they be used to ensure that the strength of the joint is not inferior to that of the bulk components? What modes of failure are characteristic of the joint? How do these failure modes differ from failure in the bulk material?

We will now review the basic criteria for *plastic yielding* and *fracture* and explore the characteristic features of failure at a joint. In the process we attempt to answer some of the above questions.

3.2.1 Yield & Fracture Criteria

Plastic yielding in engineering alloys will occur when a yielding criterion is satisfied. According to *Tresca*, it is the *maximum resolved shear stress* which determines the onset of plastic flow, and plastic shear is initiated when a critical value of this maximum shear stress is exceeded. For a volume element of the material subjected to principal stresses σ_i ($i = 1$ to 3), the maximum resolved shear stress is given by the largest value of $\sigma_i - \sigma_j$, where $i \neq j$. A second yield criterion, due to *Von Mises*, is based on a critical value for the *stored elastic energy* which can be relieved by plastic shear—the *deviatoric strain energy*. The Von Mises criterion for a triaxial stress state can be expressed in terms of the uniaxial yield stress σ_y:

$$\sigma_y = \sqrt{2 \cdot [(\sigma_1 - \sigma_2)^2 + (\sigma_2 - \sigma_3)^2 + (\sigma_3 - \sigma_1)^2]} \tag{3.6}$$

The *total stored elastic energy per unit volume* in the material, assuming *Hooke's law*, is given by:

$$U_e = \frac{1}{2}(\sigma_1\epsilon_1 + \sigma_2\epsilon_2 + \sigma_3\epsilon_3) \tag{3.7}$$

The values of strain are related to the *principal stresses* and *Young's modulus* of elasticity by relations of the form:

$$\epsilon_1 = \frac{[\sigma_1 - \upsilon(\sigma_2 + \sigma_3)]}{E} \tag{3.8}$$

and the *hydrostatic component of the stress* is:

$$\sigma_m = \frac{[\sigma_1 + \sigma_2 + \sigma_3]}{3} \tag{3.9}$$

so that the *hydrostatic component of the strain energy* is given by $1/2\sigma_m^2/K$, where K is the elastic bulk modulus: $K = E/3(1 - 2\nu)$.

3.2.1.1 Yield Surfaces

Both the *Tresca* and the *Von Mises* yield criteria can be plotted as a *failure surface* in *principal stress space*. The Von Mises criterion can be represented by a cylinder along the diagonal axis representing the hydrostatic stress component: $\sigma_1 = \sigma_2 = \sigma_3$. The radius of the Von Mises failure cylinder is then $\sigma_y/\sqrt{2}$. The Tresca maximum shear stress criterion is represented on the same diagram by a hexagonal prism whose corners touch the Von Mises failure surface (Fig. 3.5). The intercept of these failure surfaces with the plane $\sigma_3 = 0$ defines the criteria for the important case of a biaxial, *plane stress* state (Fig. 3.6). In plane stress, $\sigma_3 = 0$ while $\varepsilon_3 = -\nu(\sigma_1 + \sigma_2)/E$. Plane stress is characteristic of the stress state in *thin plates* and *shells,* in which plastic yielding is initiated on an inclined plane and leads to thinning of the plate (Fig. 3.7). A second important loading condition is that of *plane strain,* in which ε_3 is constrained to be zero. In *plane strain* no change is permitted in the through-thickness dimension, and the normal to the plane of yielding is constrained to be perpendicular to this direction (Fig. 3.8). Loading in *plane stress* is characteristic of *thick plate,* for example pressure vessels and steel armor.

Plastically constrained yielding at a free surface generally requires an applied stress that exceeds the uniaxial yield stress. As noted previously (Section 3.1.2), slip line field theory leads to a *maximum* yield stress for a fully constrained situation of approximately $3\sigma_y$. This *triaxial yield stress* is characterisic of the applied pressure required to initiate yielding in a *hardness test,* but it is *also* the critical stress at which yielding will occur at the tip of a *ductile crack.*

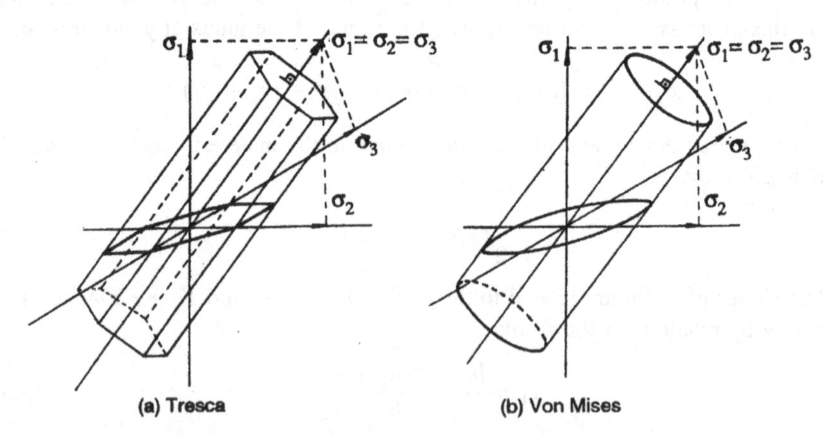

Fig. 3.5. The Tresca and Von Mises yield surfaces in stress space are a *hexagonal prism* and a *cylinder* respectively, whose common axis corresponds to the hydrostatic stress condition $\sigma_1 = \sigma_2 = \sigma_3$. Reproduced from *Introduction to Materials Engineering*, Brandon, Eylon, Nadiv, Rooen. Michlol-Publishing House, Techniar Haifa.

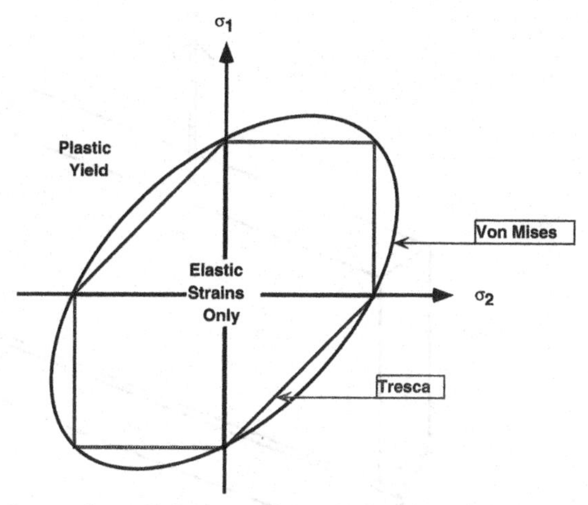

Fig. 3.6. For a *biaxial stress* ($\sigma_3 = 0$) the Von Mises criterion for *isotropic* yielding projects as an ellipse on the σ_1–σ_2 plane which touches the Tresca condition along the axes $\sigma_1 = 0$ and $\sigma_2 = 0$.

3.2.1.2 Ductile Failure

Ductile failure in a tensile test is usually initiated by debonding or microcracking at hard, brittle *inclusions* which are present in the ductile matrix (carbide particles in a steel, or oxide particles in a copper alloy). The brittle *microcrack* then initiates the formation of a *microvoid* by plastic shear of the matrix and this microvoid eventually links up with other, neighbouring microvoids to form a *dimpled*, ductile fracture surface. In a standard tensile test a ductile tensile crack will initiate and grow in *plane strain* from the *centre* of the test bar, where the plastic constraint is a maximum. As the crack approaches the edge of the specimen the stress condition changes to *plane stress*, and the plane of the fracture changes. A shear crack links the ductile tensile crack to the free surface by *shear lips* to form the characteristic ductile failure morphology (Fig. 3.9) in which the '*cup and cone*' tensile fracture reflects the plastic ductility of the matrix. The shear lips, which are formed in plane stress, become increasingly prominent as the ductility increases, leading to 'knife-edge' failures in a very ductile aluminium foil. Conversely, more brittle materials show little evidence of shear lips, even though microscopic examination of the fracture surface may still reveal *dimples*, evidence of *ductile shear* at the tip of microcracks.

The specimen dimensions are also important. A thin sheet may fail in *plane stress* by ductile *shear* (plastic thinning), while a thick bar of the same material may exhibit a ductile *tensile* failure, controlled by the growth and linkage of microvoids in *plane*

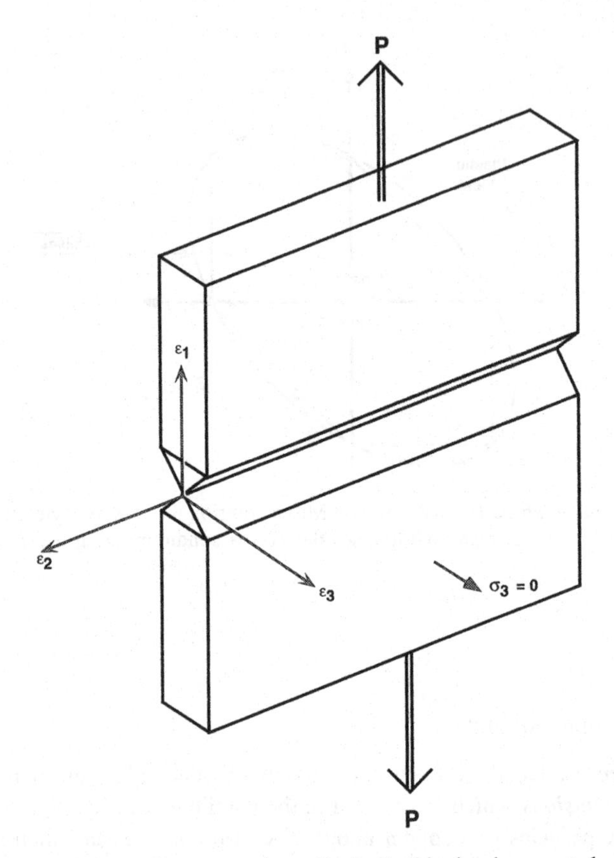

Fig. 3.7. In *plane stress* ($\sigma_3 = 0$) plastic thinning is expected.

strain. In the first case the plane of shear failure is *inclined* to the direction of loading, while in the second case the plane of tensile failure is *normal* to the loading direction.

3.2.1.3 Constraint at a Join

There is a mechanical similarity between the constraints on plastic yielding which act during the penetration of a hard punch into a ductile matrix, those that operate in the propagation of a ductile *plane strain* tensile crack and those that control the onset of tensile failure at a joint. The joining of components often involves an area of contact which may be approximated by assuming no discontinuity in material properties across the interface. The slip line field developed during yielding of this idealized 'joint' is then analogous to that developed during general yielding beneath a flat punch, as well as to that formed at the tip of a ductile crack. The three

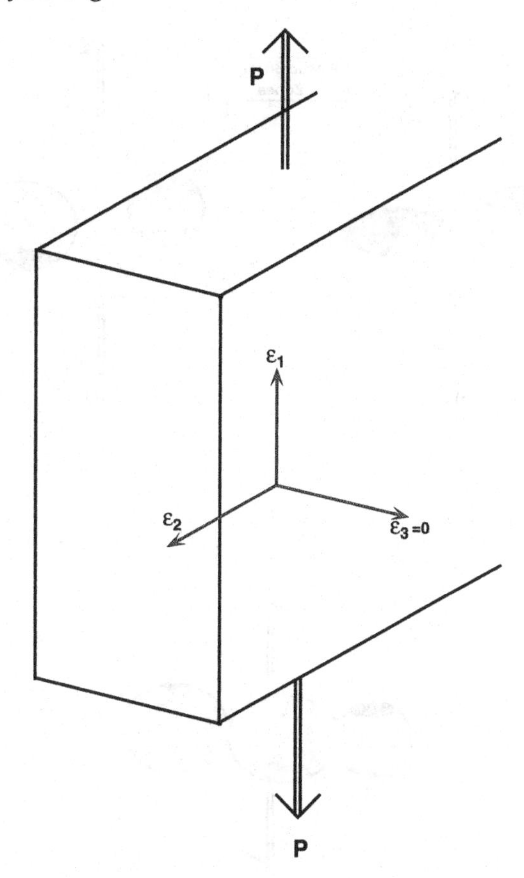

Fig. 3.8. In *plane strain* ($\varepsilon_3 = 0$) no plastic strain is allowed in the through-thickness direction.

situations are compared in Fig. 3.10. In all three cases plastic flow in the shear zone transfers material to or from the unstressed regions and relieves the applied load.

3.2.1.4 BRITTLE FAILURE

A large range of important engineering materials possess no appreciable ductility. *Concrete* structures, many adhesive joints, and a large range of thermosetting plastic components and potting compounds commonly fail in a *brittle* manner. *Glasses* and *ceramics* are completely brittle in tension over a wide temperature range, while most *cast irons* show no macroscopic tensile ductility, although they are moderately ductile in compression.

Brittle failure in crystalline materials may take place either at the grain boundaries or within the grains (*intergranular* or *transgranular* failure). In single crystals brittle

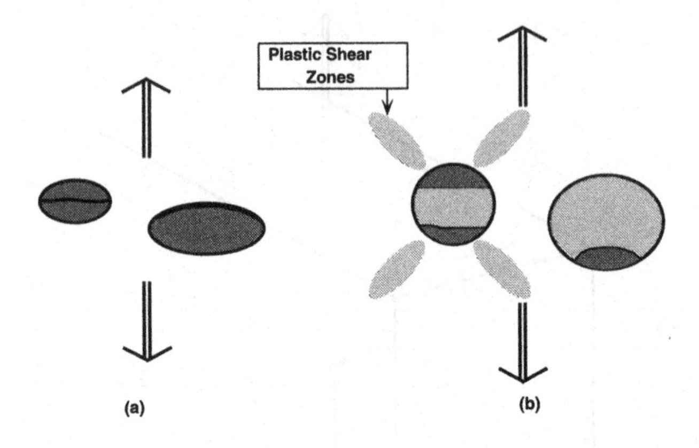

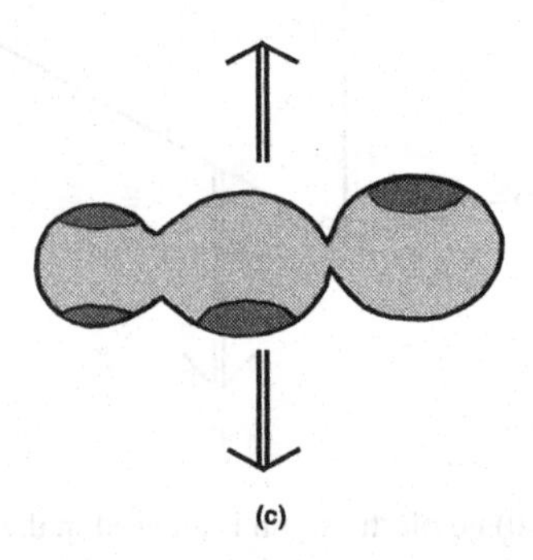

Fig. 3.9. Stages in the ductile failure of a tensile bar: (a) brittle inclusions *crack* or *debond* from the ductile matrix; (b) *microvoids* grow from the inclusion sites by plastic shear; (c) impinging microvoids form a *dimpled, ductile tensile crack*; (d) the ductile tensile crack *grows* by the nucleation, growth and incorporation of new microvoids; (e) as the tensile crack approaches the *surface* of the test bar the loading condition tends to plane stress. Final failure is in shear (plastic thinning), resulting in the formation of '*shear lips*', and the characteristic *cup and cone* tensile fracture.

failure occurs on the low energy planes of the crystal lattice and is termed *cleavage*. In many macroscopically brittle materials *mixed-mode* microscopic failures are observed—that is fracture occurs by a combination of ductile tensile failure and cleavage. In a grey cast iron the graphite flakes fail by *cleavage* on the basal plane of the graphite's hexagonal crystal lattice, but this is always accompanied by some ductile shear of the iron matrix.

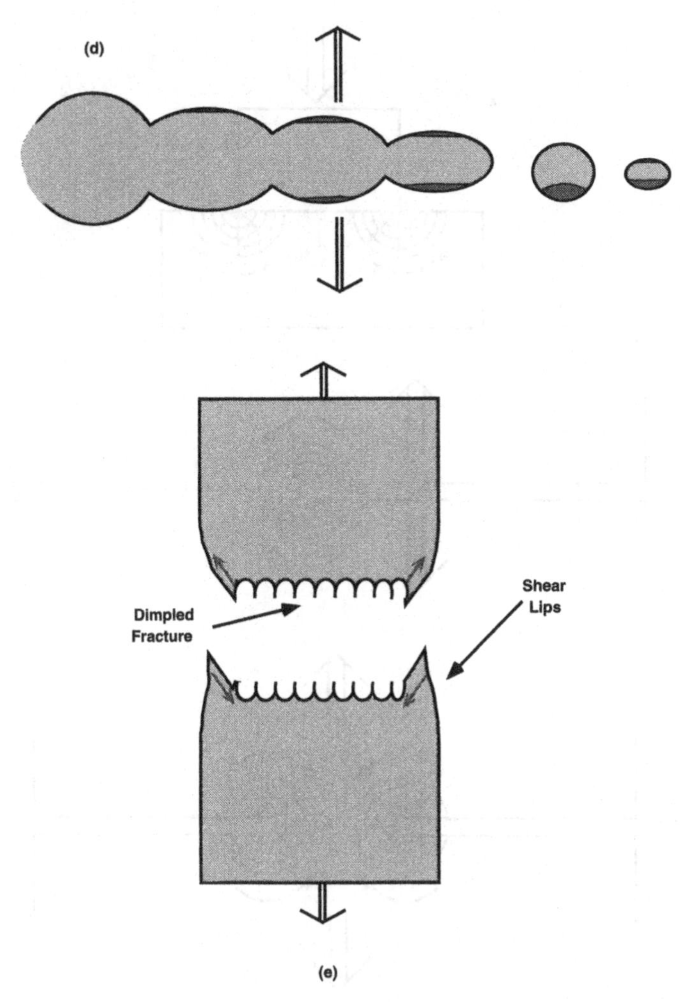

Fig. 3.9. (*continued*)

3.2.1.5 NOTCH SENSITIVITY

A material will be completely brittle if the critical tensile stress for brittle failure σ_f is less than the yield stress σ_y. In the presence of plastic constraint, such as exists at the tip of a notch, yielding may be restricted to stress levels which exceed $3\sigma_y$, so that the condition $\sigma_y < \sigma_f < 3\sigma_y$ describes a material which is *notch sensitive*. Such a material may fail *macroscopically*, in an apparently brittle manner, while the fracture surface shows tensile '*dimples*', evidence of microductility. If the critical stress for brittle failure exceeds the critical stress for *constrained* yielding, that is $\sigma_f > 3\sigma_y$,

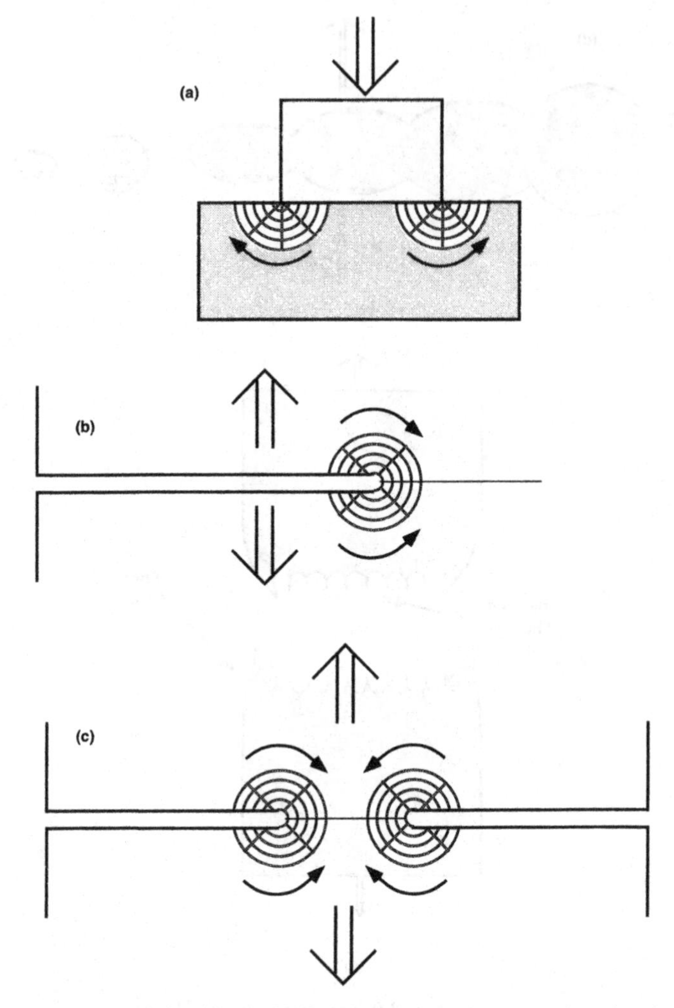

Fig. 3.10. Comparison of the 'ideal' slip line field for loading normal to the plane in three cases: (a) a *punch* of infinite stiffness; (b) the tip of a *ductile tensile crack*; (c) an ideal *plastic contact zone*.

then *general yielding* will *always* precede brittle failure, and any notch present will be *blunted* by plastic shear at the tip of the notch, and the material will be fully ductile.

3.2.1.6 Weibull Statistics

Failure of brittle components is a *stochastic* process determined by the *probability* of failure. The most successful treatment of failure probability is that described by

Weibull statistics, which gives the failure probability P_f in a volume element V as a function of the maximum tensile stress σ applied to the volume element and the integrated volume of the component V_0:

$$P_f = 1 - \exp\left[\left(-\frac{V}{V_0}\right)\left\{\frac{\sigma - \sigma_u}{\sigma_0}\right\}^m\right] \tag{3.10}$$

where the constants σ_u, σ_0 and m are considered to be material parameters; σ_u is a limiting lower stress below which failure will not occur, σ_0 is a normalizing stress and m is termed the *Weibull modulus*, and is a measure of the spread of the failure distribution (Fig. 3.11). The probability of failure at low loads is difficult to determine, since only a few samples are involved, and it is common to quote the Weibull modulus derived by assuming $\sigma_u = 0$, that is, there is always a *finite* probability of failure at *any* (positive) applied stress.

Measured values of the Weibull modulus reflect both the variability of the tensile strength of the material and the accuracy of the testing machine, as well as the mechanical tolerances in the machining and mounting of the test specimen. Values of m greater than 20 are difficult to measure, since they require an extremely accurate experimental technique. On the other hand, values of m less than 5 are not usually acceptable, since they reflect an extreme uncertainty in the engineering properties of the material. Accurate values of m for a brittle high performance

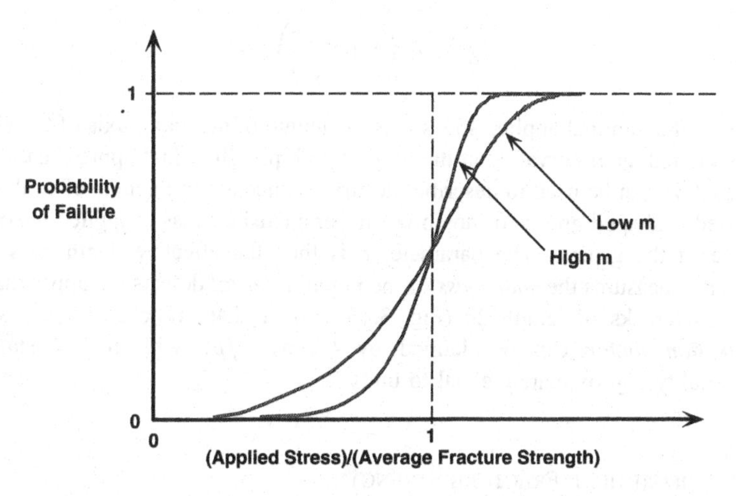

Fig. 3.11. The probability of tensile failure for brittle materials of high and low Weibull modulus, obtained by normalizing the applied stress by the average fracture strength and assuming $\sigma_u = 0$.

material are difficult to obtain. An approximate estimate can be made from the *coefficient of variance* (CV) of a set of test results: $m \sim 1.2/\mathrm{CV}$, where:

$$\mathrm{CV} = \frac{\Delta\sigma_f}{\sigma_f} \tag{3.11}$$

$$\sigma_f = \frac{1}{n}\sum \sigma_f^i \tag{3.12}$$

and

$$\Delta\sigma_f^2 = \frac{1}{(n-1)}\sum (\sigma_f - \sigma_f^i)^2 \tag{3.13}$$

In these relations σ_f^i is an individual test result, and the n test results ($n > 2$) are averaged using *Gaussian* statistics.

3.2.2 Stress Concentration

A *discontinuity* in a stressed component disturbs the stress field and there is an *increase* in the local stress at the edge of the discontinuity (Fig. 3.12). Discontinuities may take the form of *defects* at a joint between two components (inclusions, second phase particles or microcracks), or regions of *imperfect contact* or *misalignment.* When a sample is loaded normal to the plane of a defect, the *stress concentration* at the tip of an *elliptical defect* in a *two dimensional* sample generates a maximum stress given approximately by:

$$\sigma^* = \sigma_a\left(1 + 2\sqrt{\frac{c}{\rho}}\right) \tag{3.14}$$

where σ_a is the nominal applied stress, c is the length of the major axis of the ellipse and ρ is the radius of curvature at the tip of the ellipse. To a first approximation the same equation can be used to describe the stress concentration generated at the root of *any* wedge-shaped groove or notch whenever a tensile stress is applied *normal* to the plane of the groove. The parameter c is then the effective depth, while the parameter ρ measures the *sharpness* of the notch. *Internal* defects are approximated as elliptical cracks of length $2c$ (Fig. 3.13). For a sharp notch, $c \gg \rho$, a *stress concentration factor* can be defined by $k = \alpha\sqrt{(c/\rho)}$, with the constant of proportionality approximately equal to unity.

3.2.2.1 Theoretical Fracture Strength

For extremely sharp notches (such as an *atomically sharp* crack) we can assume a lower limit for ρ, the radius of curvature at the tip of the notch, of the order of the

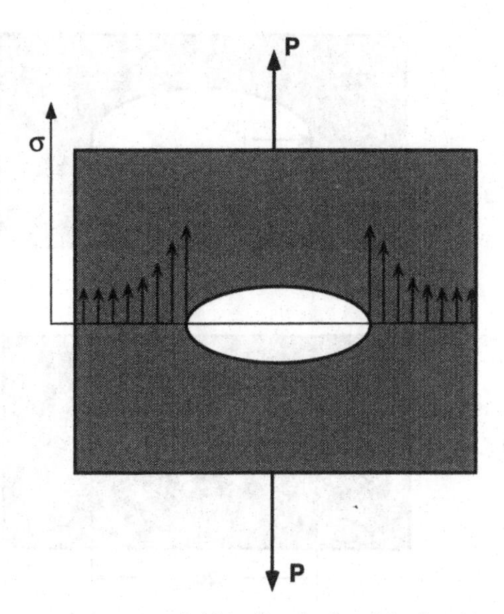

Fig. 3.12. At a *discontinuity* the stress field is disturbed and the local stress at the edge of the discontinuity is amplified.

separation of the atomic planes b, and derive a value for the *theoretical fracture strength* of a brittle solid σ_0:

$$\sigma_f \approx \alpha \cdot \sigma_0 \sqrt{c/b} \tag{3.15}$$

If a *sinusoidal* stress–displacement curve $\sigma(x)$ is assumed for the response of an ideal crystal subjected to an applied tensile stress (Fig. 3.14) then:

$$\sigma(x) = \sigma_0 \sin\left[\frac{2\pi(x-b)}{a_0}\right] \tag{3.16}$$

The *slope* of this curve at its origin (zero applied load) is determined by the *tensile modulus*: $d\sigma/dx = E/b$, where b is the equilibrium separation of the atomic planes. On the other hand, the *area* under the curve should integrate to give the *surface energy* per unit area of the two cleavage surfaces formed by the propagating crack, 2γ. The first condition yields the relation:

$$\frac{\sigma_0}{a_0} = \frac{E}{\pi b} \tag{3.17}$$

while the second leads to:

$$\sigma_0 a_0 = 4\pi\gamma \tag{3.18}$$

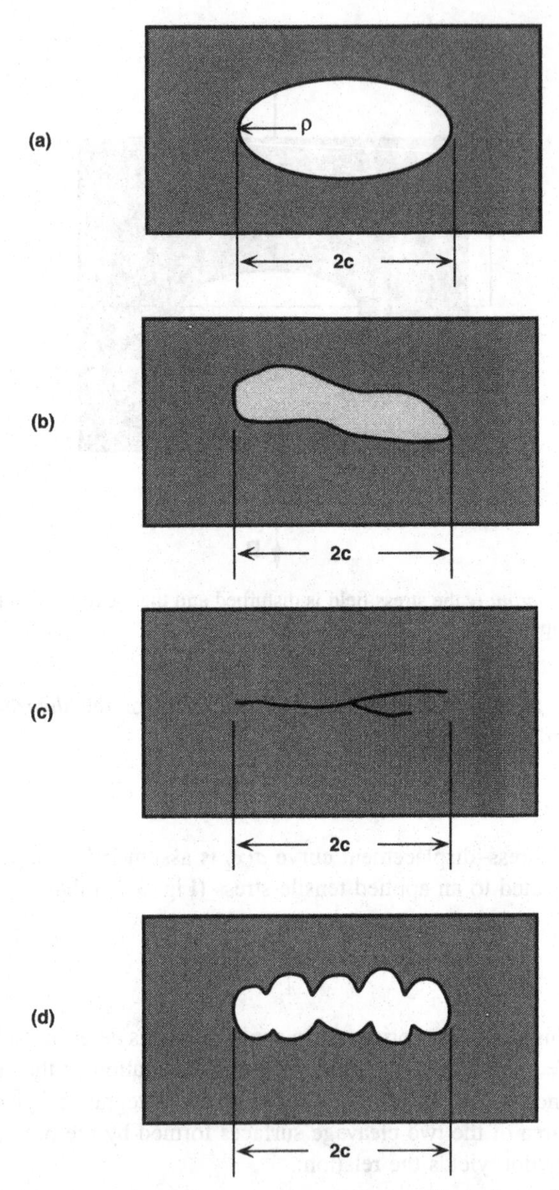

Fig. 3.13. Any internal defect can be approximated by an *elliptical crack* of length $2c$ and 'sharpness' ρ. (a) The equivalent elliptical defect. (b) An inclusion. (c) A microcrack. (d) Porosity.

The *range of the atomic forces* a_0 is unknown, so we eliminate a_0 to obtain the *theoretical fracture strength*:

$$\sigma_0 = \sqrt{\frac{4E\gamma}{b}} \tag{3.19}$$

Clearly, the *theoretical fracture strength* depends on the *elastic stiffness*, the *surface energy* and the *atomic dimensions*, and the only parameter missing is the effect of *stress concentrating defects*. Substituting for σ_0 (3.19) in the previous relation for the fracture strength of a completely brittle solid (3.15) yields the equation:

$$\sigma_f = \beta\sqrt{\frac{E\gamma}{c}} \tag{3.20}$$

where β is a geometrical constant that includes both the approximations of the model and the elastic constraint conditions (that is, *plane strain* or *plane stress*).

3.2.2.2 Crack Propagation

In the case of a macroscopically brittle but microscopically ductile fracture (the *notch sensitive* regime) plastic work is done during the growth and coalescence of the microvoids needed to propagate the ductile tensile crack (Fig. 3.9). Even in a totally brittle failure, the crack propagates under conditions which are far from thermodynamic equilibrium, so that the work of fracture γ_f must exceed the *equilibrium surface energy* of the solid. In fact $\gamma_f \gg \gamma$ for all structural materials, the extent of the inequality depending on the contribution of ductile shear and adiabatic failure to the fracture process. In brittle materials catastrophic, uncontrolled crack propagation is often observed and the crack velocity may approach (but *never*

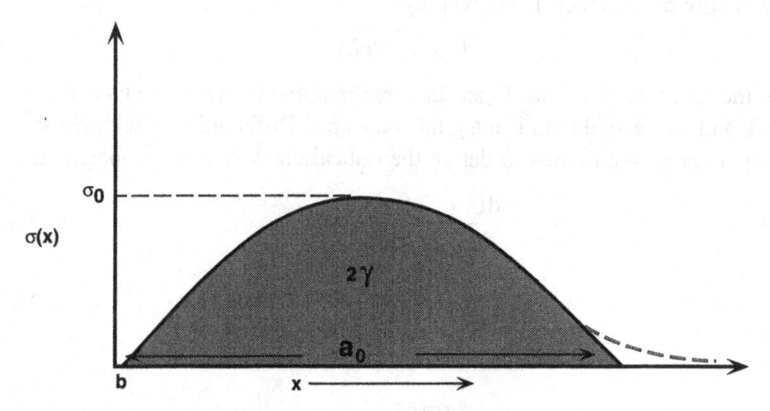

Fig. 3.14. The sinusoidal stress-displacement curve assumed in estimating the *theoretical fracture strength* of a brittle solid.

exceed) the speed of sound in the material. In practice other features of the crack propagation process normally limit the crack velocity to a small fraction of the longitudinal sound-wave velocity. In brittle materials crack *bifurcation* occurs at high stress levels, while in materials exhibiting some ductility plastic *microshear* and cavitation at the root of the propagating crack restrict the crack velocity.

3.2.3 Fracture Toughness

In general, the fracture strength is *not* a true material parameter, since it depends on the *size* of the stress-concentrating defects. We distinguish between *inherent* (and acceptable) *flaws* in the material, which limit the 'ultimate' fracture strength, and *process defects*, which are detected by suitable *non-destructive evaluation* of the components during manufacture and assembly. The design engineer possesses an additional tool for evaluating fracture probability in that branch of mechanics termed *fracture mechanics*. The foundations for fracture mechanics were laid by *Griffith* who pointed out that the fracture energy required to extend a brittle crack must be supplied by the strain energy released during crack propagation.

3.2.3.1 Griffith Theory of Fracture

We assume that strain energy U is released in a cylindrical volume of material whose radius is proportional to the crack size c:

$$U = \left(\frac{\sigma_a^2}{2E}\right)(\pi c^2 L) \tag{3.21}$$

where U is the strain energy per unit volume stored in a material of elastic modulus E under an applied stress σ_a, and the second term is the volume of the cylinder of material surrounding a straight-through crack in a sheet of thickness L. The total surface energy of the crack Γ is given by:

$$\Gamma = 2\gamma(2cL) \tag{3.22}$$

Here the first term is the work per unit area required to create the two surfaces of the crack and the second term is the total crack area. Differentiating with respect to c and then equating both terms to derive the critical crack size for propagation c^*:

$$\frac{dU}{dc} = 2\left(\frac{\sigma_a^2}{2E}\right)(\pi cl) \tag{3.23a}$$

$$\frac{d\Gamma}{dc} = 4\gamma(2L) \tag{3.23b}$$

Hence:

$$\frac{4\pi\sigma_a^2 c^*}{2E} = 8\gamma \tag{3.24}$$

Figure 3.15 plots the two opposing contributions to the internal energy of the component, *strain energy* U and *surface energy* Γ, and presents the above argument graphically. Defects smaller than c^* are stable at the applied stress level σ_a, while cracks larger than c^* will grow. This relation leads directly to the fracture strength of a defect-containing component derived from the theoretical fracture strength, $\sigma_f = \beta\sqrt{(\gamma E/c)}$. The constant β has been calculated accurately for several loading situations: either a *straight-through* or a *half-penny shaped crack* propagating in either *plane stress* or *plane strain*.

3.2.3.2 The Stress Intensity Factor

Irwin was the first to demonstrate that the *Griffith model* of crack propagation and the model based on the *stress concentration* at a crack tip were equivalent, and that an applied stress was amplified by a *stress concentration factor* at the tip of a defect. This leads directly to the concept of a *stress intensity factor* K, defined as

$$K = \frac{\sigma}{\sqrt{c\pi}} \tag{3.25}$$

and the *fracture mechanics* condition for brittle fracture in the presence of a defect becomes $K = K_c$, where the parameter K_c is termed the *fracture toughness* of the material. In terms of the Griffith derivation of the fracture strength, the fracture

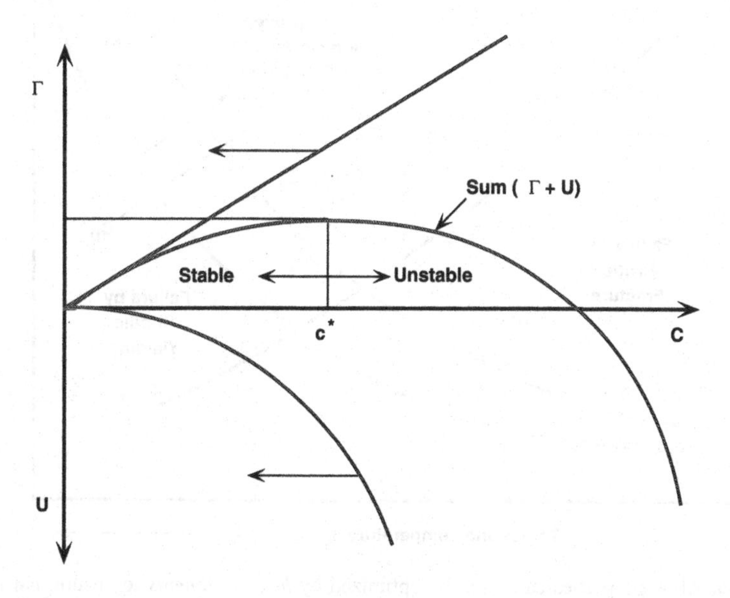

Fig. 3.15. *Strain energy released* U and *surface energy of fracture* Γ as a function of crack size c estimated from a simplified Griffith model. Cracks *smaller* than c^* are stable under the applied stress; those *larger* than c^* will grow.

toughness is given by the relation $K_c = \alpha\sqrt{(\gamma E)}$, which should be independent of the size of any defects in the material, and is therefore a *material parameter* which can be measured using a number of standard engineering tests. Once the fracture toughness K_c is known, the *critical defect size* c^*, which will lead to uncontrolled crack growth and catastrophic brittle failure, can be calculated for any specific level of the applied stress σ_a: $\pi c^* = (K_c/\sigma_a)^2$.

Many engineering alloys are susceptible to *heat treatment*, and the mechanical properties of a component can often be optimized by varying the heat treatment schedule. Thus carbon steels can be *tempered* at increasingly higher temperatures to improve the fracture toughness, but at the cost of reducing the yield strength. Selecting the *tempering temperature* to optimize the resistance to both general yielding and catastrophic fracture can be done for any allowable size of defect (Fig. 3.16). Assuming that c_0 is the minimum size of a processing defect which can be detected by *non-destructive evaluation* (NDE), then the optimum heat treatment will be that which results in ductile yielding and brittle fracture at approximately the same stress level, that is, when the condition: $\sigma_y = K_c/\sqrt{\pi c_0}$ is satisfied. Tempering

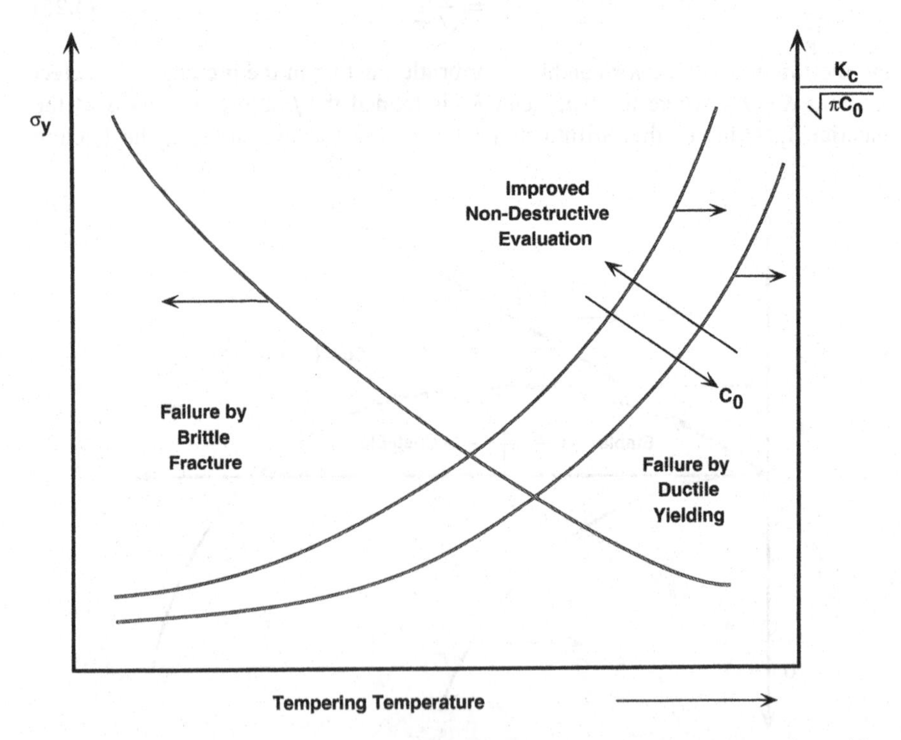

Fig. 3.16. Alloy properties can often be optimized by *heat treatments* to ensure that neither brittle fracture nor generalized plastic yielding dominate the failure process. The optimum treatment is dependent on c_0, the minimum size of *processing defect* which can be *detected*, and hence on the methods used for non-destructive evaluation.

at *lower* temperatures to obtain a higher yield stress will only result in *brittle failure* at a *lower* stress level, while tempering at *higher* temperatures to develop a higher fracture toughness will result in *plastic yielding*, again at a *lower* stress level. On the other hand, improving the resolution for NDE, and hence *reducing* c_0, will permit *higher* stress levels to be reached, since the maximum size of an undetected processing defect is then reduced.

The fracture toughness of an engineering material is dependent on the stress state during crack propagation. There are three limiting cases which are referred to as *modes I, II* and *III*, corresponding to the stress intensity factors K_I, K_{II} and K_{III} (Fig. 3.17). The mode I stress state corresponds to a tensile crack propagating in plane strain, and yields the K_{Ic} *fracture toughness*, which is a material parameter independent of the specimen dimensions (providing the initial crack length is sufficient). In mode II crack propagation is parallel to the direction of shear displacement, while in mode III the crack propagates in *transverse shear*. There is no evidence that failure can occur in pure mode II or III, since fracture requires crack opening, and hence some *tensile* component of the applied load, but *mixed mode* fracture is common. Aluminium foil tears by a ductile shear failure in *plane stress* which is very nearly pure mode III (Fig. 3.18) but involves shear on an inclined plane. In mode I fracture plastic shear on the planes of maximum shear stress lead to plastic *blunting* of the crack tip, while in mode III *transverse* plastic shear stress leads to crack tip sharpening. The commonly observed cup and cone tensile fracture of a high strength alloy (Section 3.2.1) is an example of a *mixed-mode* failure which *initiates* in mode I (the *tensile* crack at the centre of the cross-section) but *terminates* in mode II (the *shear lips* forming the edge of the fracture cone). As in the case of conventional stress analysis, the components of the stress intensity factor are additive, so that the total stress intensity factor can be expressed as: $K = K_{\mathrm{I}} + K_{\mathrm{II}} + K_{\mathrm{III}}$.

3.2.4 Failure Modes

If mismatch of the elastic moduli of the components can be ignored, then the *failure modes* which may occur at a joint can be treated using the concepts developed above. That is, the joint can be modelled as a *notch* and the geometry of loading can be modelled as combinations of *modes I, II* and *III*.

Diffusion-bonded joints between components which have been manufactured from the *same* material may show only minor, localized variations in joint composition. Such an interface may be free of defects and show no detectable changes in properties. An example is the co-sintering of green-bonded silicon nitride components. The polymer-bonded, *green* (unsintered) components are first bonded together by holding them in contact and heating above the softening point of the polymer binder. The binder is then removed by further heating (burn-out) and the weakly-bonded components are then sintered. The properties and microstructure in

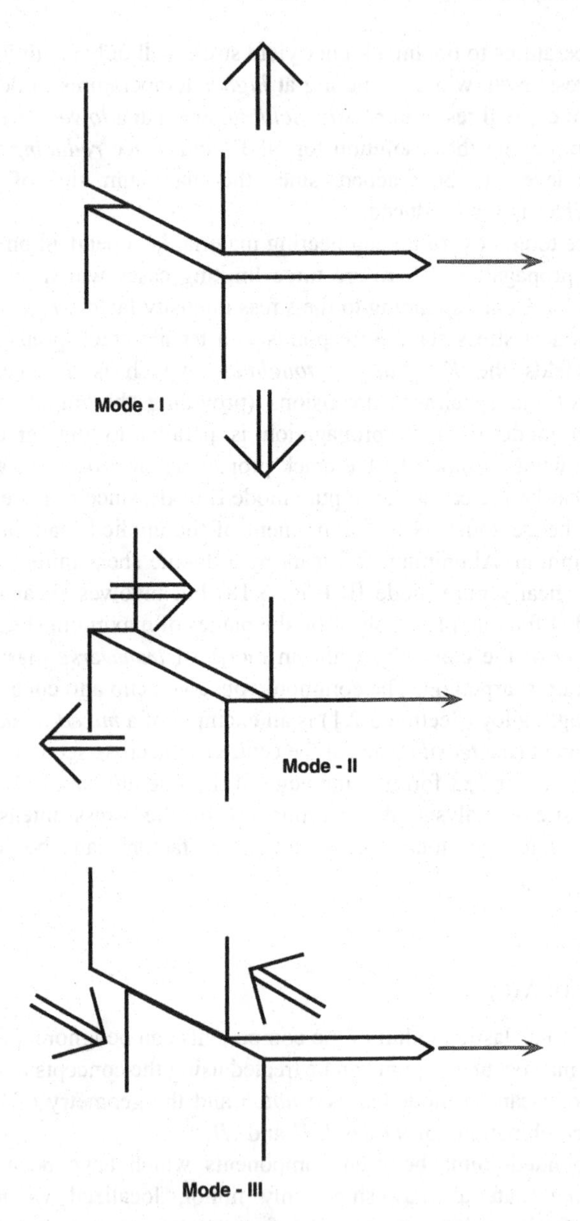

Fig. 3.17. The three *loading modes* for crack propagation result in three independent *stress intensity factors* K_{I}, K_{II} and K_{III}, and the corresponding *fracture toughness* K_{Ic}, K_{IIc} and K_{IIIc}.

the region of the final, fully-sintered interface are almost indistinguishable from those of the bulk material.

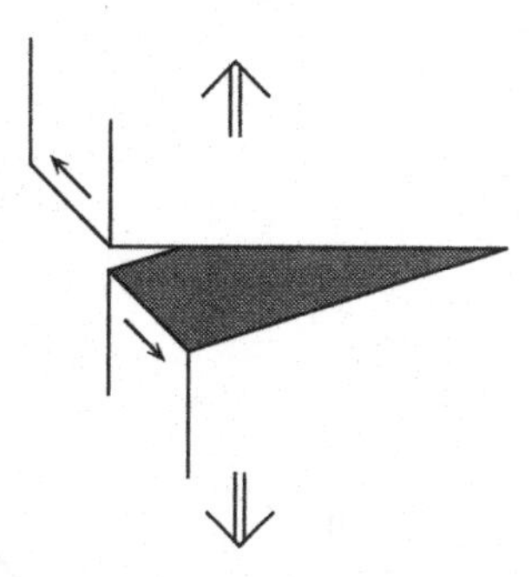

Fig. 3.18. Shear on an inclined plane in mode II leads to *plastic thinning* and subsequent ductile tearing.

However, for most joints there is a more or less sharp discontinuity in the microstructure and physical properties at the interface. The standard bulk concepts of *plastic yielding* and *fracture mechanics* must then be treated with caution. It is useful to distinguish two limiting cases. In the first case the properties of the bulk material on both sides of the joint are the same, but the properties in the immediate region of the joint are different. Examples are the ground glass joint illustrated in Fig. 1.3, as well as many glued joints. *Welded joints* constitute a major class of join for which the modification of the properties near the interface extends well into the bulk material, leading to an interface region termed the *heat-affected zone* (HAZ). Some caution is necessary when the bulk properties are *anisotropic* (directional): wooden *laminates* have directional properties which can be manipulated by varying the orientation of the layers. The properties of the laminate layers are identical but anisotropic, and the abrupt change in orientation across the interface between the layers introduces a sharp discontinuity in the properties.

The second class of joints is those between materials whose structure and properties are *dissimilar*. Load transfer across the join is then a function of the mismatch in the material properties of the two components being joined. The commonest sources of mismatch are in the *elastic modulus* and the *thermal expansion coefficient*. A tensile load *normal* to an interface separating regions of high and low modulus generates a *shear stress* parallel to the interface (Fig. 3.19a), and a tensile crack no longer propagates along the interface in pure mode I, but will have an element of mode II associated with it (Fig. 3.19b).

Thermal expansion mismatch will result in *residual stresses* in the components if a bond is formed at a high temperature and then put into service at a lower temperature (below the temperature at which *stress relaxation* is possible). In many cases *thermal stresses* may be of the same order of magnitude as the operating stresses of the system. This is especially the case in *ceramic* assemblies, where the combination of high *sintering temperatures* and high *elastic modulus* can result in exceptionally large stresses. Thermal stresses can sometimes be put to good use, as in the mounting of a heated iron hoop on a wooden carriage wheel. The iron hoop shrinks onto the wheel, placing the wood in compression and scorching away any

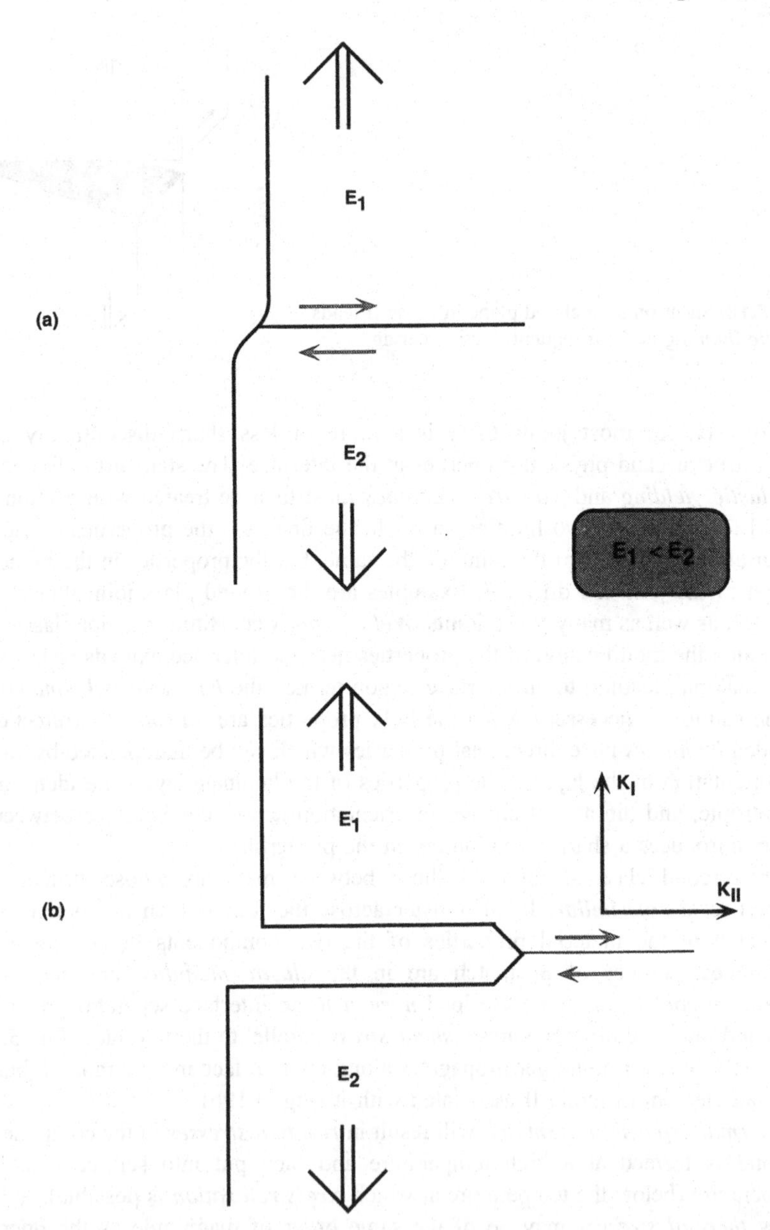

Fig. 3.19. In the presence of an elastic modulus mismatch, normal loading across the interface produces a shear stress *parallel* to the interface (a). Crack propagation *along* the interface then occurs under mixed-mode loading conditions (b).

high points. The tightly-fitting iron band then presents a wear-resistant surface to the road.

In most engineering designs a joint is an area of *weakness*, either as a result of a loss of *ductility* (and consequent *embrittlement*), or as a result of *stress concentration*. Various strategies may be used to protect against failure at a joint. Some *design strategies* are illustrated in Fig. 3.20. The joint may be *spliced*, to increase the area of the joint and hence reduce average stress at the interface. The joint may be *reinforced* by increasing the cross-section in the joint region. External *compressive* stresses may be introduced through suitable wedges.

Failures associated with a joint may be complex. While failure is frequently *initiated* in the plane of the join, mismatch of the *elastic compliance* across the interface may result in *deflection* of the fracture plane (usually into the stiffer of the two components). In *welded* joints the HAZ may be a plane of weakness, leading to failure, not in the weld itself, but rather in one of the HAZ regions either side of the weld. *Brazing* alloys generally have sufficient ductility to prevent brittle failure, and the same is true of many *thermoplastic* adhesives. *Ductile failure* in the plane of an adhesive may also be complex, depending on whether the voids generated in the course of a (predominantly) mode I failure originate at *filler particles*, in the adhesive, or at one of the two interfaces between the adhesive and the bonded components. If the *microvoids* are closely spaced compared to the thickness of the adhesive layer, then ductile crack propagation will resemble that in a constrained, mode I tensile failure, but if the microvoids are well separated it may be difficult to characterize the fracture mode. Some of these features of the failure processes at a joint are illustrated in Fig. 3.21.

Good *adhesive* joints commonly fail by *decohesion* involving *cavity nucleation* and *growth* in the ductile, viscoelastic adhesive, but this requires strong *adhesion*, which is determined by the *chemical interaction* between the adhesive and the surfaces of the components. Rigorous *cleaning* and preparation of the surface is essential to eliminate *contamination* during the bonding process.

3.3 TEST METHODS

Joints are generally designed to support a load, and must be tested to evaluate their *load-supporting capabilities*. However, it is also important to evaluate, not the joint, but rather a specific *joining capability*. Hence, many tests are made in order to evaluate the competence of the *welder* rather than the weld, while many adhesive joints are prepared to test the properties of the *adhesives* rather than the join (especially their *shelf life* or environmental sensitivity). The following section looks at some of the common methods of evaluating the mechanical performance of a joint.

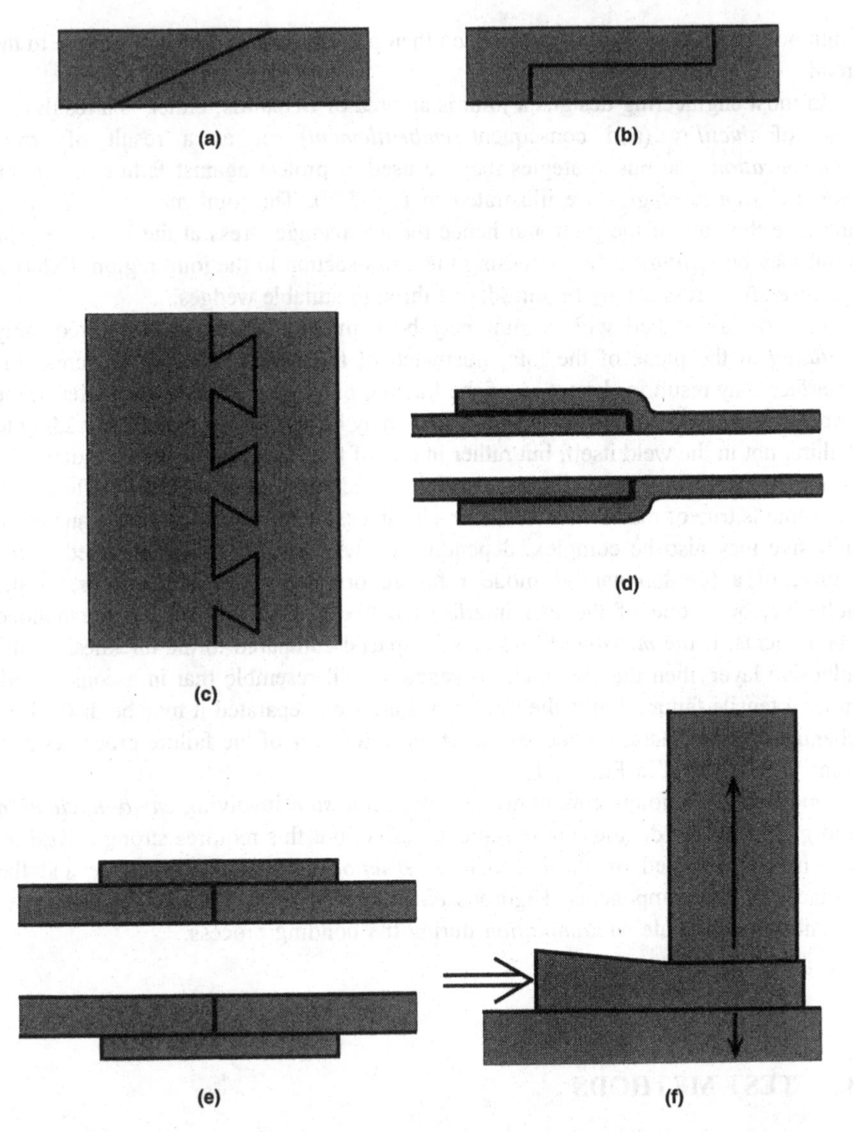

Fig. 3.20. Some design strategies for improving the integrity of a joint: (a) splicing; (b) lapping; (c) tounge and groove; (d) male and female; (e) sleeve; (f) wedge or tapered pin.

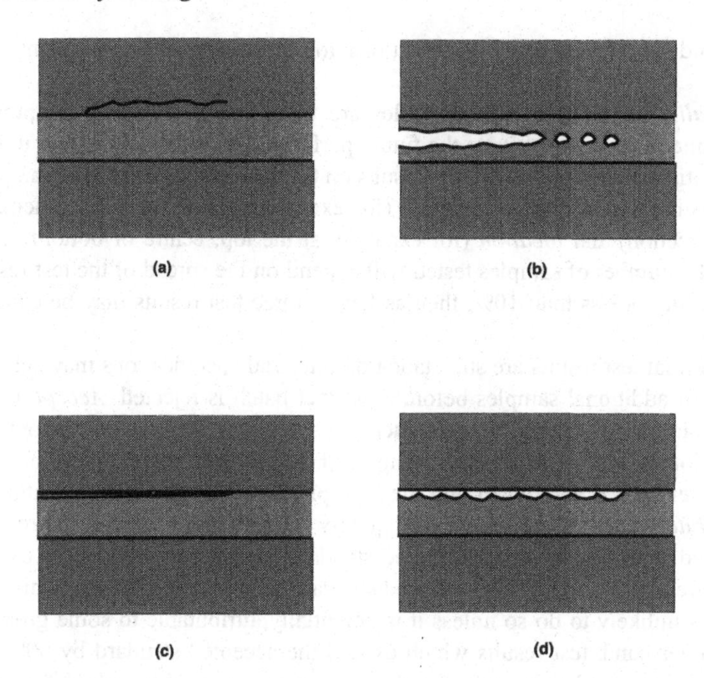

Fig. 3.21. Some failure modes at a join: (a) brittle failure in the elastically stiffer component parallel to the join; (b) ductile failure (*decohesion*) within the bonded region; (c) *loss of adhesion* at the interface between an adhesive and one component; (d) *ductile* failure of the adhesive bond at one interface.

3.3.1 Test Geometry

The design of a mechanical test method for bulk materials must consider the *sampling procedure*. This includes both the number of tests required to ensure a statistically reliable result, as well as the position of the sample in the component and its orientation with respect to the coordinate axes of the production process (the direction of solidification in a casting or the rolling direction in plastically formed sheet), or the loading in service (the axes of the principal stresses). Four types of mechanical tests are important, classified according to their objective:

1. *Process control samples.* Separately cast test coupons are prepared during the casting of components in a foundry. Welded tab specimens are specified in order to check the reliability and performance of a welder. The results of mechanical tests on this type of specimen are not a measure of the mechanical performance of the finished *product*. Rather, they are a check on the production *process*. Compliance with a *production standard* is their objective, and the relation between the test

results and any future loading conditions for the assembled component is very indirect.

2. *Quality control samples*. Samples are taken from an actual component or product line in order to evaluate the future performance of the component in service from a statistical analysis of the test results on the selected samples. The samples are chosen to have a specific *orientation* (for example, parallel or perpendicular to a rolling direction) and *location* (for example, at the top, centre or bottom of a cast ingot). The number of samples tested will depend on the spread of the test results. If the variability is less than 10%, then as few as three test results may be considered sufficient.

If the initial test results are sub-standard, industrial specifications may call for the retesting of additional samples before a product batch is rejected. *Acceptance tests* for raw materials or intermediate products are a frequent form of quality control test. It is obviously better to reject a component at an early stage in the production process, before it is assembled into the final product. No significance is attached to the *absolute values* of the results of a quality control test. In effect, the results are normalized with respect to an accepted standard, which they will either exceed or fail. Failure by, say, 10% of the required standard may justify retesting, while failure by 50% is unlikely to do so unless it is potentially attributable to some gross error. On the other hand, test results which exceed the accepted standard by 50% would probably be regarded as a symptom of *over-design*, and lead to rethinking of either the material requirements or the design specifications.

3. *Testing for design and development.* There are many situations in which it is necessary to predict, as accurately as possible, the way in which a component will perform in service. This is true of *design data*, which are to be used to model the engineering system. But it is also true of the *post-mortem* assessment of engineering failures. It is also true of *prototype* development. Simulative testing of a prototype design is often employed to replicate and exacerbate the performance requirements in order to pin-point design weaknesses. For these purposes, the data must reflect accurately the *actual properties* of the material (yield stress, fatigue limit, stress rupture life and so on). That is, *absolute* values of the properties are required, and relative values are insufficient.

4. *Proof testing*. In the event that the capacity of a component to fulfil its engineering function remains in doubt, it is common to resort to *proof testing*. In a proof test the *component* is subjected to a test regime which simulates the actual service conditions, but at some agreed values of the test parameters which exceed those expected in service. Thus testing may be performed at a temperature 50°C *above* the maximum expected service temperature, or at an *overload* 20% above the maximum service load. Proof testing is performed in the expectation that a component or assembly which survives the proof test is unlikely to fail in service.

Proof testing eliminates the weaker components from a product population, but it may also damage (and hence weaken) some of those components which *pass* the proof test. It follows that there is no guarantee that a proof-tested component will not fail in service; the most that can be said is that the *probability* of failure has been greatly *reduced* by proof testing.

3.3.1.1 TEST CONFIGURATIONS

A *uniaxial load* can be applied by a standard testing machine to create tensile, compressive or shear stresses (Fig. 3.22). *Combined* modes of testing are also common, for example in a three- or four-point *bend test*, in which the stresses vary with position in the sample (Fig. 3.23). The applied load may also be varied during the test, as in a *mechanical fatigue test*, or held constant at elevated temperatures while the displacement is monitored, as in a *creep test*. If the specimen is loaded to a given *strain* and the load is then monitored as a function of time at this strain, then the process of *load relaxation* can be followed. Tests performed in a controlled environment are used to investigate susceptibility to failure by *stress corrosion* (crack propagation at constant stress in a chemically-active environment), *corrosion fatigue* (environmentally exacerbated failure by mechanical fatigue) or *radiation damage* (degradation of the mechanical properties associated with irradiation in a nuclear reactor or some other radiation source).

Impact loading of the specimen is the characteristic feature of the standard Charpy test, in which a notched bend sample is impacted at low velocity by a swinging hammer. The Charpy and other, related impact tests were developed to evaluate *notch sensitivity* to brittle failure. The energy absorbed in a Charpy test has *no* design value (since it clearly depends on test geometry), but it *is* a clear indication of a possible loss of ductility in the presence of plastic constraint.

3.3.1.2 WELD TESTING

Tensile specimens are normally cut from welded metal alloy sheet with the line of the weld normal to the loading axis. A satisfactory test will be one in which tensile failure occurs neither in the weld metal, nor in the heat-affected zone (HAZ), but rather in the cross section of the bulk metal (Fig. 3.24). In such a case it is commonly stated that the tensile 'strength' of the welded joint exceeds that of the bulk metal. However, when a soldered joint, which has the same basic geometry as a weld, fails in the region of the solder, it does so in a region of *maximum plastic constraint*, and it is incorrect to regard the *nominal* failure stress (the failure load divided by the cross sectional area of the joint) as a measure of the strength of the solder. The measured strength is a function of the thickness of the soldered joint and the surface finish of the components (Fig. 3.25), increasing as the thickness of solder decreases, with the surface roughness limiting the minimum thickness of solder.

The *thermal history* of a welded joint has a large influence on its ductility. An effective quality control test is used to check for loss of ductility. The welded joint is bent about an axis parallel to the weld line on a mandrel of standard radius. If a U-

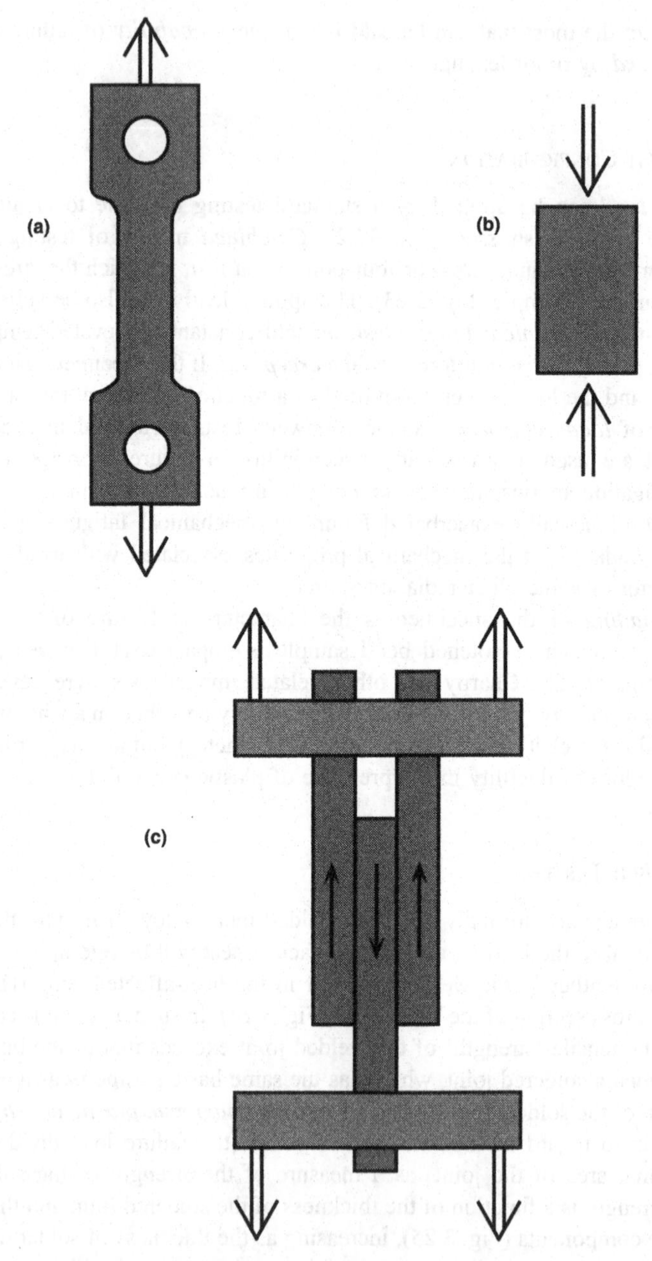

Fig. 3.22. A uniaxial loading machine can be used to test in: (a) *tension*; (b) *compression*; (c) *shear*.

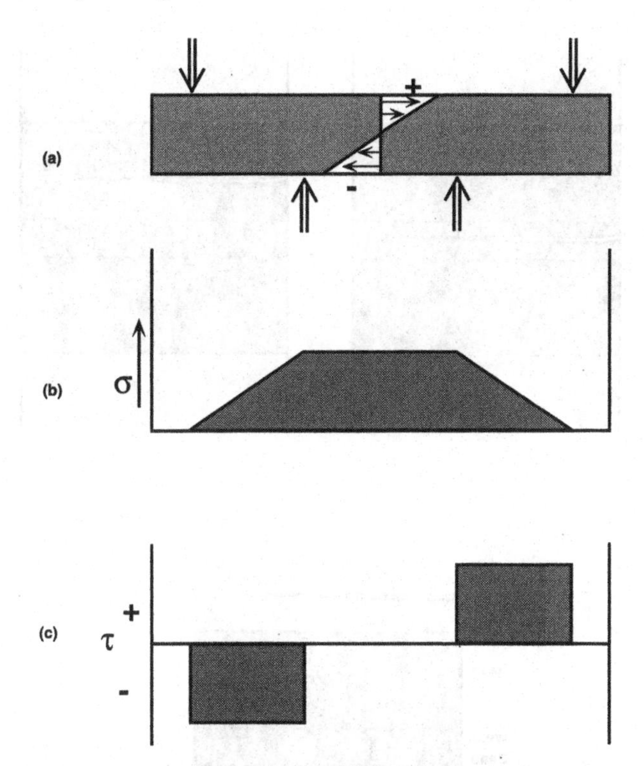

Fig. 3.23. A *four-point bend* test generates a controlled stress pattern. (a) In the *central region* the stress *parallel* to the specimen axis is a maximum at the surface and varies linearly through the thickness from *tensile* to *compressive*, passing through a *neutral axis* at the mid-line. (b) The *axial stress* at the surface is *constant* in the central region but *falls linearly* between the top and bottom loading points. (c) A *constant shear stress* is generated in the through-thickness direction between the top and bottom loading points, but the shear stress is zero elsewhere.

bend can be produced without any sign of visible cracking in either the weld bead or the HAZ, then the weld is deemed satisfactory (Fig. 3.26).

3.3.1.3 MIXED-MODE TESTING

Welded lap-joints are tested in *shear*, although offset of the tabs either side of the weld introduces a considerable component of normal stress. The test conditions must be standardized and the results have little design significance. Seam welds are a typical example of the lap-joint configuration, but spot-welds are also tested in this geometry. A satisfactory test result is one in which tearing of the welded sheet occurs outside the spot-weld zone, that is, holes are torn in the welded sheet.

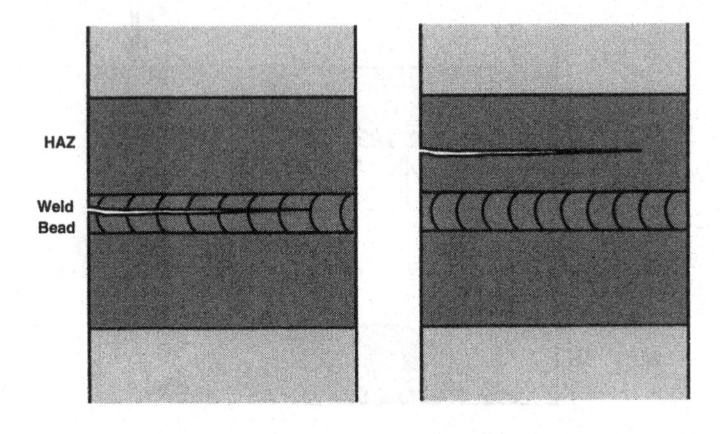

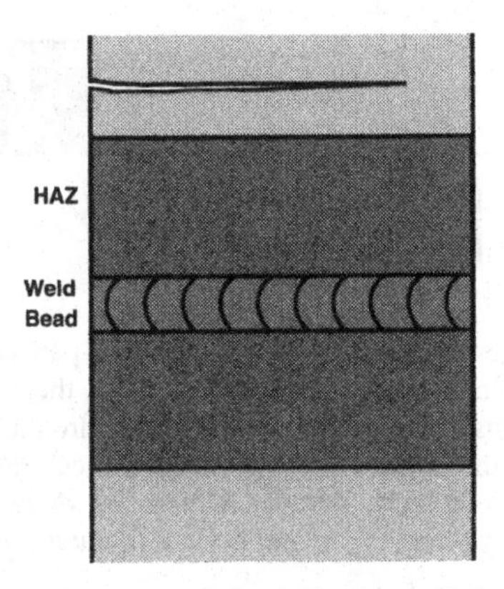

Fig. 3.24. A tensile weld specimen may fail within the weld bead, in the heat-affected zone (HAZ) or in the bulk metal.

Peel tests are not dissimilar in configuration and are used to estimate the *adhesive strength* of a glued joint. They are best supplemented by a double-lapped shear test (Fig. 3.22c) in which the applied stress approximates pure shear and the test configuration corresponds to a *mode II* fracture toughness test. There is a continuous spectrum of test configurations, running from *tensile* tests on a glued joint (analogous to a *mode I* fracture toughness test), to a *peel* test and a *single-lapped shear* test (both *mixed-mode* tests, but with different mode mixities), and finally to the *double-lapped shear* test. These options are illustrated in Fig. 3.27.

Where tests are performed on joints between *unlike* materials, *mixed-mode* failure is to be expected at the joint, although the loading angle can be varied with respect to

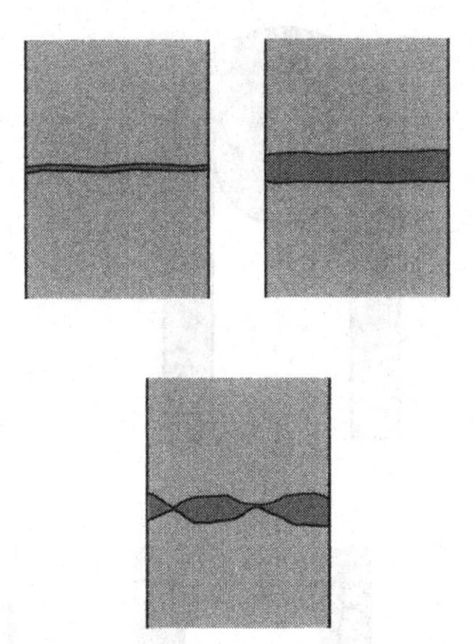

Fig. 3.25. The strength of a soldered joint depends on the *thickness* of the solder layer and the *surface finish* of the components, and not only on the tensile strength of the solder.

the direction of failure propagation, in order to compensate for a K_{II} component of the stress intensity factor at the advancing crack tip. Theoretical and experimental work along these lines has been reported by Evans and his colleagues, but the complexity of the test system is not justified in normal engineering practice.

3.3.2 Coating Adhesion & Strength

The adhesive strength of *coatings* is a special area of interest. Several methods for evaluating the strength of the bond between a substrate and a coating have been reported. Among the more exotic, *piezoelectric excitation* has been used to generate compressive acoustic waves in the bulk sample, which are reflected from the coating interface as a tensile wave. At a sufficiently high input energy, the coating is *spalled* away from the surface. Of more practical importance are a number of *qualitative* tests, in which the substrate is bent about a mandrel and the critical surface tensile strain for the onset of cracking and flaking of the coating is determined. Such tests are sensitive to the position of the *neutral axis* of bending with respect to the coating interface, and in analysing test results it is usual to assume that the coating is *thin* with respect to the distance from the neutral axis and experiences a uniform tensile stress.

By testing the substrate in *tension* and monitoring the separation of the cracks formed in the coating as a function of the tensile strain in the substrate, it is possible to obtain a more *quantitative* estimate of both the *tensile strength* of the coating and

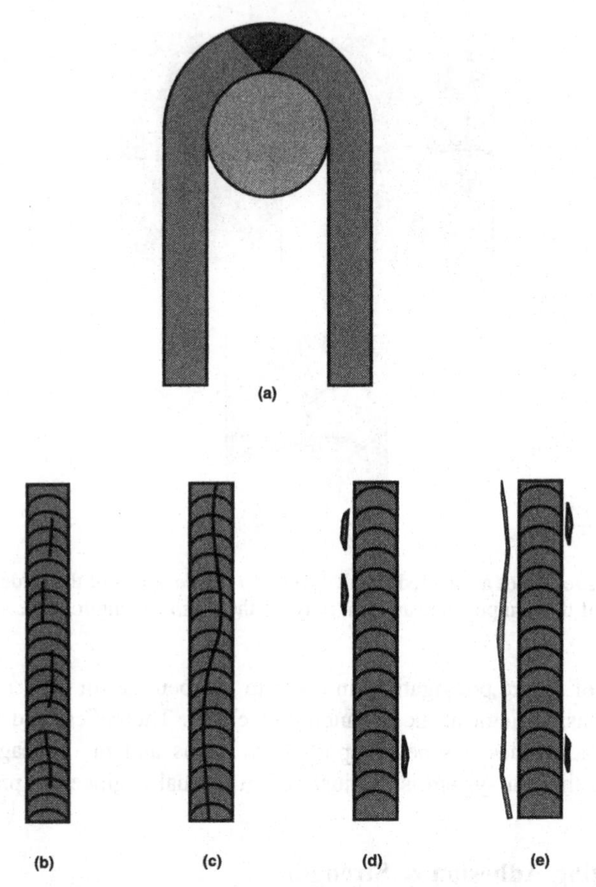

Fig. 3.26. (a) A welded strip is bent about an anvil of standard radius. (b) Incipient cracking in the weld bead. (c) Failure of the weld bead. (d) Incipient cracking in the HAZ. (e) Failure in the HAZ.

the *adhesive strength* at the interface. Failure occurs in two stages, the first being tensile failure of the coating. At this stage *tensile* cracks propagate through the coating thickness *normal* to the applied stress, with a crack-opening displacement which reflects the through-thickness elastic relaxation. *Shear* failure at the interface between the coating and the substrate then marks the onset of *adhesive* failure (Fig. 3.28). As the (predominantly mode II) shear failure propagates along the interface, the coating then peels and flakes off.

3.3.2.1 Bond Strength in Composite Materials

Similar failure processes to those observed at joints are also found in *composite materials*. High strength reinforcing fibres deflect cracks in a brittle matrix, which

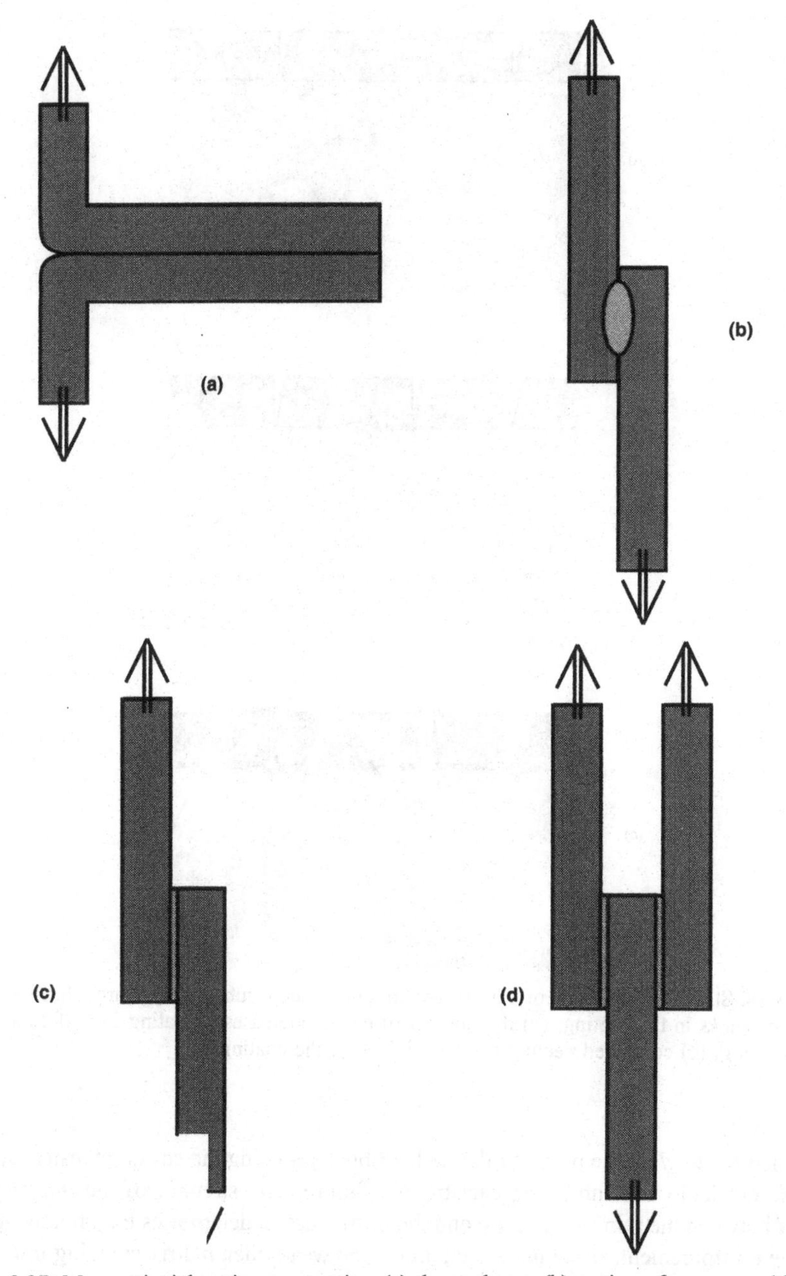

Fig. 3.27. More uniaxial testing geometries: (a) the peel test; (b) testing of a seam weld; (c) testing a glued, single lap-joint; (d) testing a double lap-joint.

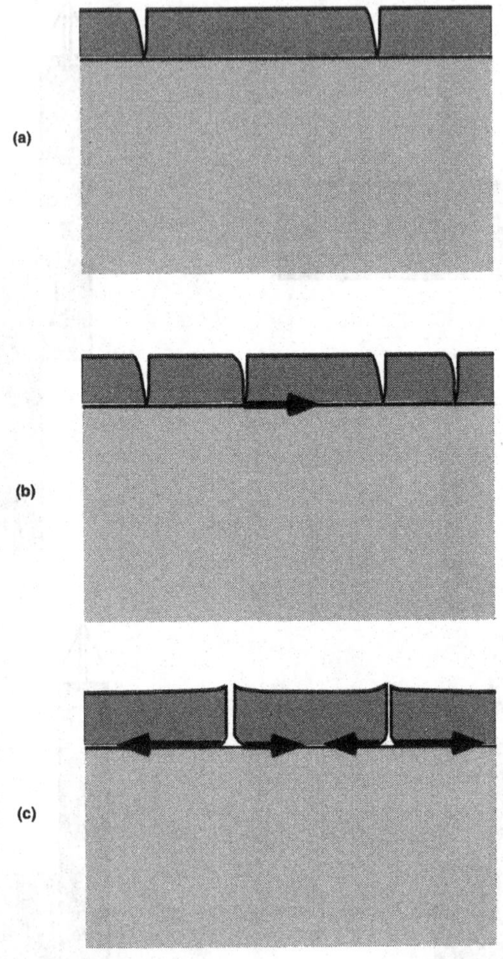

Fig. 3.28. Stages in the failure of a brittle coating on a ductile substrate: (a) through-thickness *tensile cracks* in the coating; (b) the number of cracks increases, initiating loss of adhesion and *peeling*; (c) continued peeling leads to *flaking* of the coating.

are then *bridged* by the partially-debonded fibres, reducing the stress intensity factor at the crack tip and inhibiting catastrophic failure. The strength of the *interfacial bond* between the reinforcing fibre and the brittle matrix determines the effectiveness of the reinforcement. If the interface bond is too weak, then matrix cracking initiates at very low loads and only the fibres carry the load. If the interface bond is too strong, then the matrix crack generates sufficient stress concentration to fracture the fibres before debonding can occur, and the material fails by catastrophic brittle fracture. In the intermediate range the fibres *partially* debond and *bridge* the matrix cracks, increasing the effective toughness of the composite.

A major tool for evaluating the mechanical design of a joint is *finite element analysis*, in which the joint is modelled by a *mesh* of volume elements. Self-consistent stress analysis, using iterative computer codes, is used to derive the *stress distribution* in and around the joint. *Computer modelling* of an engineering component requires an accurate analysis of the operating parameters and boundary conditions for the system, as well as a knowledge of the appropriate material parameters. However, *no* model can take complete account of the detailed structure and processes which occur in real joints. Engineering skill, knowledge and intuition are the only reliable guides to the design, evaluation and application of mechanical tests needed to ensure the performance of a joint in service.

Summary

When two solid bodies are brought together initial contact is established at only three points. The total, integrated contact area increases with the normal load transferred across the contact surface, accompanied by an increase in the number of contact points. While the initial material response is elastic, the stress concentrations developed at the contact points eventually lead to plastic flow in the contact areas. At this stage the increase in contact area is linear in load, but the true contact area will only approach the nominal area of the contact surface when the average compressive stress due to the applied load approaches the yield stress of one of the contacting bodies.

Rolling and sliding contacts involve relative movement across the contact surface and this leads to wear, which depends on the applied load, the surface roughness and the relative velocity of the contacting bodies, as well as on their elastic response and their hardness.

It follows that both the static and dynamic response at a contact surface depend on the mechanical properties of the bulk materials. The load response at the contact points is triaxial and yield in the contact areas involves triaxial yielding, similar to the loading that occurs in a hardness test or beneath a flat punch. The condition of loading at the contact surface is then one of constrained plastic flow. If yielding cannot occur, as in the case of very hard or brittle materials, then microcracking and fracture will lead to local crushing at the contact points.

A characteristic of many joints is their notch sensitivity: semi-brittle failure associated with the combination of the increase in yield strength associated with plastic constraint and the stress concentration associated with the presence of a notch. The failure stress is not a deterministic property, whose value is uniquely determined for a specific combination of material and testing parameters, but rather a stochastic property statistically distributed, usually according to Weibull statistics, about a mean value. Crack propagation, on the other hand, can generally be described using fracture mechanics, and the fracture toughness of a material is a good indication of its resistance to crack propagation, as well as of the sensitivity of the material to the presence of inclusions and other defects.

Fracture toughness is a mechanical property of an engineering material only for the case of pure tensile, mode I, loading: the K_{Ic} fracture toughness. At a joint the loading is typically mixed mode and it is not easy to define the crack propagation resistance at the joint interface. This is true of a continuous joint, such as that formed by a weld, but is especially the case for an adhesive joint, which is commonly loaded in shear, or a bolted or rivetted joint, which is discontinuous. Joints between dissimilar materials pose special problems associated with mismatch of the mechanical and thermal properties, especially the elastic constants and the thermal expansion coefficients.

The integrity of a joint can only be guaranteed if adequate test methods are available to simulate the operating conditions. The test objectives vary from simple process and quality control to carefully designed tests with specific engineering development objectives. Proof testing is a reliable way of evaluating the performance of a joint between actual components and eliminates the weaker members of the product population.

While tensile tests are used wherever possible, shear tests are necessary to evaluate the properties of either welded or glued lap-joints, and in practice most tests involve some degree of mixed-mode loading. Coatings present a particular problem, and standard tests have been developed to ensure that the coating adheres adequately to the substrate. Similar problems arise in composite materials, in which the bonding between the reinforcing fibres or layers must ensure load transfer to the matrix, but must not be so strong as to promote failure of the matrix in the event of failure of a fibre. The mechanical principles which determine the adhesion of coatings and and the performance of composites are no different from those that control the strength of a joint between two bulk components.

Further Reading

1. A. Higdon, E. H. Ohlsen, W. B. Stiles, J. A. Weese and W. F. Riley, *Mechanics of Materials*, John Wiley & Sons, New York, 1976.
2. S. P. Timoshenko and J. N. Goodier, *Theory of Elasticity*, Third Edition, McGraw-Hill Book Company, London, 1985.
3. G. E. Dieter, *Mechanical Metallurgy*, Third Edition, McGraw-Hill Book Company, New York, 1986.
4. M. F. Ashby, *Materials Selection in Mechanical Design*, Pergamon Press, Oxford, 1992.
5. R. W. Hertzberg, *Deformation and Fracture Mechanics of Engineering Materials*, Third Edition, John Wiley & Sons, New York, 1989.
6. W. A. Backofen, *Deformation Processing*, Addison-Wesley, New York, 1972.
7. S. Suresh, *Fatigue of Materials*, Cambridge University Press, Cambridge, 1992.
8. Metals Handbook, Ninth Edition, Vol. 8: *Mechanical Testing*, ASM, Metals Park, Ohio, 1988.

Problems

3.1 Describe the difference between *true* contact area and nominal contact area. Assuming elastic contacts only, list in order of increasing true contact area joints between two flat surfaces of:
a. Diamond polished (1/4 μm) aluminium.
b. Iron sand-castings.
c. As-sintered alumina plates.
d. Diamond sliced alumina.

3.2 Describe how the *true* contact area between two ductile metals changes with:
a. Surface roughness.
b. Applied load.
c. Yield stress.
d. Elastic modulus.

3.3 A machine part is loaded in biaxial stress such that: $\sigma_1 = 120$ MPa, $\sigma_2 = 90$ MPa, and $\sigma_3 = 0$ MPa. The material from which the part is made has a tensile (and compressive) uniaxial yield stress of $\sigma_y = 360$ MPa. Calculate the design safety factor according to the Von Mises criterion.

3.4 A rod made of brass has a uniaxial yield stress $\sigma_y = 210$ MPa. Calculate the design safety factor according to the Von Mises criterion for the following loading conditions:
a. $\sigma_x = 70$ MPa, $\sigma_y = 30$ MPa.
b. $\sigma_x = 70$ MPa, $\tau_{xy} = 30$ MPa.

3.5 A semi-infinite plate of low carbon steel ($\sigma_y = 215$ MPa) is loaded under a tensile stress of 107 MPa. The plate contains an elliptical hole of major and minor axes 5.0 cm and 3.0 cm respectively.
a. Calculate the stress intensity factor.
b. Will the plate fail under the above stress?

3.6 A stainless steel plate ($\sigma_y = 459$ MPa) has a width $W = 25$ mm and thickness $t = 5$ mm and contains a hole located 8.75 mm from one end of the plate. The hole diameter is $d = 5$ mm. Estimate the maximum tensile load which can be placed on the plate. Include a safety factor of 2.0 in your calculation.

3.7 Which of the following statements refer to a ductile and which to a brittle fracture?
a. No plastic deformation prior to fracture.
b. The crack velocity approaches the velocity of sound.
c. Failure is initiated by pore nucleation and growth.
d. Cracking at low temperatures and high strain rates.
e. Relatively little energy is required for crack formation and propagation.

3.8 Silica glass has an elastic modulus $E = 70$ GPa and a surface energy $\gamma = 6 \times 10^{-5}$ J cm^{-2}. A silica glass fibre with a length of 200 mm and diameter of 0.052 mm fails under a load of 0.22 N. Estimate the length of the critical crack responsible for failure assuming that it lies perpendicular to the load direction.

3.9 Calculate the critical crack length for brittle fracture of cast-iron assuming:

$$\sigma_a = 90 \text{ kN cm}^{-2}.$$
$$\gamma = 1.2 \times 10^4 \text{ J cm}^{-2}.$$
$$E = 2.05 \times 10^8 \text{ N cm}^{-2}.$$
$$\pi = 0.1 \text{ J cm}^{-2}.$$

3.10 A thick steel plate contains a random array of cracks which range in size from 0.5 to

1.0 mm. Under what tensile stress is failure expected, and what will be the probable mode (type) of failure?

$$K_c = 53 \text{ MPa} \cdot \text{m}^{1/2}.$$
$$\sigma_y = 400 \text{ MPa}.$$

3.11 A thick-walled pressure vessel is seam-welded. The weld defects present include porosity, irregular inclusions and cracks. Using the information provided below and assuming *worst* case geometries:

a. What is the minimum crack size which will cause failure?
b. Estimate the maximum allowable crack size for a crack located adjacent to a spherical pore.
c. Estimate the maximum allowable crack size for a crack located adjacent to an inclusion.

Given:

$$\sigma_a = 800 \text{ MPa}.$$
$$K_c = 99 \text{ MPa} \cdot \text{m}^{1/2}.$$
$$b/a = 3 \text{ (for inclusions)}.$$

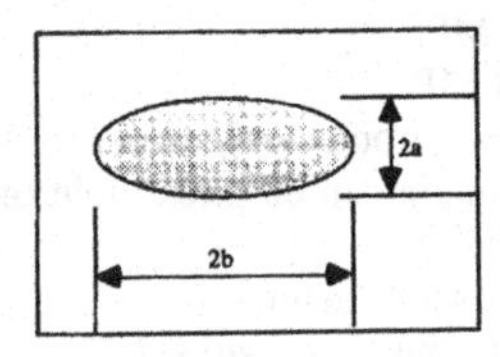

Part II

JOINING METHODS

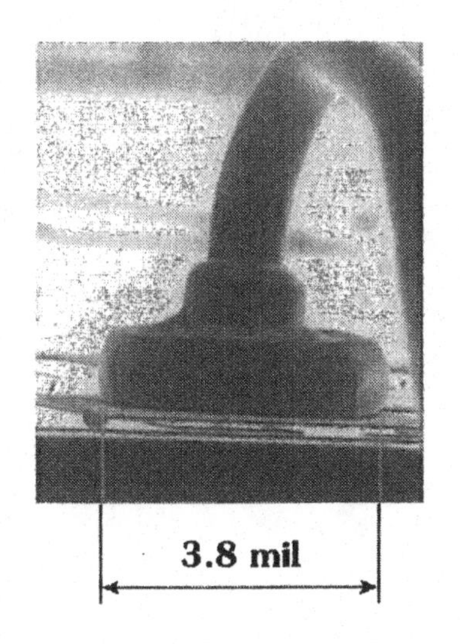

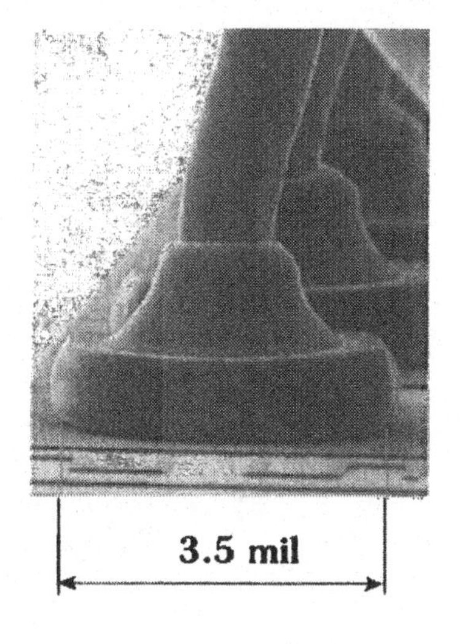

4

Mechanical Bonding

Nails, screws, rivets, bolts, clamps and fasteners are ubiquitous in engineering systems and constitute by far the most versatile class of joining methods. These range from the nailing of two wooden planks to the bolting in place of an engine head, and from the rivetting of boiler plate to the zip fastener of a winter jacket. Cup and cone ground glass seals, described in the Introduction, are mechanical joints. So are the *velcro*, hook and felt, fabric fasteners which have effectively replaced the metal 'pop-fastener'. Figure 4.1 illustrates some of this wide variety, and in this chapter we will examine some of the principles involved in the *design* and *fabrication* of mechanical bonds and joints.

In particular, we will expand the discussion of *stress concentration* in mechanical assemblies and the role of *residual stress* in preserving the integrity of an assembly. We will analyse the requirements for *seals* and *gaskets* and describe some ways in which these requirements are met in practice. The role of *friction* as a positive factor in stabilizing a joint will be discussed, together with some negative aspects of friction. The role of *coatings* in controlling frictional properties will be described, as well as some of the possible *corrosion* problems associated with mechanical joining processes.

Examples will be given of both *permanent* and *demountable* mechanical joining systems, ranging from ultra-high vacuum seals to orthopaedic implants.

4.1 STRESS CONCENTRATION & RESIDUAL STRESS

There are always regions of *stress concentration* present in a material, associated with either the microstructure itself or the presence of *processing defects* in the material (inclusions, porosity or microcracks). It is convenient to distinguish between three different length scales when discussing stress concentrations and residual stresses, micro, meso and macro. On the *microstructural* scale, variations in stress may be associated with elastic or plastic *anisotropy* of the individual grains. If

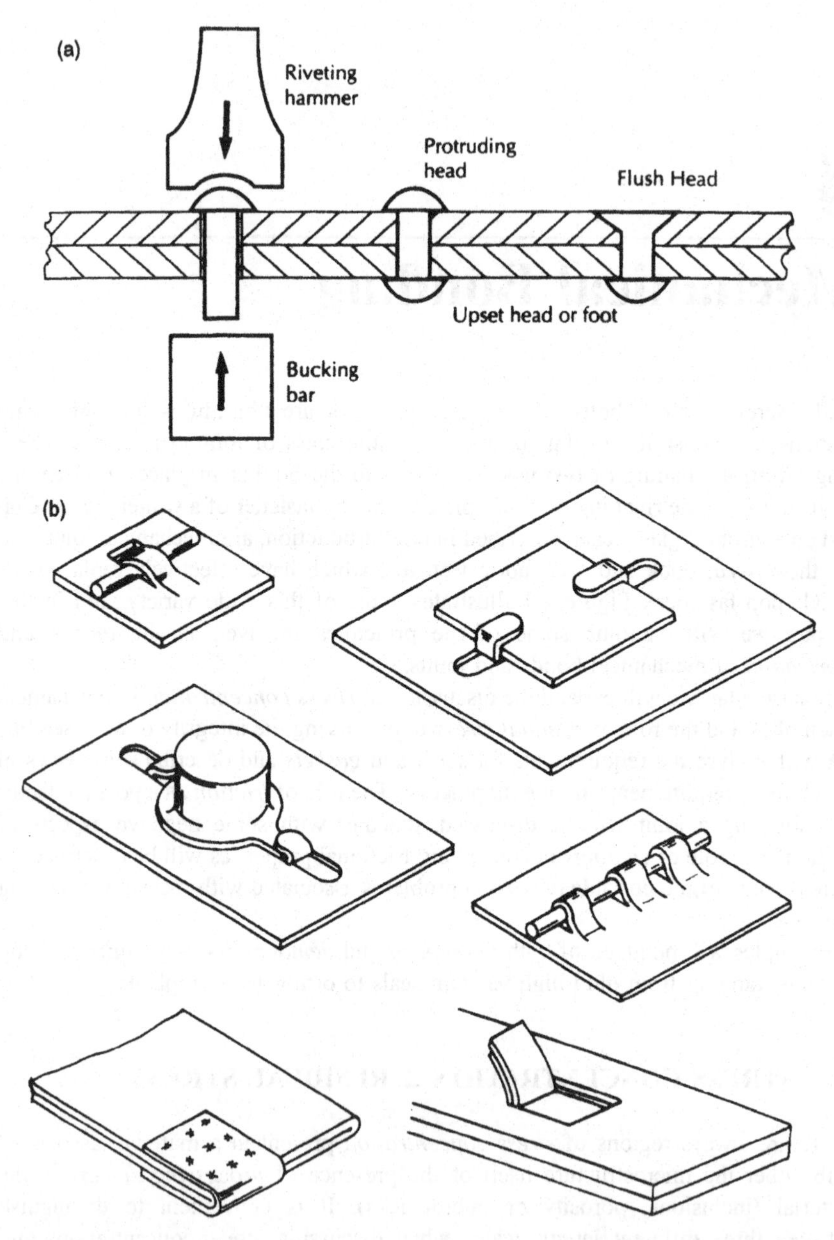

Fig 4.1. Some examples of mechanical fastening: a. Rivetting, b. Tabs, c. A zip fastener. Reproduced with permission from (a) Parmley, *Standard Handbook of Fastening and Joining 2nd Ed (1989)*. The McGraw-Hill Companies. (b) *Smithells, Metals Reference Book*, Brandes and Brock (1992) Butterworth-Heinemann Ltd. (c) From How Things Work: The Universal Encyclopedia of Machines Vol. 1 (1972), Paladin/Granada Publishing Limited.

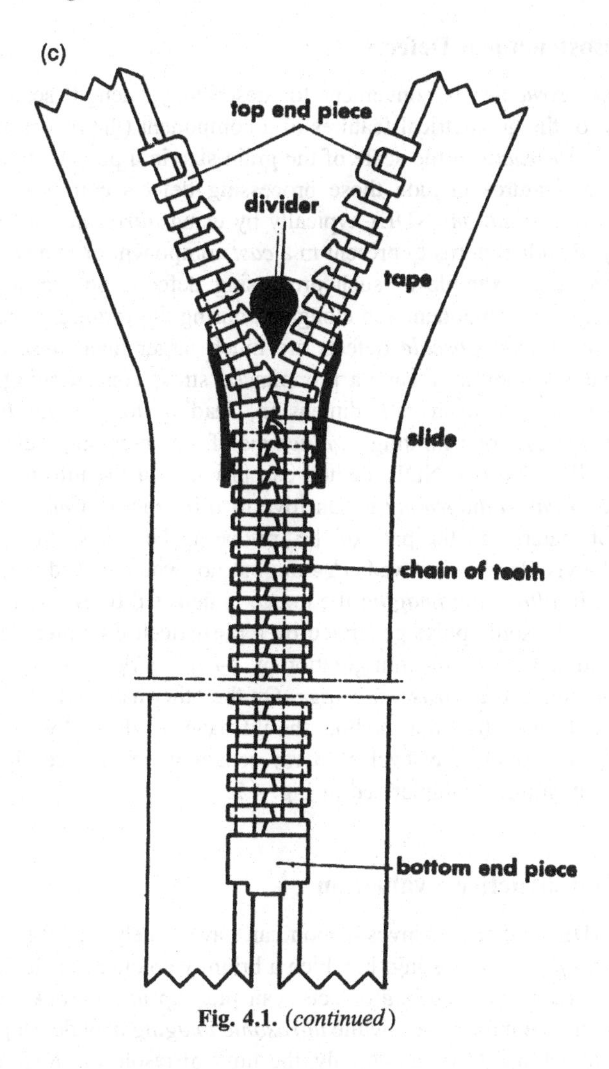

Fig. 4.1. (*continued*)

the normal to a grain boundary corresponds to a direction of *high stiffness* (high elastic modulus) in one grain, but to a direction of *low stiffness* in the neighbouring grain, then, when a load is applied *normal* to the boundary, the elastic discontinuity will give rise to *shear stresses* at the edge of the boundary facet and parallel to the boundary. Hard, *second phase particles*, in a precipitation- or dispersion-hardened alloy, are also sites of stress concentration in the matrix phase, and in such regions the stress in a uniaxially-loaded specimen is *triaxial* rather than *uniaxial*.

4.1.1 Mesostructural Defects

The term *mesostructure* is convenient for describing a length scale intermediate between that of the geometrical features of a component (the *macrostructure*), and that of the *microstructure* (the scale of the grain size in a polycrystalline material). Mesostructural features include those processing defects commonly observed in *non-destructive evaluation* (NDE), typically by using *ultrasonic imaging* or *X-ray radiography*. Residual porosity present in a *cast* component or in a *sintered* powder compact is a good example of such processing defects. So are slag inclusions, retained in a forged component and elongated during the forging process. Since the dimensions of these *mesoscale* defects are much larger than those of the microstructural features, they may have a pronounced stress concentrating effect, especially if they are small in one dimension, leading to a *notch-like* structure. Evaluating the stress concentrating *significance* of the processing defects is a major objective of NDE. No one NDE method can provide *all* the information required. For example, *X-ray radiography* is sensitive to differences in *mass thickness* (the total mass of material in the path of the irradiating beam), so that an ellipsoidal defect may be visible when viewed *edge-on*, but not when viewed *perpendicular* to the ellipsoid. In *ultrasonic imaging* the image is detected by *reflection* from a free surface of the ultrasonic pulse generated by a piezoelectric emitter. The ellipsoidal defect detected in the *edge-on* configuration by *X-ray radiography* may be below the limit of resolution in the *ultrasonic image*. On the other hand, when viewed *normal* to the ellipsoid, the ultrasonic method should have no difficulty in detecting the defect which was below the level of X-ray sensitivity in this configuration. This hypothetical situation is summarized in Fig. 4.2.

4.1.2 Non-destructive Evaluation

In general, NDE methods of investigation can have widely varying resolution and sensitivity. *Dye penetrant* testing, in which a brightly coloured or fluorescent dye is allowed to penetrate into *surface* defects, can pick up *microcracks* of length less than 0.1 mm in favourable cases, while *ultrasonic imaging* may detect planar defects down to 20 or 30 μm. More commonly, the limit of resolution for the length of a *planar crack* is of the order of 0.2 mm, while *volume defects* need to have their minimum dimension greater than 0.1 mm to be detected with any certainty by ultrasonic methods. Since it is the difference in mass thickness that determines contrast in X-ray radiography, it is not only the size but also the density of the defect which limits its detectability by X-rays.

In a joint the same considerations apply. A *mesodefect* may be a region of *debonding* in a composite laminate, a *crack* at the edge of a fastener or a *scratch* on the sealing surface of a cylinder head. The methods used to detect these defects are those conventionally employed in NDE, with the addition of *leak testing* (see Chapter 10), and other indirect methods.

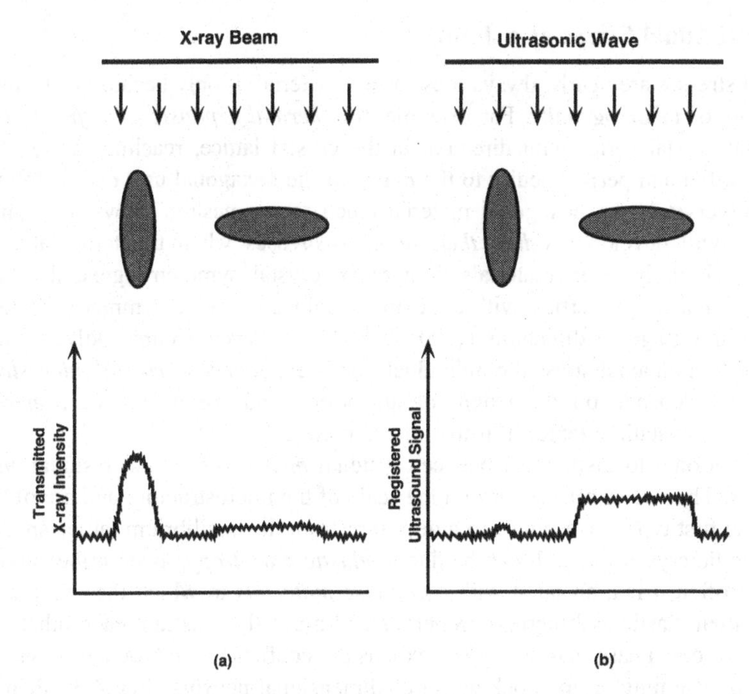

Fig. 4.2. An elliptical defect is more readily detected by *X-ray radiography* when edge-on to the incident beam (a) but will show up better in an *ultrasonic inspection* when the defect is perpendicular to the incident radiation (b).

4.1.3 Meso- & Macroscale Stress Concentrators

Stress concentrations are unavoidable at *reentrant angles* in a joint: the root of a screw thread, the head of a bolt, or the edge of a lock seam. They are also a major factor limiting design of *brazed* and *soldered* joints, since the *elastic* properties of the filler metal commonly differ from those of the components being joined. *Adhesive* joints employ organic polymers as sealing agents, and these have elastic moduli which may be less than 1% of that of a metal or ceramic. They are an extreme example of elastic mismatch and stress concentration in the region of the joint.

Although the distinction between the *macro-* and *mesoscale* of a structure is no less artificial than that between the *meso-* and *microscale*, it is useful to restrict the term macroscale to the *design* features of the joint (the *radius* of a seam or the *pitch* of a screw thread), while using the term mesoscale for those features which are at the limit of the *production tolerances*: scratches, imperfectly bonded regions, large inclusions, small cracks and other imperfections.

4.1.4 Residual Stress at a Joint

Residual stresses are nearly always present in a material, if only because of material *anisotropy* or *inhomogeneity*. For example, the *thermal expansion coefficient* of a hexagonal crystal varies with direction in the crystal lattice, reaching its extreme values parallel and perpendicular to the *c*-axis of the hexagonal unit cell. It follows that a polycrystal of a hexagonal material (such as magnesium alloys or alumina ceramics) will develop *residual thermal microstresses* when the temperature is changed. Similarly, even materials with *cubic* crystal symmetry generally have anisotropic *elastic* properties, with the extreme values for cubic symmetry being in the cube and diagonal directions (⟨100⟩ and ⟨111⟩). When a cubic polycrystal is subjected to a uniaxial stress the individual grains experience a *triaxial microstress* field which depends on the extent of anisotropy and the degree of *preferred orientation* (crystalline texture) in the microstructure.

It is important to distinguish between residual *microstresses* and residual *macrostresses*. The *microstresses* vary on the scale of the microstructure and are of two types. The first type corresponds to a deviation from the equilibrium lattice spacing for a specific crystal plane. It can be due to *elastic anisotropy*, as described above, but more often it is associated with *two-phase materials* in which the two phases differ in their elastic and thermal properties and mutually constrain each other. For example if one phase has a *higher* expansion coefficient and both phases are *continuous* (forming an interlocking, three-dimensional network, Fig. 4.3), then on changing the temperature one phase will place the other in *compression*, while this thermal compressive stress is balanced by a hydrostatic *tension* in the second phase.

The second type of *microstress* most commonly arises as a consequence of the *inhomogeneity* of *microplasticity*, which places some crystal planes in compression, while the same family of planes, in the same orientation, is placed in tension elsewhere in the material. In consequence the X-ray diffraction peak corresponding to this family of planes is *smeared*, indicating a *spread* of lattice spacings, even though the *average* lattice spacing of these planes remains unchanged.

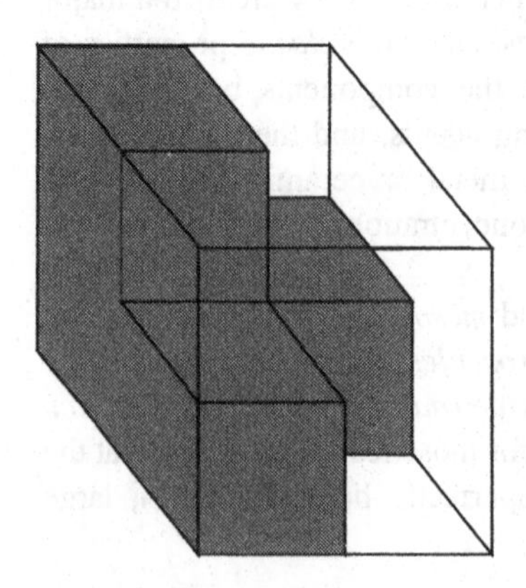

Fig. 4.3. In *three* dimensions, two *isotropically-distributed, continuously-interconnected* phases are dominated by *hydrostatic* constraints that plane one phase in compression and the other in tension. The microstructure can be approximated by a three-dimensional rectangular grid whose unit cell is illustrated here.

Macrostresses exist on a scale which corresponds to the geometrical features of the component: the *width* of a weld bead, the *pitch* of a screw thread or the *diameter* of a rivet. They can also be detected by *shifts* in the apparent lattice spacing of the crystal planes, as observed by X-ray or neutron diffraction.

The role of *residual stresses* in determining the integrity of a mechanical joint is crucial. The case of the nail used to hold together two planks of wood is typical (Fig. 1.1). The nail is forced into the wood by hammer blows that place the nail in compression and overcome the frictional resistance of the wood fibres. Ignoring the anisotropy of the wood, the increment of *shear force* resisting penetration of the nail is given by: $dF_s = 2\pi r \mu \sigma_n dz$ where σ_n is the normal *radial stress* resisting penetration of the nail and μ is the coefficient of friction between the nail and the wood, while the axis of the nail defines the z direction. This shear force is balanced by the increment of *compressive force* applied to the nail: $dF_c = \pi r^2 d\sigma_c$. To simplify the problem, we assume that the compressive force is applied in *dead-loading* rather than by hammer blows. The *total* force required to push the nail into the wood to a depth L is then given by: $F_c = \pi r^2 \sigma_c^0$, where σ_c^0 is the maximum compressive stress applied at the head of the nail: $\sigma_c^0 = \int_0^L 2(\mu\sigma_n/r)dz = 2\mu\sigma_n(L/r)$. As the load on the head of the nail (and the contra-force supporting the base of the wood) is released, the stress is *redistributed*. As the compressive stresses on the nail and the wood are released, and reversed sliding initiates at both the head and the tip of the nail, the frictional shear at the interface *reverses* at the head and tip of the nail, placing the wood in tension at the head and the nail in tension at the tip (Fig. 4.4a). When the external compression has been *completely* relaxed the distribution of *residual stress* will approximate that shown in Fig. 4.4b. Near the head of the nail there is a residual *compressive stress* in the nail, which reverses to a residual *tension* near the tip. In practice the wood is much more *compliant* than the steel, and reverse sliding initiates at the tip of the nail, so that the frictional force at the *tip* of the nail is reversed *first* to *oppose sliding* at the wood/nail interface. In the limit of an *infinitely stiff* nail the frictional stress reversal at the *head* of the nail is negligible. Figure 4.4c, d show what happens as the compressive load is reduced to zero in this case. The *sign* of the axial stress now reverses *throughout* the nail from compressive to tensile. At zero load the nail is in *residual tension*, with the frictional forces acting in *oppsite directions* on the top and bottom halves of the nail with the maximum stress at the mid-point. The *maximum residual tensile stress* in the nail for this case is just *half* the stress required to force the nail into the wood (as opposed to 1/4 of the maximum compressive stress for the case of *equal* elastic moduli). Clearly, the balance of the shear forces at the interface requires that the wood remain in *axial compression*, but this compressive stress falls off with radial distance from the interface. Note that if the stiffness of the nail were *less* than that of the wood, then the reversal of frictional resistance would initiate at the *head* of the nail, placing the *wood* in residual tension. That is, the successful implementation of a nailed joint depends on having a nail of *higher modulus* than the material to be joined.

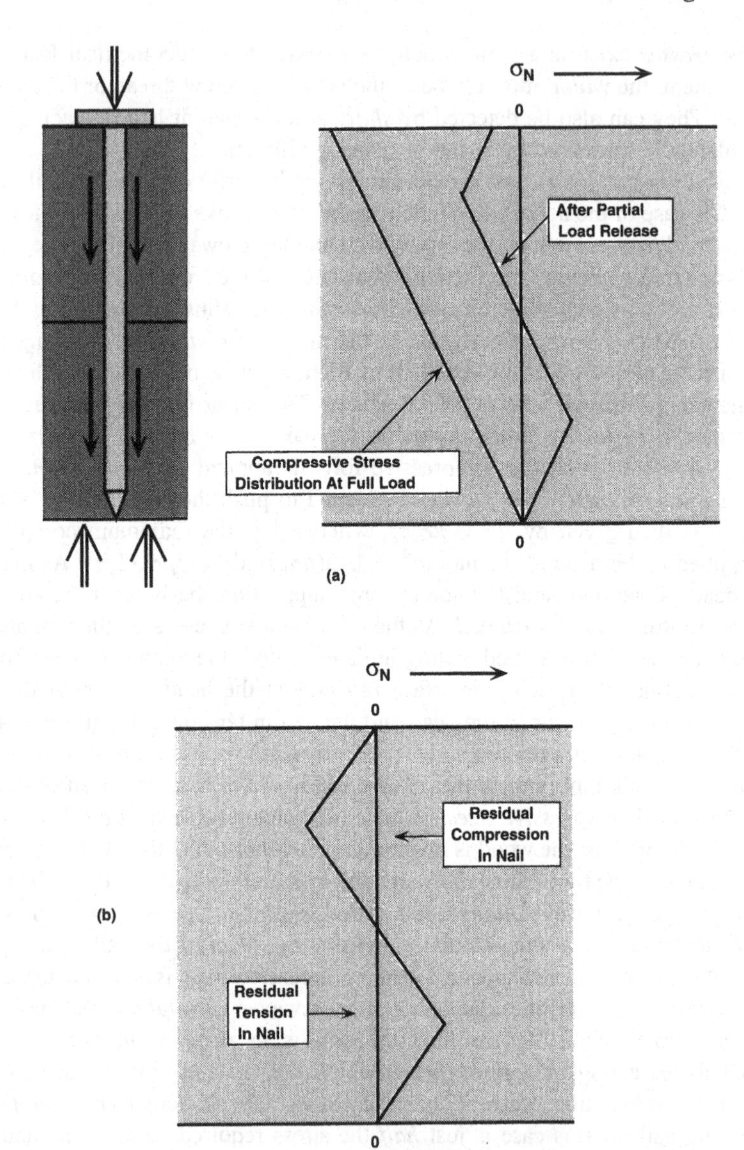

Fig. 4.4. A nail is forced into two wooden planks in contact, and the compressive force applied to the nail is balanced by a compressive contra-force applied to the bottom surface of the two planks. (a) When the force is released and the elastic moduli of wood and nail are assumed identical, sliding starts at both the tip and head of the nail and the frictional forces are reversed at both ends. (b) The final residual stresses in the nail for this case are compressive near the head and tensile near the tip. (c) If the elastic modulus of the nail is assumed infinitely large in comparison with the wood, then reverse sliding only occurs from the tip of the nail. (d) For this case the final residual stress state in the nail is entirely tensile, with the peak stress at the mid-length.

(c)

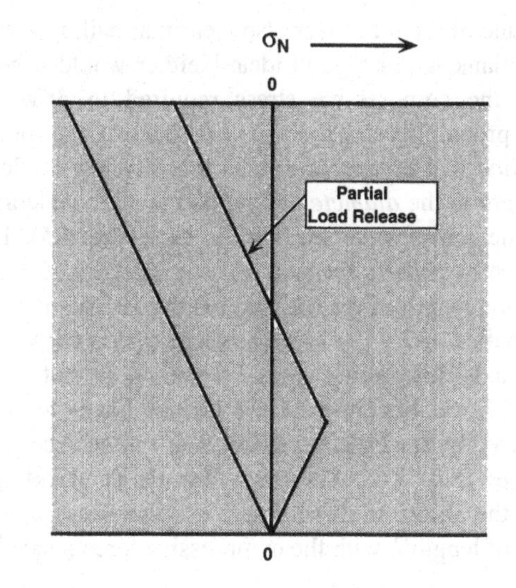

(d)

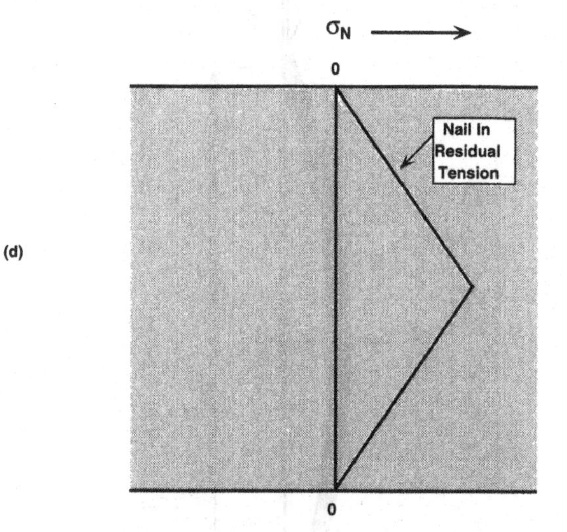

Fig. 4.4. (*continued*)

We can place the interface between the two wooden planks anywhere we wish, but clearly the *optimum* position, at least for the case $E_W \ll E_N$, is at the mid-point of the nail, to coincide with the region of maximum residual compressive stress in the wood. Furthermore, the nail should penetrate into the lower plank to a *depth* equal to the *thickness* of the upper plank, in order to achieve the maximum compressive stress in the plane of the joint. It is also clear that nailing a thicker, upper plank, to a thinner, lower plank is *not* a good idea. Neither would it be useful to use a longer nail: although the compressive stress required to drive in the nail is greater (increasing the probability that the nail will buckle!), the compressive stress acting at the joint *interface* will be unchanged, so that nothing would be gained. Finally, the ratio of the *length* to the *diameter* of the nail is of considerable importance: the nail will *buckle* if the compressive stress is too high (Fig. 4.5). The condition for elastic buckling is given by *Euler's* formula: $F_b = \pi^2 EI/L^2$, where E is the elastic modulus, L the unsupported length of the nail and I is the second moment of area of the cross-section of the nail, $I = 2\int_{-r}^{+r} y^2 b(y)\mathrm{d}y$, where $b(y)$ is the width of the chord parallel to the bending axis. Integrating gives $I = \pi r^4/4$, so that: $F_b = \pi^3 Er^4/4L^2$.

In practice, *plastic* buckling of the nail is likely to occur before the *elastic* buckling predicted by the Euler equation, and will be accompanied by the formation of *plastic hinges* (Fig. 4.6). Assuming that elastic buckling is responsible for an upper limit on the ability to drive a nail into the wood, then equating the buckling force for a nail of length L with the compressive force needed to insert the nail to the

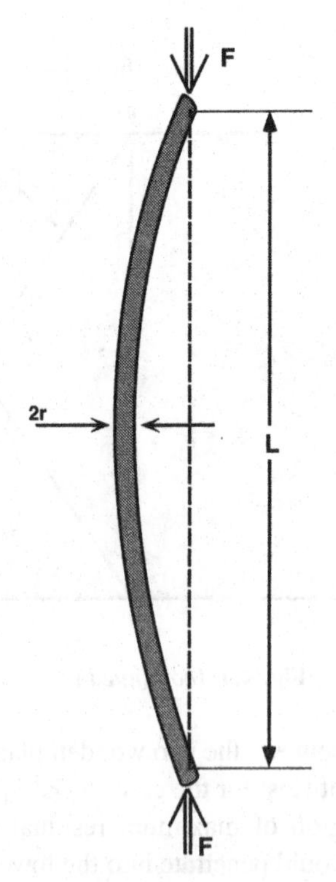

Fig. 4.5. Elastic buckling under a compressive axial load.

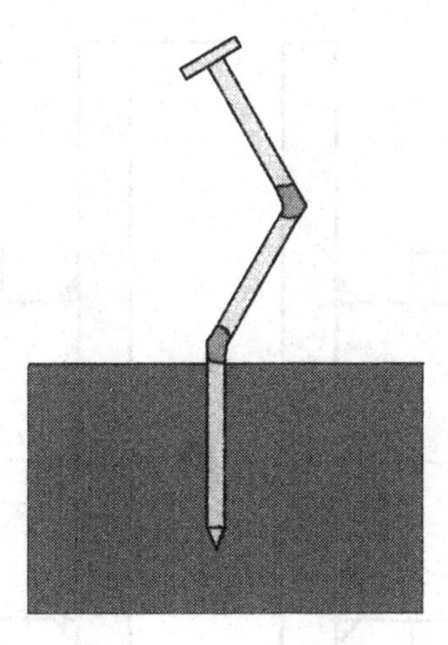

Fig. 4.6. Plastic buckling and the formation of plastic hinges in a partially encastrated nail.

same depth L (and ignoring complications associated with encastrating the nail in the wood), we find that the ratio of L/r is related to the *ratio* of the *modulus* of the nail to the *frictional stress* at the wood/nail interface: $(L/r)^3 = k(E/\mu\sigma_n)$, where k is a constant of order unity. The shear stresses resisting penetration of the nail into the wood are typically of the order of 10 MPa, while the elastic modulus of iron is 60 GPa. It follows that the limiting value of L/r is about 15. That is, a nail of radius 1 mm (2 mm diameter) should be no more than 15 mm long if buckling is to be avoided.

The above treatment has been given in some detail in order to illustrate the importance of *residual stresses* in mechanical joints and to demonstrate some of the mechanical factors which control both the ability to develop a suitable residual stress field and the optimization of the residual stress distribution.

A second example, also extremely simple, is that of a *bolted flange* (Fig. 4.7). The tensile stress in the bolts can be adjusted to any value, up to the yield stress of the metal, by tightening the nuts, but the stress concentration at the root of the screw thread will initiate yielding at a reduced stress, below that required for *general yielding* of the complete cross-section. If the bolts are used to attach a flange which must seal against an *internal pressure*, then the tensile force in the bolts must generate sufficient compressive stress in the flange to ensure that the *compressive stress* exceeds the *internal pressure* so that the seal can remain leaktight. This internal pressure *adds* to the tensile stress in the bolt developed during assembly of the flange. These are *major* engineering requirements for the joining system which must be taken into consideration when specifying the maximum *torque* applied when tightening the nuts. We will return to this example below.

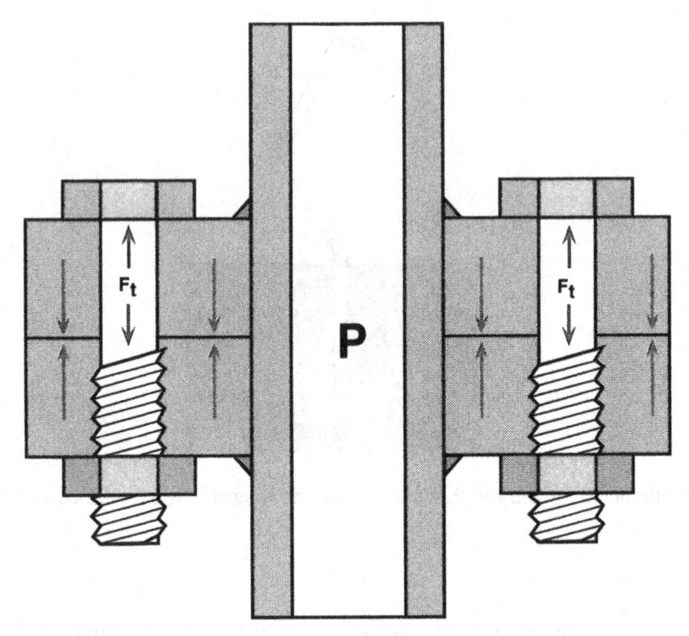

Fig. 4.7. The bolts holding a bolted flange must be tightened to a sufficient tensile load F_t to ensure that the compressive force on the flange exceeds the internal operating pressure P.

4.1.5 Measurement of Residual Stress

There are two alternative methodologies for determining the residual stresses in an engineering system, either *destructive* or *non- destructive.* In general, the destructive methods are both more accurate and more informative, even though they result in permanent damage or destruction of the component.

The commonest *non-destructive* test for residual stresses in *polycrystalline* materials involves measuring the distribution of the crystal lattice spacings by *diffraction*, most often X-ray diffraction, although neutron diffraction is also used where available. Changes in *position* of the diffraction peak which is associated with a specific family of lattice planes are directly proportional to the average elastic strain *perpendicular* to these reflecting lattice planes, while *broadening* of the diffraction peak corresponds to *variations* in this lattice strain from one region to another within the irradiated volume. The average residual *stresses* can be derived if the *single crystal elastic constants* are known or can be calculated (as is usually the case). Care must be exercised in the interpretation of the *peak broadening,* since broadening can also be a consequence of a reduction in the *size* of the coherently diffracting regions, for example, in submicron, *fine-grained* materials.

Residual stresses can also be measured *destructively* by removing a portion of the material and measuring the *shape changes* which then occur. Extruded tube usually contains residual *radial* strains which can be observed by *slitting* the tube (Fig. 4.8a). The residual radial strain is proportional to the change in *tube diameter* observed on slitting: $\varepsilon_r = \Delta d/d_0$. *Longitudinal* residual strains in the tube can be observed by cutting a narrow *tongue* at one end (Fig. 4.8b). The strains in a cold-rolled sheet can be analysed by a similar process. The sheet is clamped at both ends, a known *thickness* is then removed (by machining or, better, by chemical etching), and the *radius of curvature* is then measured after releasing the clamps (Fig. 4.9).

Strain gauge rosettes can be used to monitor the residual *biaxial* strains in the surface layers of a component. A hole is drilled at the centre of the rosette to relieve the *biaxial surface stresses* and the strain change associated with stress relief is monitored by the three strain gauges of the rosette, placed at 0, 90 and 45° to a principal axis of the component (Fig. 4.10).

The smallest available strain gauge rosettes contain 2 mm strain gauges, which is sufficient for measurements in and around many *welds*. X-ray beams can also be collimated to similar dimensions in order to monitor the residual stresses around a weld by X-ray diffraction, while *microbeam* X-ray sources are also available to limit the irradiated area even further. However, diffraction methods rely on the *statistical sampling* of a large number of grains by the diffracted beam. For most engineering alloys this statistical limitation restricts the useful spatial resolution of X-ray diffraction residual stress measurements to about 1 mm.

4.1.6 Assembly & Joining Stresses

When a system is *assembled* additional stresses may be introduced into the components which will affect the performance of the engineering system. The successful operation of the system under mechanical loading depends on the *sum* of all the stresses acting on the components of the system. These stresses arise from *three* sources: *residual stresses* arising during the manufacture of the components, stresses associated with the *assembly* and *joining* of the components, and stresses due to the *operation* of the system in service. In this section we concentrate on the stresses associated with *assembly* and *joining*.

Major factors in determining the distribution and magnitude of the *assembly stresses* are the dimensional *tolerances*. These may be of two types: those associated with *surface finish* and those associated with the *dimensional accuracy* of component manufacture. As described previously, the *surface finish* and *surface roughness* can have a major influence on the *wetting* of a surface by an adhesive, solder or braze metal, as well as on the *mechanical strength* of the bond formed at a joint. The extent of the *misfit* between two components which are to be joined is not always subject to control, and in a large system can reach large values. A railway line is a good example. The *coefficient of thermal expansion* of mild steel is approximately 12×10^{-6} °C^{-1}. Assuming a 30 m length of rail and a temperature differential of 20°C between night and day (easily attained in most climates), this leads to a very appreciable displacement of 7.2 mm which has to be accommodated

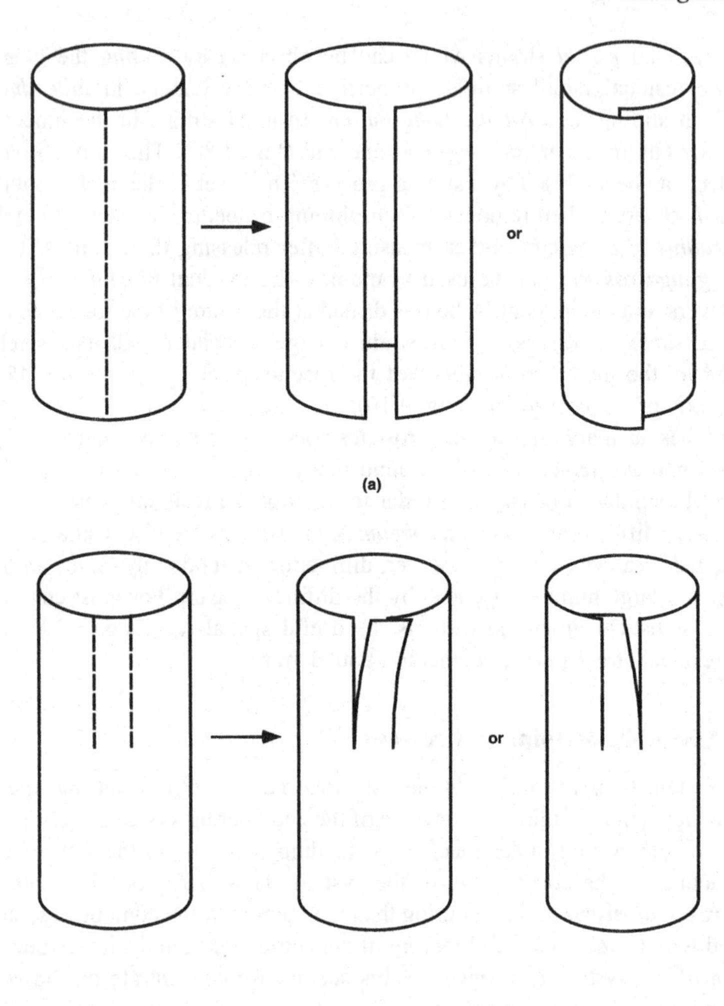

Fig. 4.8. (a) Sitting a tube longitudinally reveals both the sign and magnitude of the residual radial stresses in the tube wall, as measured by the change in diameter of the tube. (b) Cutting a narrow tongue longitudinally in one end of the tube reveals the sign and magnitude of the longitudinal stresses in the tube wall by the sign and magnitude of the curvature of the tongue.

at the rail joints. For a *welded* rail the only possibility is an *axial elastic strain*, in this case 2.4×10^{-4}, corresponding to compressive or tensile stresses of approximately 15 MPa.

Welded assemblies may give rise to residual stresses associated with the *heating cycle* involved in the welding process. As an example, consider the steel support

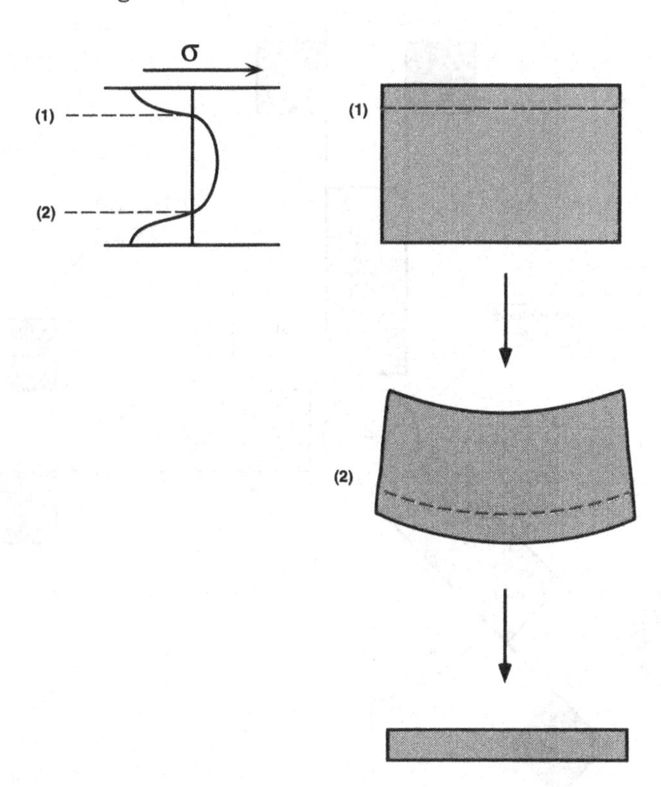

Fig. 4.9. A residual stress distribution in cold-rolled sheet can be determined by machining away successive layers of the clamped sheet and then measuring the curvature of the sheet after releasing the clamps.

strut welded to the bracket shown in Fig. 4.11. If the length of the region heated during welding is 4 cm, with the temperature linearly distributed up to the melting point, then the change in length on cooling is $1/2\alpha T_m \times 40$ mm, where T_m is the melting point of the steel and α is the *thermal expansion coefficient.* Inserting the value for α given above and putting T_m as 1650°C, the length change is approximately 0.4 mm. Distributed over a 20 cm strut length, this corresponds to a tensile strain of 2×10^{-3}. For a rigid bracket and a strut tensile modulus of 60 GPa, this results in an *assembly stress* of 120 MPa, which is an appreciable fraction of the yield stress for structural steel.

The bolted flange described previously (Fig. 4.7) must be compressed by the bolts to a stress which is sufficient to prevent leakage at the proposed operating pressure. Let us assume an internal pressure of 100 MPa acting on a flange of 10 cm internal diameter. The total force acting on the flange is then 250π kN. If this is carried by four bolts, then each bolt must be tightened by a *torque* of at least $250\pi/4$ kN. The radius of the bolt must be sufficient to withstand the applied torque without yielding.

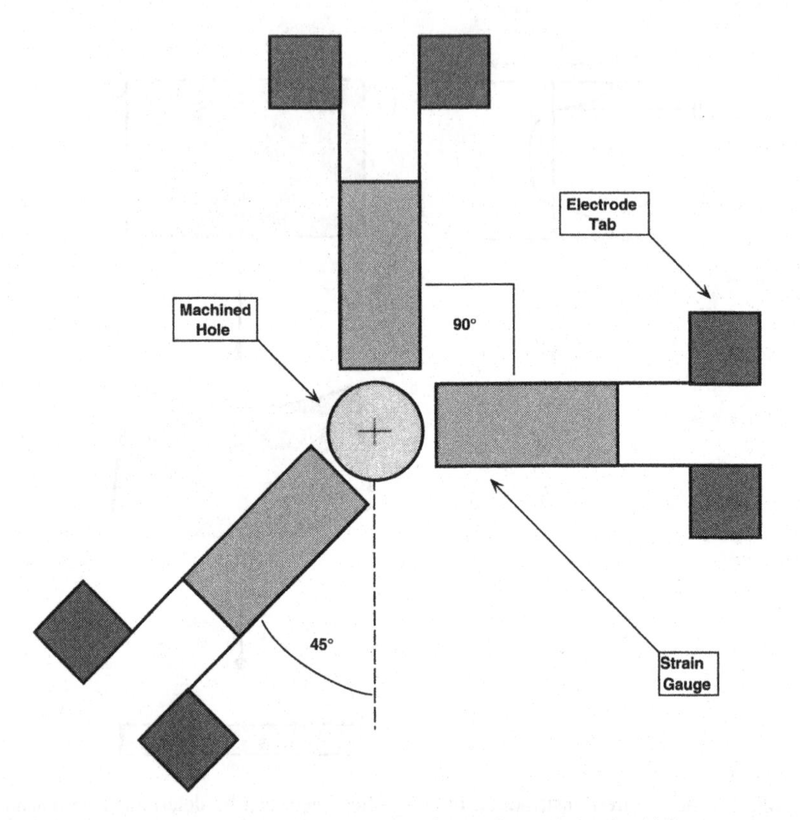

Fig. 4.10. The distribution of *biaxial stresses* at and near the surface of a thick component can be measured by drilling a hole at the centre of a strain gauge rosette and monitoring the changes in the strains as a function of the depth of the hole.

If the yield stress of the steel is 500 MPa, then the required condition is: $\pi r^2 \times 500 > 250\pi/4 \times 10^3$, where r is given in mm. The limiting value for r is just over 11 mm—a fairly massive bolt! If the number of bolts is increased, then the radius can be reduced. If the bolts are threaded to the head, then the increased risk of failure due to the *stress concentration* at the root of the threads must be taken into account. In any event, it is necessary to exert a torque that will keep the flange sealed *under the operating conditions of the system.*

As a final example we can consider the lock seam illustrated in Fig. 1.2. Wrapping the sheet around the mandrel to form a tube introduces a pattern of residual stress in the thickness of the sheet, which would be partially released (with an increase in tube radius) if the sheet were to be removed from the mandrel *before* forming the lock seam. This residual stress can be analysed by assuming that the forming process approximates pure four-point bending (Fig. 4.12). The tangential tensile and compressive stresses are a maximum in the surface fibres and are symmetrical about the

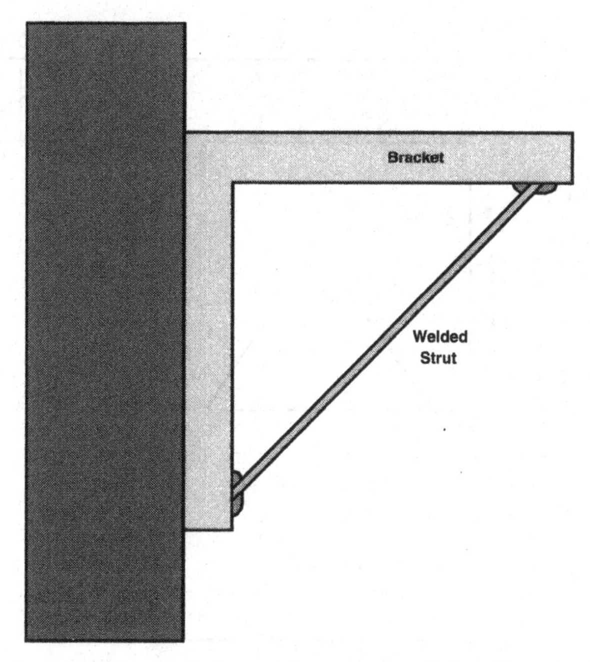

Fig. 4.11. *Residual stress* in a welded strut is a result of the dimensional changes accompanying cooling of the welded zone to room temperature.

neutral axis at the centre of the sheet. *Yielding* initiates at the top and bottom surfaces and spreads towards the neutral axis (Fig. 4.12b). On releasing the external bending moment, the stress at the surface redistributes to give a residual stress pattern in which the *external* fibres of the bent sheet are placed in *tension* and the *internal* fibres in *compression* (Fig. 4.12c), and the radial component of the *average* through-thickness stress reduces to zero by a partial relaxation of the bending strains (Fig. 4.12d). This relaxation, termed *springback*, is identical to that described above in the measurement of a *residual radial stress* by tube-slitting (Fig. 4.8a). The residual radial *tensile* stress is in equilibrium with a *compressive* stress in the lock seam which guarantees the integrity of this purely mechanical joint. Once more, the presence of the *stable* residual stress distribution is an integral component of the joining process.

4.2 CORROSION, FRICTION & WEAR

Mechanical bonding systems are frequently susceptible to *corrosion* processes associated with electrochemical potential developed in the region of the join. *Electrochemical corrosion* requires the presence of an *anode*, a *cathode* and an

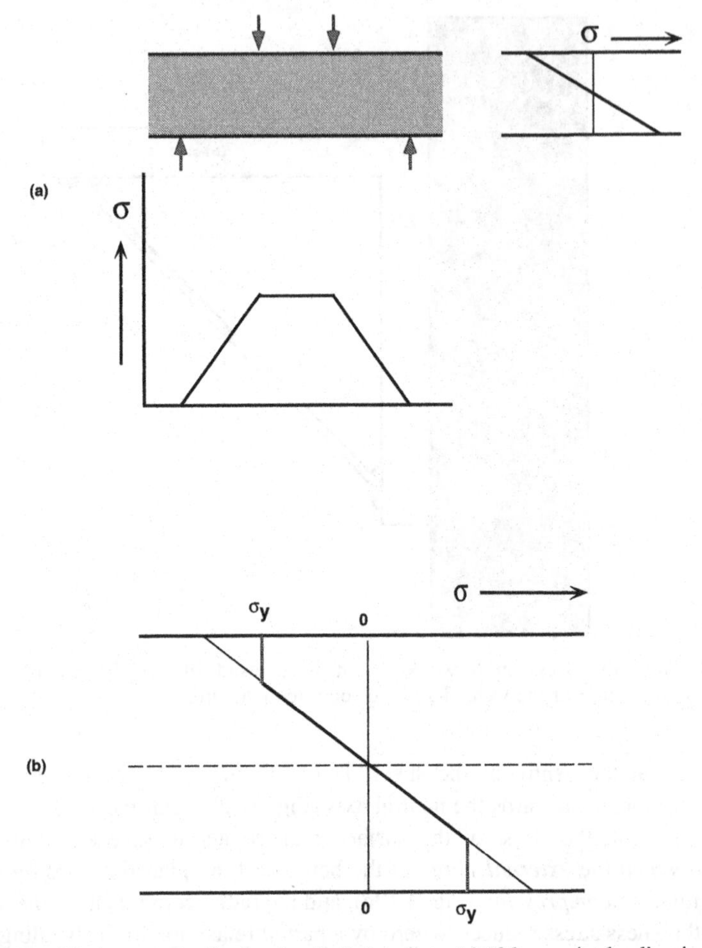

Fig. 4.12. Residual stress distribution in plastic bending. (a) If four-point loading is assumed, the elastic stress distribution in the cross-section is constant between the loading points and varies linearly through the thickness. (b) Plastic yielding initiates at the external surfaces and spreads towards the neutral axis. (c) Releasing the *external* stress reduces the *internal* stresses

electrolyte. In the growth of an oxide film, discussed in Section 2.5.2, the oxide film is itself the electrolyte and the charge carriers may be diffusing *anions* or *cations* in the crystal lattice, and *electrons* or *holes* in the conduction band of the oxide. At the *anode* the typical anodic reaction is $M \rightarrow M^{2+} + 2\varepsilon$, while at the *cathode* the typical cathodic reaction is $1/2O_2{}^{+}2\varepsilon \rightarrow O^{2-}$. In oxidation the anodic and cathodic reaction sites are separated by the thickness of the growing oxide film, but in corrosion the anode and cathode are often quite far apart, while the electrolyte is either an *aqueous solution* or condensate on the surface of the assembly. The *anodic* reaction remains

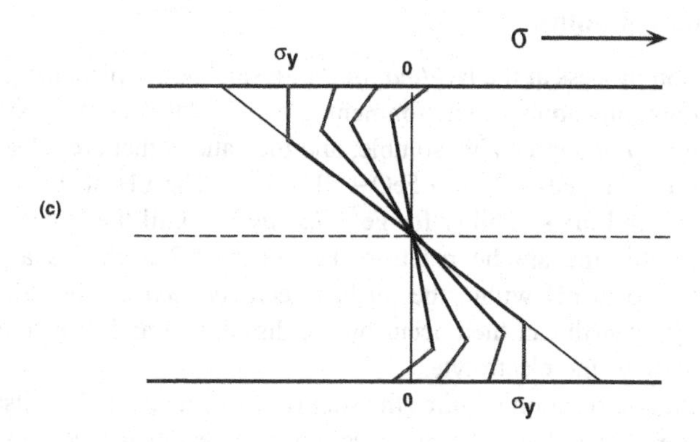

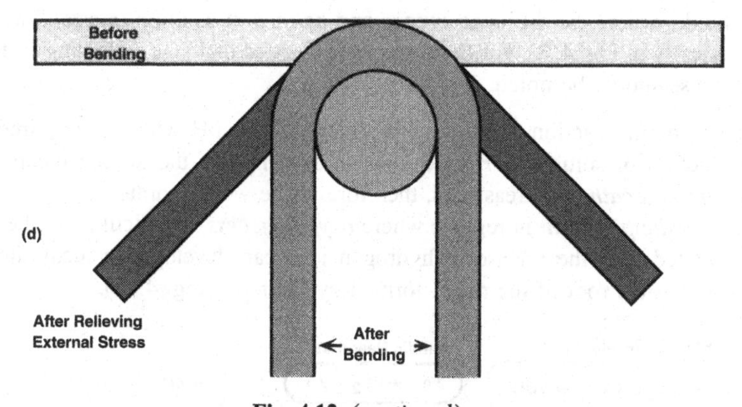

Fig. 4.12. (*continued*)

the same (with the *cations* being, at least in the first instance, dissolved in the electrolyte), but a typical *cathodic* reaction would be: $1/2O_2{}^+H_2O + 2\varepsilon \rightarrow 2OH^-$. The *anions* are taken into solution in the electrolyte, but in many corrosion reactions of practical importance subsequent *diffusion* of the anions and cations through the electrolyte leads to the precipitation of a *corrosion product*: $M^{2+} + 2OH^- \rightarrow M(OH)_2$. In the case of steels the corrosion product is *rust*, which has the approximate composition $FeO \cdot OH$. Rust has a much lower density than steel, and the resultant increase in solid volume may exert considerable pressure if the corrosion product is constrained. This change in volume associated with formation of solid corrosion products is responsible for the commonly observed *blistering* of paints and other protective coatings, and occurs when corrosion is initiated at localized sites on the component beneath the coating.

4.2.1 Corrosion at a Joint

Consider the corrosion process at the *rivetted joint* of a steel boiler plate (Fig. 4.13). We assume a salt-containing aqueous environment at pH 7 (neither acidic nor basic). At pH 7 the Fe^{2+} ion is *not* appreciably soluble, and the cations that are released by the anodic reaction are: H^+ $Fe + H_2O \rightarrow FeO + 2H^+ + 2\varepsilon$. The pH at the anode is progressively *reduced* and the solubility for Fe^{2+} *increases,* until the release of the ferrous cation dominates the anodic reaction: $Fe \rightarrow Fe^{2+} + 2\varepsilon$. This is a *pitting* reaction in which the local pH within the pit has been *reduced* by the release of hydrogen ions and pit growth can then occur by the dissolution and diffusion away of the ferrous cations into the electrolyte.

The balancing *cathodic* reaction requires the supply of oxygen, which is dissolved from the atmosphere. Since the cathodic reaction releases *hydroxyl* ions, it is accompanied by an *increase* in pH, but this is not usually very appreciable, since the cathodic reaction occurs at a free surface of the metal and is distributed over a large area.

Precipitation of the *corrosion product* occurs in the region between the anode and the cathode, where the diffusing *anions* and *cations* in solution intermix, as shown schematically in Fig. 4.13. For the case of the rivetted plate the following features of this process should be noted:

1. The cathodic reaction *suppresses* the reduction in pH which is required if the production of soluble *ferrous* cations is to dominate the anodic reaction. The *anodic* and *cathodic* areas tend, therefore, to be well-separated.
2. The *anodic pits* form in regions where *oxygen access* is difficult, and the acidity associated with the release of hydrogen ions can develop. An ideal site is the *crevice* at the root of the notch formed by the rivet (Fig. 4.13).

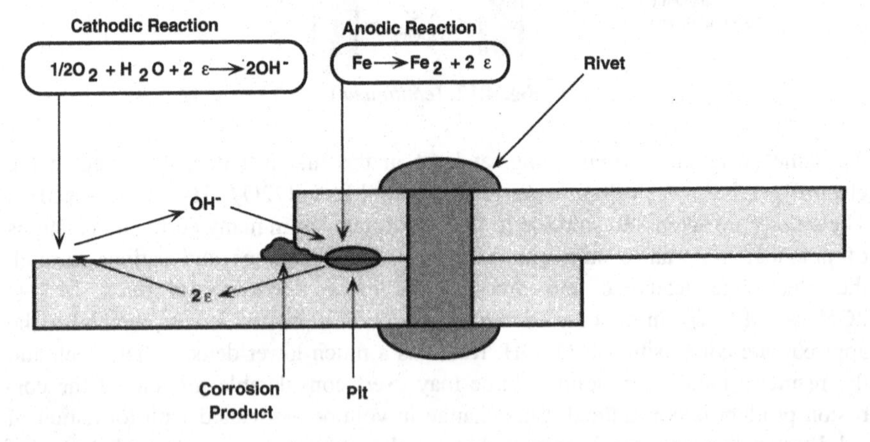

Fig. 4.13. *Crevice corrosion* at a rivetted joint is characterized by *anodic pitting* in the crevice, where *oxygen access* is restricted, and the formation of a solid corrosion product (*rust*) which exerts a pressure on the joint.

3. Since the cathodic area is *larger* than the anodic area, the highly localized anodic corrosion process (*pitting*) is not appreciably affected by polarization at the cathode. The steady state *corrosion potential*, at which the reaction proceeds, is therefore close to the *cathodic electrode potential*, and the *corrosion current* is controlled by the *polarization* of the anode (Fig. 4.14).
4. The corrosion product, *rust*, is precipitated in the crevice, where the pH is approximately neutral, and the volume increase generates a crack-opening stress on the joint. The stress grows as precipitation proceeds, and may rupture the rivet if its yield stress is exceeded.

This type of corrosion is termed *crevice corrosion*, and is extremely common in bolted and rivetted steel structures. The only effective preventative measure is to *seal* the crevice, either with a suitable protective paint, or with some gasket material at the joint.

4.2.2 Loading Response

When a joint is subjected to a varying stress it will respond with some combination of *elastic*, *plastic* and *sliding* displacement. If the response is purely *elastic*, then

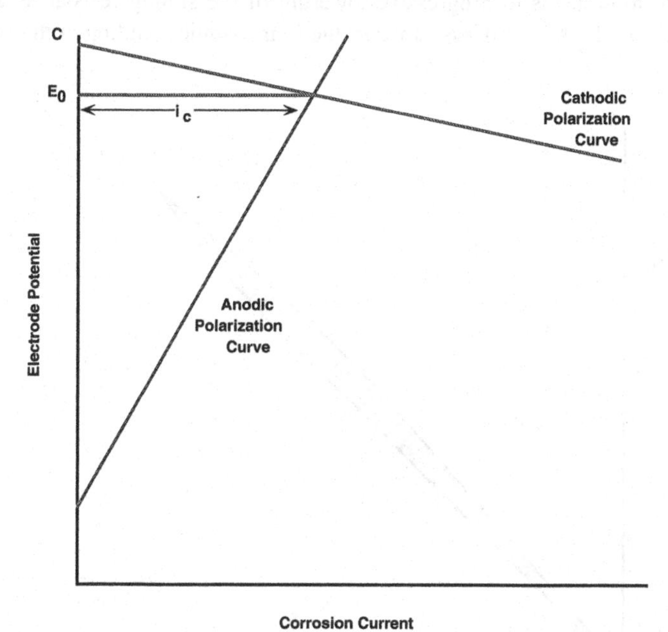

Fig. 4.14. Both the anodic and cathodic potentials are affected by *polarization* when a corrosion current flows, but the restricted anodic area in the crevice leads to much greater polarization of the anodic reaction. The steady state corrosion potential is therefore quite close to the equilibrium cathodic potential, but the corrosion current i_c is controlled by the polarization of the anode.

there is *no* nett displacement at the end of a loading cycle, although some *hysteresis loss* will occur (Fig. 4.15). During subsequent loading cycles the characteristic mechanical response curve may change as a result of *cumulative damage*, eventually leading to failure of the joint in *mechanical fatigue*, even though the response in any one loading cycle appears elastic. If a loading cycle results in some *nett displacement*, then the response is partially *plastic. Repeated loading* of a bolted joint in the region of the *yield stress* may loosen the bolts by small increments of *plastic deformation*, to the point where the joint is not sealed in compression at the maximum operating load.

In mechanical joints, the displacements induced by loading include possible *sliding* parallel to the plane of the joint. In the presence of a compliant gasket, sliding displacement is taken up by *elastic shear* of the gasket and no damage accumulates in the joint. The components may slide with respect to one another when localized shear occurs at the *interface*, leading to *friction* and *wear* (see Section 3.1.5). The *coefficient of static friction* μ and the *applied compressive force* F_n together determine the resistance to sliding F_s: $F_s = \mu F_n$. These displacements frequently combine to cause failure at a mechanical joint from incremental plastic deformation under *repeated* loading to the point where the reduction in the normal compressive force leads to progressive lowering of the sliding resistance. Damage accumulates rapidly by a process of *wear* due to mechanical sliding at the interface.

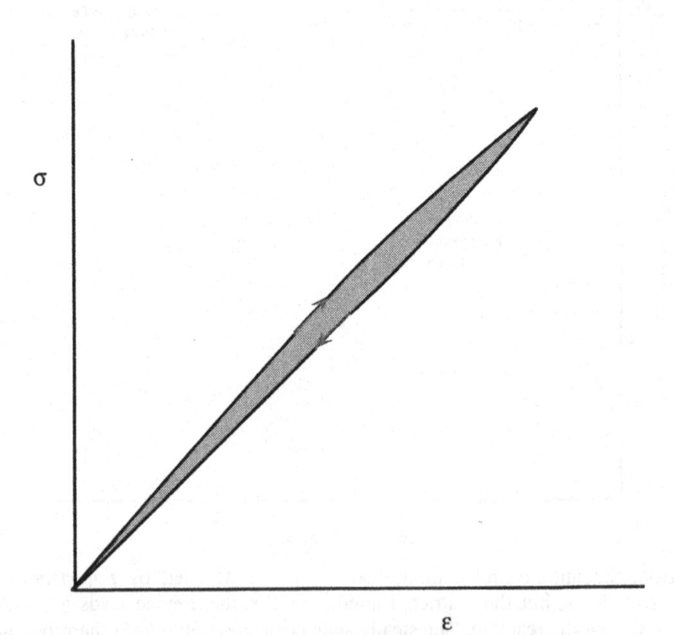

Fig. 4.15. Even an *elastic* mechanical response (no nett shape change on releasing the load) will be accompanied by some energy losses (*hysteresis*).

The breakdown of *contact points*, the accumulation of *wear debris*, and the effect of wear debris on *abrasion* at the sliding surface have been discussed earlier. The sliding *amplitude* may be limited to distances of the order of the separation between the contact points. Damage then accumulates as the result of localized, *reverse* sliding and is termed *fretting corrosion*.

4.2.3 Damage Mechanisms

The major sources of *damage* in mechanical joints can now be summarized under three headings:

1. *Corrosion-induced stresses* and stress concentrations, especially as a result of *crevice corrosion*.
2. *Fatigue cracking* associated with *cumulative damage* due to hysteresis losses in a repeated elastic loading cycle.
3. *Wear damage* associated with repeated *reverse sliding* at a joint which has been loosened by incremental *plastic relaxation* of the compressive stress.

Once the design of a mechanical joint has ensured that *overload* is not a primary cause of failure, then *failure prevention* in the joint reduces to the problem of *inhibiting* these damage mechanisms: ensuring that the joint is adequately *sealed* to prevent corrosion, avoiding overloading in the *prestressing* of bolts and fasteners, and maintaining adequate *compressive stresses* at the joint itself.

4.3 SEALS & GASKETS

A simple and effective method of improving the integrity of a mechanical joint is by the use of a *seal* or *gasket.* The distinction between these two terms is of historic rather than technical importance. A *seal* was designed to effect a closure which could not be opened without breaking the *seal*, typically for documents or deposit boxes. A *gasket* was a fibrous packing used to caulk a joint, typically in ship construction. In common technical parlance the two terms are frequently interchangeable. *Gaskets* are used for *demountable* joints while *seals* may be either *permanent* (as in elastomeric sealing compounds and thermosetting resins) or *demountable* (as in vacuum O-rings or thermoplastic waxes). They may be intended just to 'seal', or they may be intended to transmit movement—*sliding seals*. Most often they are *single-phase* materials, commonly *polymeric*, although metal seals (gold wire) and ceramic seals (silicon carbide) are also technically feasible solutions to some engineering problems. Some sealing compounds contain *filler powders* to modify their setting characteristics and mechanical properties.

4.3.1 Gasket Load Response

Gaskets are commonly organic-based fibrous compounds whose primary function is to *seal* the joint between two components without allowing the components to come into direct contact. They are a *load-spreading* element whose response to the *compressive stresses* exerted at a joint serves to accommodate dimensional mismatch. The typical *load response* of a gasket is illustrated in Fig. 4.16. The initial, *high compliance*, portion of the curve ensures that the gasket material is squeezed into any gaps in the joint, while the final, *high stiffness* portion allows the compressive loading to be controlled to the appropriate level.

The initial response is *irreversible* (Fig. 4.16), so that a gasket will *leak* if the compressive load on the joint is relaxed. Furthermore, although the gasket is demountable, the irreversible compaction means that it cannot be used a second time, but must be replaced.

4.3.2 Viscoelastic Behaviour

Polymeric materials are commonly used for a wide variety of *seals* and *gaskets*, and the properties of the polymer determine the response of the seal. Elastomeric O-rings have *viscoelastic* properties, and behave qualitatively according to a *Voigt–Maxwell* model (Fig. 4.17). The model consists of four elements, a *spring* and a *dashpot* in

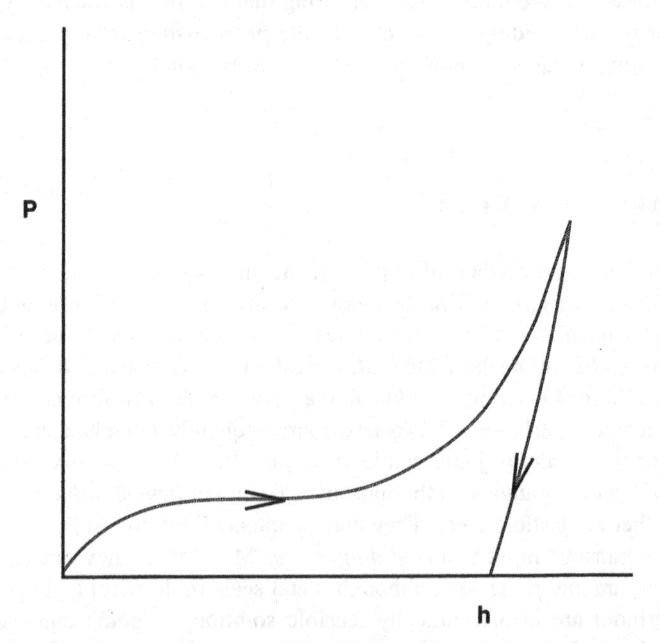

Fig. 4.16. The load-displacement response of a gasket has a an initial *high compliance region* which is irreversible, followed by a *high stiffness* region.

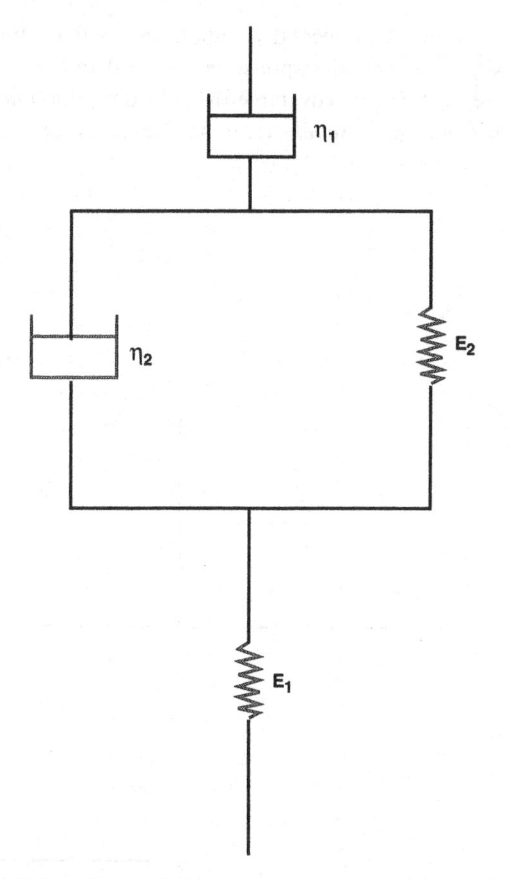

Fig. 4.17. The *Voigt–Maxwell model* for a viscoelastic material combines both a *viscous* (plastic) and an *elastic* response, with a region of time-dependent *viscoelastic* behaviour. In the simplest case *four* material constants are required to describe the behaviour: a *viscosity coefficient* η_1, an *elastic modulus* E_1, and two *viscoelastic constants* η_2 and E_2. The *ratio* of the viscoelastic constants gives a *viscoelastic relaxation time*, $\tau = \eta_2/E_2$.

series, and a *spring* and a *dashpot* in parallel. The spring and dashpot in *parallel* constitute the *viscoelastic* element: a load applied instantaneously to this element is gradually transferred from the *dashpot* to the *spring*, and the *total* limiting extension is determined by the *stiffness* of the spring. On releasing the load on the *viscoelastic* element, the *spring* exerts a *compressive* force on the *dashpot* and the system relaxes with a *time constant* which depends on the *coefficient of viscosity* of the *dashpot*. The displacement in the viscoelastic element is *recoverable* (elastic) given sufficient time, hence the term *viscoelastic*.

The spring and dashpot in *series* model an *elastic* element (the spring) together with a *plastic* element (the dashpot). The rate of plastic displacement depends on the *coefficient of viscosity* of this second dashpot, and will be linear if we assume *Newtonian viscosity*. The overall response is sketched in Fig. 4.18. In this model the *elastic* response ε_e is *fully* recoverable on unloading, the *plastic* response ε_p is *never* recoverable, and the *viscoelastic* response ε_r is one of *time-dependent recovery*.

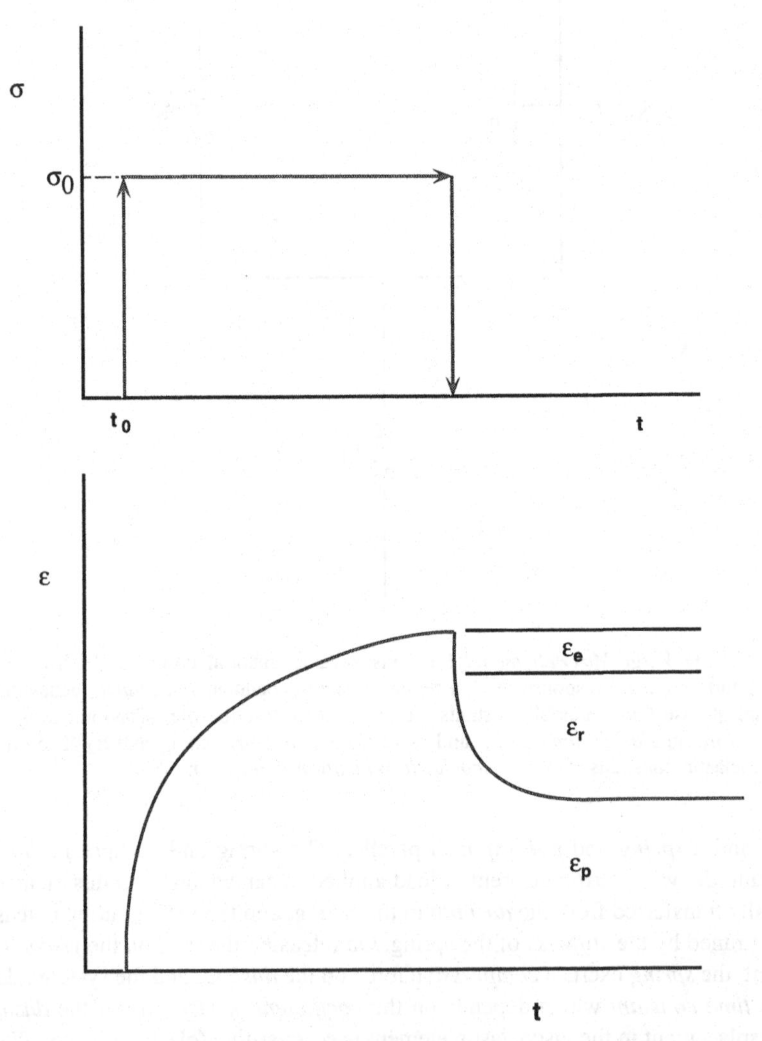

Fig. 4.18. When subjected to a stress the Voigt–Maxwell model shows an *instantaneous* elastic response, a *time-dependent* viscoelastic response, and a *constant rate* of viscous displacement. On removing the stress, *elastic* strains ε_e are instantly recovered, *plastic* strains ε_p are irrecoverable, and viscoelastic strains ε_r recover over a period of time.

4.3.3 O-Ring Design

Most *elastomeric* demountable seals show some form of *viscoelastic* behaviour. That is, the compressive load acting on a seal *relaxes* as a function of time, so that the seal may require tightening at some stage. Moreover, these seals also exhibit time-dependent, *non-recoverable* plasticity, so that they will eventually deform *plastically* to the point at which they have to be replaced. Figure 4.19 shows four possible designs for an *O-ring seal*. In the first the O-ring is not allowed any appreciable freedom to adapt to the sealing pressure, so that the stresses in the plane of the O-ring seal are very large. *Plastic flow* will rapidly distort the shape of the O-ring and the life of the seal will be rather limited. In the second design the square grooves in the two surfaces allow the O-ring to be compressed throughout its cross-section at a lower stress, limiting *permanent plastic set* and increasing the life of the

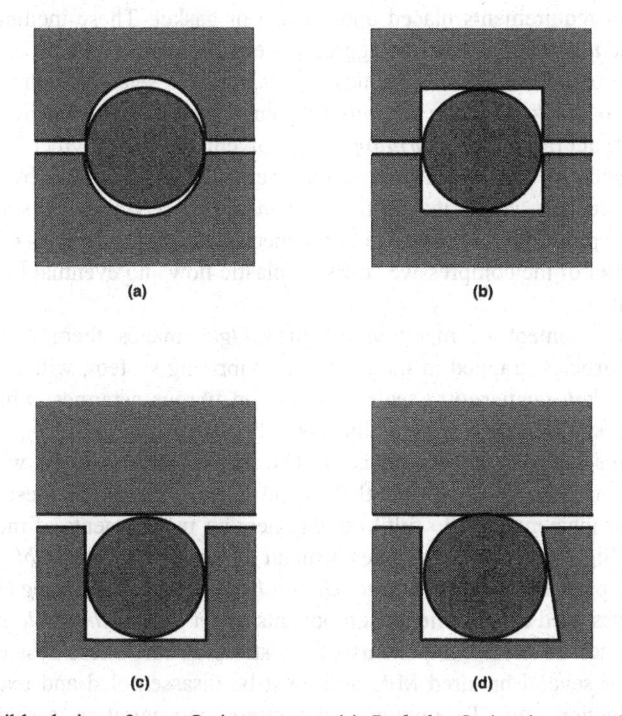

Fig. 4.19. Possible designs for an O-ring groove. (a) *Bad*: the O-ring is constrained by the machined grooves and can only accommodate an applied compressive load by being squeezed into the gap between the components. (b) *Better*: the O-ring has freedom to adapt its shape to the applied load, but will still tend to flow laterally into the gap between the components. (c) *Good*: the O-ring now makes contact with the top and bottom flanges at four points which are all well away from the gap. (d) *Best*: the trapezoid cross-section of the groove is designed to ensure that the O-ring is retained when the joint is demounted, minimizing the chances of entrapping particulate and other contamination.

seal. However, the double groove is unnecessary, and a better design is shown in the third sketch. Here the O-ring makes contact with the three sides of a *single* groove, and the seal is made with a concentrically machined flat flange. The O-ring is uniformly compressed and the three contact points make it much easier to mount the O-ring in the groove. Finally, an additional refinement is introduced in the fourth sketch, in which the *reentrant angle* of the inner radius of the groove acts to retain the O-ring regardless of the orientation of the seal. The additional machining costs are negligible and the life of the seal is maximized. Moreover, the seal can be demounted and reassembled *without disturbing the O-ring,* although some care needs to be exercised to prevent any accumulation of dirt particles or fibres.

4.3.4 Other Engineering Requirements

In addition to the *mechanical properties* of the materials selected there may be other engineering requirements placed upon a seal or gasket. These include the *thermal* and *chemical stability*. Seals for aggressive environments, such as are encountered in sewage treatment and petrochemical plants, must be selected with respect to their resistance to attack. *Organic solvents* frequently attack a seal, not by dissolution in the solvent, but rather by *absorption* of the solvent into the polymer. The seal *swells* as the strands of the polymer molecular network are separated by the absorbed solvent molecules, loses its elastic *resilience*, and acquires *plastic* rather than *elastomeric* properties. The change in the mechanical characteristics of the seal then leads to relief of the compressive stress by plastic flow and eventual loss of integrity of the joint.

The solid content of many industrial *sludges* makes them highly *abrasive.* Abrasive particles, trapped in the joints of a vibrating system, will accelerate *wear* processes. *High temperature* seals may rely on fibrous ceramics, which may have highly undesirable side effects. In the case of *asbestos*, the *carcinogenic* properties of the fibres are the primary concern. The *health hazard* posed when asbestos-sealed boiler systems are dismantled is the reason for taking these seals out of service, *not* their inability to fulfil the engineering requirements of the application.

At very high pressures it becomes difficult to design a *demountable* seal that will contain the pressure without leakage. *Hot and cold isotactic pressing* (HIP and CIP) are processes used to consolidate components under high *hydrostatic pressures* and in the absence of shear components of the stress tensor. These systems operate at pressures of several hundred MPa and must be disassembled and reassembled for each production run. To minimize the stress concentration associated with a machined screw thread, a *grooved closure* is made with a hardened steel *locking spiral* (a spring, Fig. 4.20), while the end-cap is coated with a high pressure anti-friction coating.

In some cases it is possible to use a *sealing compound* with a low melting or softening point. The seal is then *solid* at the operating temperature, but can be demounted by heating *above* the softening point. A *molten tar coating* applied to a

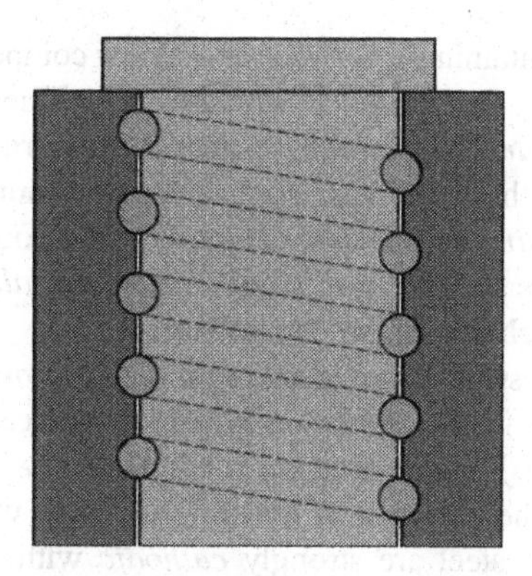

Fig. 4.20. A hardened steel spiral reduces the stress concentration in the demountable end closure of a HIP pressure vessel.

wooden fishing boat is a good example, as are the wax *sealing compounds* used for many undemanding vacuum applications. *Low melting point soldered joints*, based on *indium*, are also readily diassembled, as are some high vacuum seals that are based on *gallium metal*, which has both a low melting point and low vapour pressure.

The *surface finish* of the mating surfaces is a major factor in ensuring the integrity of many joints which have been sealed with a gasket. Concentric *machining grooves*, perpendicular to the sealing direction, are usually acceptable, while *scratches*, corresponding to the same *roughness* but with the direction of abrasion running *across* the seal may be unacceptable. Highly *polished* surfaces may be more susceptible to damage associated with fine abrasive scratches than a *machined* surface, which will then be preferred.

4.4 FASTENER MATERIALS & COATINGS

The range of fasteners available for mechanical joining systems is enormous and includes both *metallic* and *non-metallic* materials. Of the non-metallic materials *nylon* threaded screws and bolts are a generic example. 'Nylon' is the trade name of a range of DuPont *polyamide* polymers, but the term is commonly used for a range of tough, high-strength, hydrogen-bonded plastics. Polymer compositions used for fasteners usually contain a *glass* or *ceramic filler* to increase the elastic modulus. The *electrical insulating* properties of most plastics make them especially suitable for many applications in communications equipment. The term *polymer* is reserved for long chain molecules which have been formed from a monomer by a polymeric reaction. *Low molecular weight* polymers ($M < 1000$) are liquid at room temperature, while *high molecular weight* polymers are solids. The term *plastics* is used for engineering materials containing a *polymer* as a major component. Plastics may also contain *filler* powders or fibres (which increase the stiffness), *plasticizers* (which improve the ductility), *flame retardants* (to inhibit exothermic conflagration), *dyes*, *anti-fungal agents* and other additives. Plastics are by definition solids, and may be *ductile* or *brittle*, *crystalline* or *amorphous* (glassy). The polyamide, nylon family of polymers and plastics are *ductile*, *tough* and largely *crystalline* (hydrogen bonding cross-links the polymer chains in the crystal lattice).

Metal alloy fasteners suffer from a susceptibility to *corrosion* which is a consequence of differences in the electrode potential between the fastener and the components being joined. The use of *brass* or *stainless steel* nuts and bolts to secure *carbon steel* flanges in the presence of moisture is a sure way to initiate corrosion, since brass and stainless steel are strongly *cathodic* with respect to the steel. To avoid forming a corrosion cell, the *electrode potentials* of the fasteners and the components should be closely similar. Since most fasteners are required to operate in *tension* (placing the joint in *compression*), it is important that they should have sufficient *ductility*. The lower limit of ductility for high strength bolts is of the order of 4%, which is considered sufficient to ensure that yielding occurs at *stress-concentrating defects*, such as may occur at the root of a machined thread. *Self-tapping* steel screws may have lower ductility than this (and must have sufficient hardness to machine the thread into the mating component). Materials for *rivets* must be much *more* ductile, in order to accommodate the *upset forging* which accompanies the securing of the rivet. Rivets are commonly made from low carbon steels or high ductility, wrought aluminium alloys.

The performance of a fastener can often be improved by *coating*. Metallic coatings serve a dual purpose, both providing some *corrosion protection* and preventing mechanical *seizure*, by reducing the *coefficient of sliding friction*. Metal coatings for bolts and screws are termed either *anodic* or *cathodic*, depending on whether the coating is anodic or cathodic with respect to the material of the fastener. A coherent anodic coating will raise the electrode potential of the fastener, preventing corrosion at the lower cathode potentials. *Nickel-coated* bolts are anodic with respect to carbon steel, but any *defects* (such as microcracks in the coating) can lead to intense *localized corrosion* of the steel beneath the coating. The coating will eventually *spall* and peel off the surface as a result of the added volume of the corrosion products. *Chromium-coated* bolts form an adherent chromia film which makes them *cathodic* with respect to the steel. *Defects* in the coating are then immune to corrosion, since they remain cathodic. Chromium coatings can be extremely *hard*, reducing the area of plastic contact and hence improving resistance to *wear*. *Cadmium-coated* bolts are also cathodic to steel, and the oxide film formed on cadmium is not very protective. A steel bolt is therefore *cathodically protected* by cadmium, and the large surface area of the coating ensures that no localized attack is likely. However the cadmium also serves a quite different purpose, in that the basal plane of the *hexagonal crystal structure* of cadmium provides an easy shear path for the breaking of sliding contacts. Cadmium-plated nuts and bolts can therefore be tightened with minimal frictional losses in the threaded contacts and are therefore less likely to *seize*.

4.5 MECHANICAL JOINING SYSTEMS

The simplified mechanical joining systems described below are intended to illustrate some of the factors which determine the selection of an assembly method for a mechanical joint.

4.5.1 Lock Seams

Lock seams made from sheet metal are ubiquitous. An example was given in Chapter 1 and the residual stress field necessary to ensure the integrity of the lock seam was described in Chapter 3. To be completely effective the lock seam must be sealed, and this may be done either with *solder* (as in a paint can) or by an *organic resin* (as in some long-life food storage containers, especially in low pH applications, such as preserves and jams).

4.5.2 Screws & Bolts

In spite of their poor ductility and their tendency to promote corrosion by acting as strong cathodes, *brass screws* and *bolts* continue to find diverse applications, for example as major components of domestic electrical equipment. The primary reason is their good *machinability*, which is also a consequence of their limited ductility. Alternative methods of mechanical assembly are being substituted for brass nuts and bolts, even though nickel plating reduces corrosion susceptibility by reducing the electrode potential with respect to steel. One reason is the *cost* of the alloy, but another is the susceptibility of brass to environmentally-assisted crack growth: *stress corrosion cracking*. This phenomenon is associated with the development of either *intergranular* or *transgranular* microcracking at the site of a *stress concentration* and in the presence of *moisture*.

Steels provide by far the widest range of screws and bolts, including high tensile strength, heat-treatable alloy steels, and the *precipitation-hardened*, high strength stainless steels (the PH steels) for corrosion resistant applications. *Electroplated* steel bolts are a cheaper alternative to the PH steels, if they are capable of surviving the application without *localized corrosion* or *spalling* of the coating. As in other engineering applications, the primary advantages of steel are its *cost effective strength* and *toughness* and these have to be set against its primary disadvantage–susceptibility to degradation by *corrosion*.

4.5.3 Rivets

A *rivet* is pushed into a locating hole in the two components to be joined and *cold-headed* (upset-forged) to form the join. Sufficient *compressive* elastic energy must be stored in the components to ensure that the rivet is placed in *tension* by stress relaxation when the compressive forging pressure is released. The rivetting process is illustrated schematically in Fig. 4.21. The quality of the final rivetted structure depends sensitively on the preparation of the hole and the control of the punch pressure cycle. A wide range of rivet designs is available, in which a compromise is sought between the *strength* of the assembled joint, the required *ductility* of the rivet material and the *control* of the forging process.

Assuming the rivet to be elastic to the yield stress, to have the same compliance as the material being joined, and no workhardening beyond the yield stress, then the compressive load required to forge a rivet of radius r is just: $F_c = \pi r^2 \sigma_y$. If the compressive load is increased beyond this point the elastic energy is stored in the joint assembly (Fig. 4.21). The situation can be analysed approximately using similar assumptions to those made previously (Section 4.1.4). As the compressive load is released, the sign of the stress in the rivet is reversed, placing it in *tension*. If the forging force exceeds $3F_c$, then the relaxation process will place the rivet under a tensile stress which exceeds its *yield stress*, and *reverse* plastic flow will occur. The residual tensile stress in the rivet increases linearly from zero to σ_y as the forging force increases from F_c to $3F_c$. The tensile strength *perpendicular* to the rivetted joint will be a maximum when the forging force is the *minimum* required for general yielding, $F = F_c$, so that the rivet experiences *no* prestress, but the tensile strength *parallel* to the joint is *improved* by the tensile residual stress in the rivet, since the *frictional force* at the interface reduces the stress concentration at the rivet by assisting *load transfer* to the components (Fig. 4.22). The optimum cold-heading conditions thus depend on the expected stresses in service, and will be somewhere in the range $F_c < F < 3F_c$. A more accurate analysis of the mechanical performance of a rivetted joint is outside the scope of this text.

The importance of rivetted structures dates from the earliest days of the industrial revolution, in the construction of boilers for steam engines and, later, in the construction of steel-hulled ships. *Welded* structures to a large extent superseded *rivetted* plate in the ship-building industry after the 1939–45 World War, but for many years *rivetted assemblies* have dominated the airframe industry, in which the material of first choice is a wrought aluminium alloy. Today aluminium rivetted airframe structures are being challenged in some areas, by *adhesive lap-joints*. In many applications a rivetted join is in competition with a modified lap-joint, which combines the *lock-seam* design with an *adhesive*. This combination both seals the join and prevents excessive stress concentration, by distributing the load over a larger contact area. The main limitation on the lock-seam remains the requirement for *ductility* of the sheet (which must be bent to a radius of the order of the thickness).

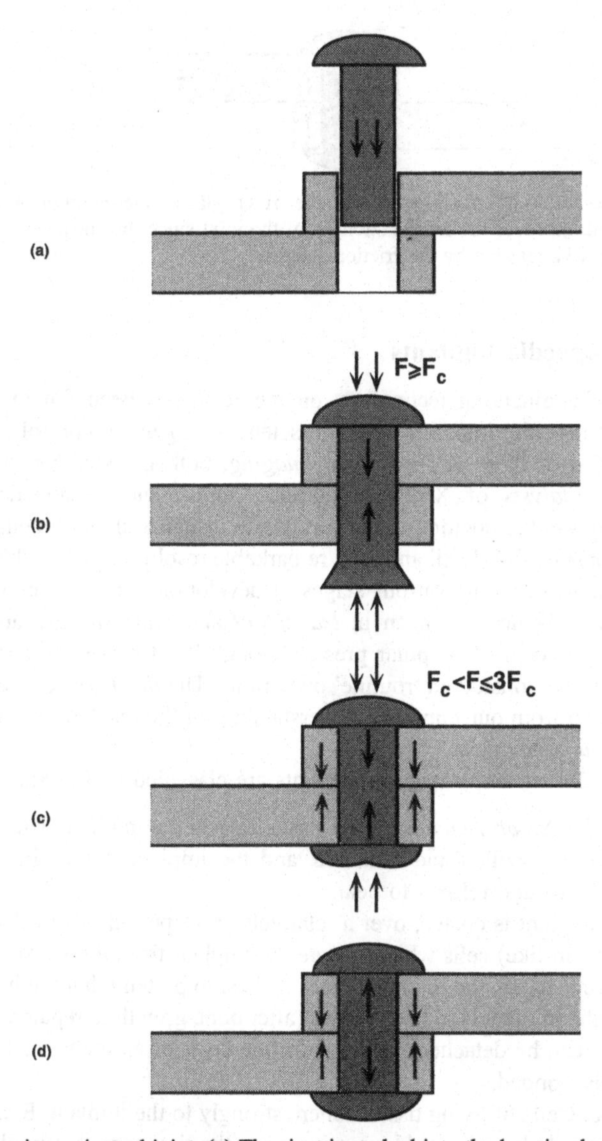

Fig. 4.21. Forming a rivetted joint. (a) The rivet is pushed into the locating holes in the sheet components. (b) Upset forging of the rivet is initiated when the applied force exceeds that required for general yielding. (c) The compressive force used to forge the rivet also places the components in compression. (d) Provided that the compressive force is not excessive, the residual tensile stress in the finished rivet will be below its yield stress.

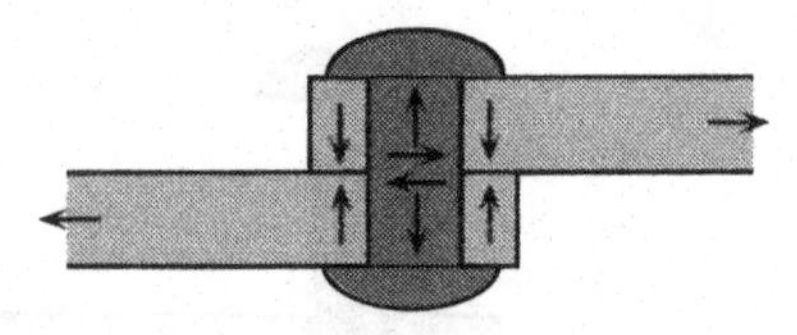

Fig. 4.22. The residual tensile stress in the rivet places the components in residual compression and improves the tensile strength of the joint since the shear load at the joint is distributed over a larger area by the frictional forces.

4.5.4 Orthopaedic Implants

The impact of engineering technology on medicine has been felt in three areas, which complement the impact of the life sciences in genetic control and pharmacology. The *first* of these is *biomedical imaging*, both invasive and non-invasive. *Tomographic analysis* of X-ray radiographic and nuclear magnetic resonance (NMR) data allows the doctor and the surgeon to visualize three-dimensional tissue structure in considerable detail and with remarkable resolution, while the *ultrasound imaging* of the foetus in its various stages of development is a commonplace. The *second* area of advance has been in *transplant* and *microsurgery* and has been extensively reported in the popular press, although heart bypass surgery and heart transplants are now considered 'routine' operations. The *third* area concerns us here, and has followed from our improved understanding of the reaction of a living tissue to an *implanted component.*

Tissue reactions to implanted components are classified under three headings:

1. *Incompatible*: An *ab-reaction* of the tissue rejects the implant. The surrounding tissues become swollen and inflamed, and the implant is the site of pain and oedema. The wound refuses to heal.
2. *Inert*: The implant is coated, over a relatively short period of time, by a layer of *epithelial* (skin-like) cells which *isolate* the implant from the surrounding living tissue. Once isolated, the organism does its best to pretend that the implant is not present. If the implant is to be removed (after bone growth to repair a fracture, for example) it can be detached readily from the layer of epithelial cells, which are very loosely bonded.
3. *Compatible*: Cells of living tissue adhere strongly to the implant. Removal of the implant can only be achieved by cutting away the tissue, so that implant removal is accompanied by *tissue damage.*

Implant surgery introduces two *new* engineering requirements for any bonding process:

1. The time-dependence of *post-surgical healing* and the influence of the healing process on bond formation must be considered. A compromise is usually necessary between the *immobilization* of the patient following surgery and the

need for suitable *muscular stimulation* to complete the healing process (such as the regrowth of a broken bone).

2. The need to avoid *tissue incompatibility*. This is the inverse of the usual requirement for the *component* to be stable in the environment (that is, it must exhibit sufficient corrosion or oxidation resistance). We now require the *environment* (the biological tissues) to be stable in the presence of the component!

We confine the present discussion to *orthopaedic* implants intended to *repair* the bony skeleton. The *setting* of a broken bone is achieved by fixing the two ends in position and encouraging *natural* bone growth processes to make the repair. The growth of bone is stimulated by *stress*, and loading the bone in *compression* encourages growth. However, if the applied load is *too* high, *dissolution* will occur (the bone will shorten). Dissolution of bony tissue will also occur if the bone is *unstressed* (*calcium loss* is a major problem for astronauts held in weightless orbit for extended periods of time). It follows that a 'working window' of compressive load exists for bone healing to occur.

The *mechanical* problem is then to fix the bone in position, but allow for sufficient *compressive* stress at the contact between the two fractured ends in order to promote growth of new bone. The commonest method of achieving this is by the use of *metallic plates* and *screws*, typically made from a high quality stainless steel or a titanium alloy. Both these materials are *biochemically inert*, and so can be removed relatively easily in a subsequent operation, once healing is complete, should this be considered medically desirable. However these metals also have a very much higher elastic modulus than bone, so that relaxation of the compressive load on the broken bone can occur at very small displacements (Fig. 4.23). In consequence, *dissolution* of the bone may replace bone *growth* at the fracture. To prevent this considerable attention has been given to the design of an *isocompliant* implant, with elastic properties approximating those of bone. Some plastic materials appear suitable, but have not yet gained wide acceptance.

The setting and fixing of broken bones is a task for *traumatic* (unplanned or emergency) surgery. The replacement of defective joints is undertaken in *elective* (planned) surgery. One of the commonest operations in *orthopaedic* elective surgery is that of *hip-joint replacement*. This is usually indicated if the *meniscus* which cushions the load applied to the hip-joint is damaged or destroyed. The operation involves replacing the joint with an artificial *cup*, set into the pelvic girdle, and a matching *ball and shaft*, inserted into the head of the femur (Fig. 4.24). Until recently the preferred materials were a high density *polyethylene* cup and a highly polished *stainless steel* ball mounted on a stainless steel stem. Today the stainless steel ball has been largely replaced by an alumina ball with a conical hole which fits over the stainless steel or titanium alloy shaft. The cup is cemented in place with a *methacrylate resin* which is formulated to retain its dimensions after molding (many such compounds shrink as cross-linking process proceeds). Unfortunately, the cross-

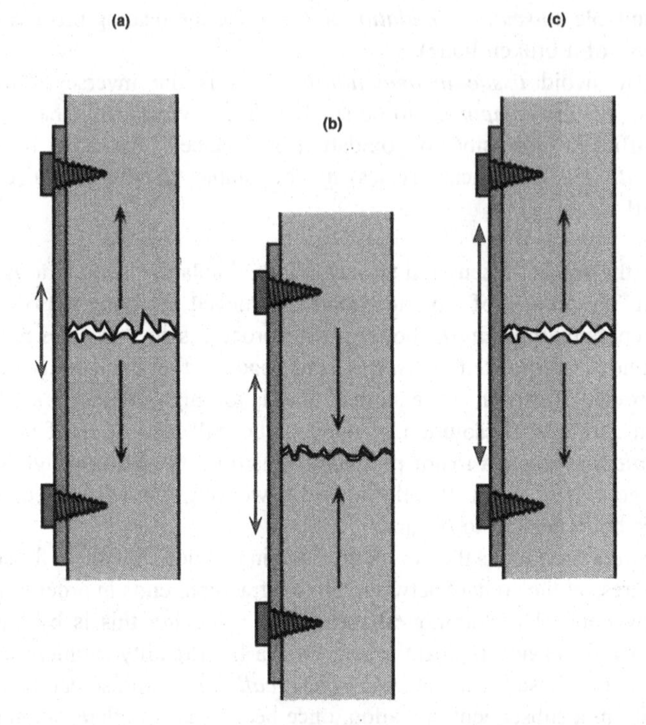

Fig. 4.23. A metal bone plate fixes the two halves of a fractured bone in contact until new bone has formed to repair the break. (a) If the *tensile* stress in the plate is insufficient, then the *compressive* stress in the bone will be insufficient to promote growth and bone *dissolution* may occur. (b) There exists a window of compressive stress in the fracture contact area within which the *growth* of new bone is stimulated. (c) If the tensile stress in the metal plate is excessive, then the *compressive* stress in the contact area will again result in bone *dissolution* rather than new growth.

linking of methacrylate is an *exothermic* chemical reaction and the heat evolved must be dissipated. Moreover, the reaction is *accelerated* if the methacrylate does heats up, and the heat evolved can kill the surrounding living tissue (*necrosis*).

The *stem* of the metal ball is forced into the *shaft* of the femur (the thigh bone) and held in place by a residual *compressive radial stress*. Here the problem is once again the high elastic modulus of the metal compared to the bone and its limited biocompatibility. Very little dissolution of the bony tissue is required before the compressive stress on the stem of the implant is relaxed. At the same time, the metal is *inert*, rather than *biocompatible*, leading to a weak bond at the metal–tissue interface. The cyclic *bending stresses* which act on the hip-joint during walking can result in rotation of the cup out of position or, worse, loosening of the stem of the implant in the femur shaft and eventual failure of the joint. One possible solution to

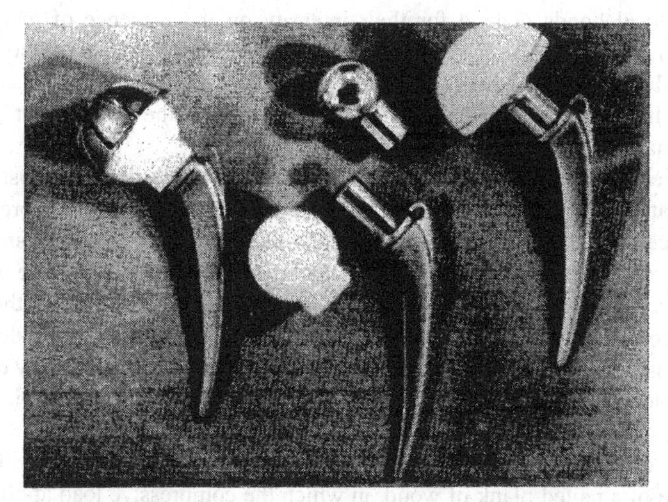

Fig. 4.24. An artificial hip joint has two components: the *cup*, which is set into the pelvic girdle; and the *ball*, whose stem is inserted into the head of the femur (the thigh bone). Reproduced with permission from Plenum Press, from Park, *Biomaterials: An Introduction* (1979)

this problem which has been clinically tested and approved is the development of a stem with a *porous layer*, formed by partial sintering of either packed metal fibres or a coarse metal powder (usually a titanium alloy) to the stem of the implant. The interstices in the porous surface layer are large enough to allow the *ingrowth* of bony tissue, together with the capillary blood supply needed to keep the tissue alive. Of course, such a joint is now strongly bonded to the bone, and replacement of the stem would be a major surgical undertaking.

A major problem in an artificial hip-joint is the *wear* which can occur at the interface between the ball and the plastic cup, especially if there is some misalignment of the joint. Both the *wear damage* and *frictional losses* are reduced by the combination of a polymer cup with an alumina ball and a metal stem. It is worth recognizing that surgical procedures have to be performed on soft tissues of complex shape and in the presence of rather unstable viscous fluids. The dimensional tolerances which can be achieved are strongly dependent on the individual skills of the surgeon, and perfect alignment of the implant components is not easy.

Summary

Mechanical bonds are ubiquitous in engineering systems. They are enormously versatile, permitting assembly of components made from diverse materials and often allowing for ease of disassembly of the components and their eventual replacement.

All mechanical bonds depend for their integrity on the presence of macroscopic residual compressive stresses between the components which maintain the intended geometrical constraint of the component within the engineering system during its operation. In most cases mechanical bonds also rely on the presence of frictional forces to maintain the compressive constraint in service.

Since mechanical bonding depends on the pressure of residual macrostresses, it follows that the performance of mechanical bonds is sensitive to the presence of stress concentrations associated with discontinuities or defects. Such stress concentrators are generally mesostructural, that is, they exist on a scale which is intermediate between that of the microstructure of the components and the macrostructure corresponding to the design features of the assembly. Non-destructive evaluation is the principle tool used to monitor the mesostructural integrity of a joint, often supplemented by an analysis of the macroscopic and microscopic residual stresses.

The role of friction in maintaining the integrity of a mechanical joint is illustrated by the case of a nailed plank of wood, in which the compressive load applied to the nail leads to a normal residual compressive stress between the two wooden components which is balanced by a residual tensile stress in the nail. The load is transferred from the nail to the wood by the frictional shear stresses at the interface between the two.

Similar conditions apply to a bolted flange, the tension in the bolts balancing the compressive load on the flange and the frictional forces being generated at the contact surfaces of the screw thread. The successful performance of these mechanical assemblies relies on the sum of the assembly stresses and the stresses developed during operation of the system being less than the stress leading to plastic relaxation or failure (generally the yield stress of one or other of the components).

Residual stresses can be measured either destructively or non-destructively, but the interpretation of the results may be difficult, especially when the primary concern is to evaluate stress concentrations at the mesostructural level. In addition the macrostresses associated with assembly may be quite sensitive to geometry and scale, as well as to relatively small changes in temperature. In many cases these stresses are associated with thermal expansion mismatch or mismatch in the elastic moduli of the components.

Mechanically bonded joints are often sensitive to electrochemical corrosion associated with variations in either the electrochemical potential across the joint, or the supply of oxygen in the joint region. Crevice corrosion, associated with anodic attack in a notch or crevice, can lead to the build up of low density corrosion products with a high specific volume. The corrosion products may exert sufficient pressure on the joint, already weakened by the corrosion process, to cause failure.

In the presence of a variable load, repeated plastic relaxation may lead to a loss of joint integrity, while small repeated shear displacements may lead to fine-scale wear and fretting corrosion. The commonest forms of damage in mechanical joints are due to overloading during assembly, corrosion and corrosion-induced stresses, cumulative plastic damage associated with repeated loading (mechanical fatigue) and wear accompanying reversed shear displacements.

Gaskets and seals are used for two primary purposes: to redistribute the load at the joint and reduce stress concentrations, and to prevent leakage of gas or liquid across the joint. Many seals are required to operate under extremes of pressure and temperature, or to accommodate relative movement between the components. Most sealing materials are to some extent viscoelastic: that is, they relax as a function of time under load. Most seals and sealing compounds are required to be sufficiently elastomeric—rubber like—to ensure that all minor displacements across the joint can be accommodated without loss of contact between the seal and the components, that is, without leakage.

Mechanical fasteners may be metallic or non-metallic and cover an enormous range of designs and applications. All are intended to be used in tension, maintaining the joined components in compression. Coatings for fasteners may be intended to afford corrosion protection or to reduce frictional stresses, or both.

Among the many mechanical joining systems are screws, bolts, nails and rivets, as well as those systems in which the components are constrained to form the join, as in a lock seam. Orthopaedic implants provide a variety of examples of mechanical joining systems whose single aim is to repair the integrity of living skeletal tissue. The selection of materials for the implants used in orthopaedic surgery introduces two additional engineering requirements: the need to account for the time dependence of bone growth, and the need to avoid tissue incompatibility, that is the adverse response of the living tissue to the presence of the implant.

Further Reading

1. J. E. Shigley, *Mechanical Engineering Design*, First Metric Edition, McGraw-Hill Book Company, New York, 1986.
2. R. W. Messler, Jr., *Joining of Advanced Materials*, Butterworth–Heinemann, London, 1993.
3. I. C. Noyan and J. B. Cohen, *Residual Stress–Measurement by Diffraction and Interpretation*, Springer-Verlag, London, 1987.

Problems

4.1 Sketch the cross-section in the plane of loading for a beam in three-point bending. Show the stress distribution across the beam, in the plane of maximum stress and parallel to the axis of loading. Now assume some plastic deformation has occurred during bending: show the location and state of residual stress across the beam *after* the external load has been removed.

4.2 Describe *in words only* the production process by which a paint can is produced by lock seaming.
   a. What are the limitations of the materials which can be used in such a process?
   b. In which area of the can would you expect to find regions of stress concentration?

c. Describe the location and type (tensile or compressive) of residual stresses which result from the process.
d. What can be done to 'relieve' the state of residual stress, either during or after the process has been completed?

4.3 Explain the differences in the final stress distribution associated with joining by bolts and rivets.

4.4 What are the geometric differences between wood screws and bolts. How are these differences related to the joining method and how do they affect the residual stress distribution in the component after joining?

4.5 The head of a wood screw has failed in brittle fracture during installation.
a. Make two reasonable guesses at the cause of failure.
b. Suggest possible solutions to prevent the failure.

4.6 Nuts frequently come loose from the bolt during service, especially if the system is subjected to mechanical vibration. How can this be prevented?

4.7 How does the low elastic modulus of elastomer and fibre affect joint design and assembly?

4.8 Two tube sections are joined by bolting together their end flanges with four equally spaced 5 mm diameter bolts. The tubes (and hence the retaining bolts) are then loaded in tension.
a. Failure of the joint eventually occurs at one or other of the bolts. What are the probable sites of failure along the bolt?
b. Estimate the stress concentration factor at a screw thread by making reasonable assumptions about the geometry of a 5 mm diameter bolt.
c. During assembly, the bolts are prestressed in tension to $1/3$ of the yield stress of the material (200 MPa) in order to ensure a leak-tight joint. Using your own estimate of the probable stress concentration factor for the thread geometry, what is the expected failure strength of the bolted joint? Explain your assumptions.

4.9 A hard rubber gasket is used to ensure a gas-tight seal in a pump. Assuming that the seal is leak-free when the true contact area equals the apparent cross-section of the seal, derive an expression showing how the required closure pressure depends on the surface finish of the ground metal sealing surface. In general, leakage occurs when a critical over-pressure is reached in the pump. Qualitatively, how is this over-pressure related to the initial closure pressure? (The radius r of a Hertzian contact area is given by: $r = 3\sqrt{(P/E_b \cdot R/2)}$, where P is the applied force, E_b is the biaxial modulus and R is the radius of curvature of the contacting surfaces.) Explain your assumptions.

4.10 Given the following three designs for a rubber vacuum seal, select the most appropriate. Explain your answer in detail.

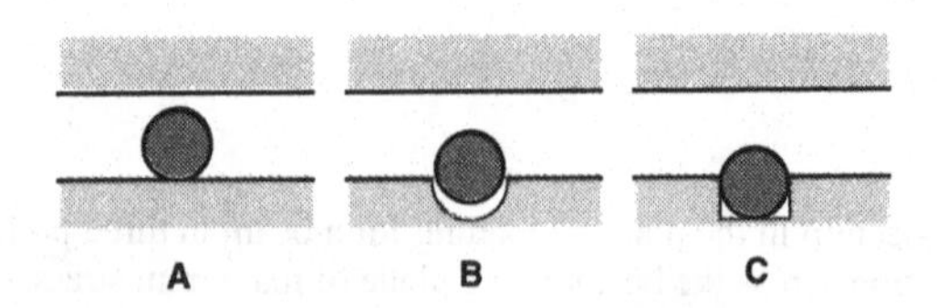

4.11 A bolt and nut assembly is used to joint two plates. The bolt and nut are made of brass ($K_c = 20$ MPa $\cdot$ m$^{1/2}$). Given that the ratio of the thread separation to the thread-root radius of curvature is $C/r = 10$:
a. Calculate the size of a critical flaw located at the thread base.

b. What is the critical thread separation which results in failure? Explain your assumptions.

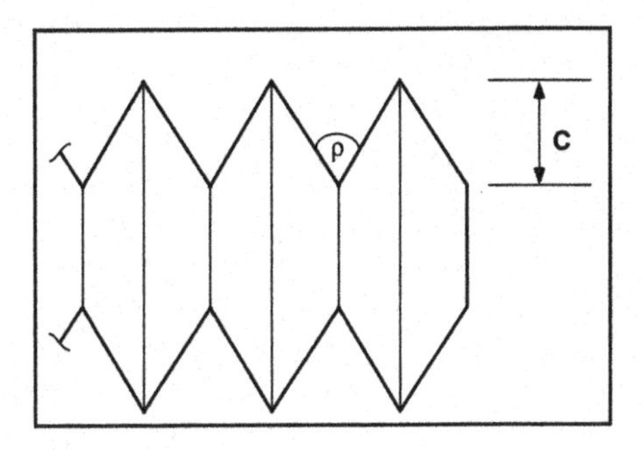

5

Welding

In *welding* two components are joined by heating the region at the interface above the melting point of one or other of the components. Welding is distinguished from *soldering* and *brazing*, in which a low melting point filler metal is used to make the join, as well as from *diffusion bonding*, in which the temperature at the interface is kept below the melting points of *all* the phases present.

In addition to welding as a term to describe bonding by localized melting, the term *solid state welding* is used to describe at least two processes in which the melting point of the components is *not* exceeded at the interface. The first of these is *friction welding*, in which bonding relies on the heat generated by friction at a sliding interface subjected to compression. The frictional heating raises the temperature at the interface to the point at which extensive plastic flow occurs to relieve the compressive stress. The second process is *explosive welding*, in which a flyer plate travelling at high velocity impacts a stationary component to generate hydrostatic pressures which exceed the *Hugoniot elastic limit*, the pressure above which dynamic plastic flow accompanies a ballistic shock front. Hugoniot pressures are typically in the range of 5 to 100 GPa, at least an order of magnitude greater than the static yield stress. The bonded interface formed in both friction welding and explosive welding is a region of intense turbulent flow, where surface defects and contamination are *extruded* from the bonded zone by the high plastic strains resulting in a strong joint of high integrity. Both processes are characterized by a high gradient of *plastic strain* at the joint, which limits the width of the zone affected. Both processes are also limited in their applications by some fairly severe restrictions on joint geometry.

5.1 WELDING SCIENCE

In the present chapter we are primarily concerned with the *scientific principles* of welding, in particular the basic *transport* processes that determine *mass* and *heat transfer* during welding. We also outline some engineering solutions to problems of *weld design* and describe the more important sources of heat for welding technology. Since an *electric arc* is one of the commonest of these, we describe the physical characteristics of an electric arc. Finally, we outline the nature and characterization of *welding defects*.

It is as well to appreciate why, despite the high temperatures and large power inputs required, *welding* remains one of the most versatile and effective means available for the assembly of individual components into larger modules for both large and small engineering systems. At one extreme, the outer shell of a welded reactor pressure vessel may have a thickness of the order of 0.5 m. Welded armour plate for a battle-tank is also massive (at least 0.2 m). At the other extreme, ultrasonically welded conducting wires for microelectric assemblies may have a diameter of less than 15 μm, and submicron device technologies are attempting to reduce this. The *range of linear dimensions* for welded structures thus covers over 4 orders of magnitude.

Welded assemblies are able to carry loads similar to those supported by the individual components from which they are constructed, without requiring the addition of appreciable mass or volume to the overall assembly. It is the high *strength to weight ratio* (at reasonable processing costs) that compensates for the high processing temperatures and environmental hazards associated with welding. No mechanical bond can compete in its strength to weight ratio with this load-carrying efficiency of a welded joint. No adhesive bond can match the tensile and shear strength of a welded joint. If disassembly is not a requirement, then welding is very often the joining method of first choice.

5.1.1 Weld Geometry

Any *weld design* must aim at ensuring the integrity of the weld and, effectively the same thing, minimize the effects of *welding defects*. There are two major considerations. Firstly, the control of the dimensions and thermal history of the molten metal in the *weld pool*. Secondly, the analysis of the *geometrical constraints* imposed by the system and the effect of these constraints on the development of *residual stresses* during the welding cycle.

Welding generally approximates one or other of two limiting cases of *heat transfer*. In the first case welding stresses and heat transfer are *two-dimensional* (Fig. 5.1a). The temperature is assumed constant in the through-thickness direction and varies only in the plane of the components being welded. The component is in a state of *biaxial stress* throughout the welding cycle, with no stress component in the through-thickness direction. That is, welding is accomplished in *plane stress*. The welding of a thin sheet in a single weld pass approximates this situation.

In the second limiting case the components being welded behave as though the through-thickness dimension were *infinite*, the components filling the whole of the half-space below the heat source and the weld pool (Fig. 5.1b). In the *multipass*

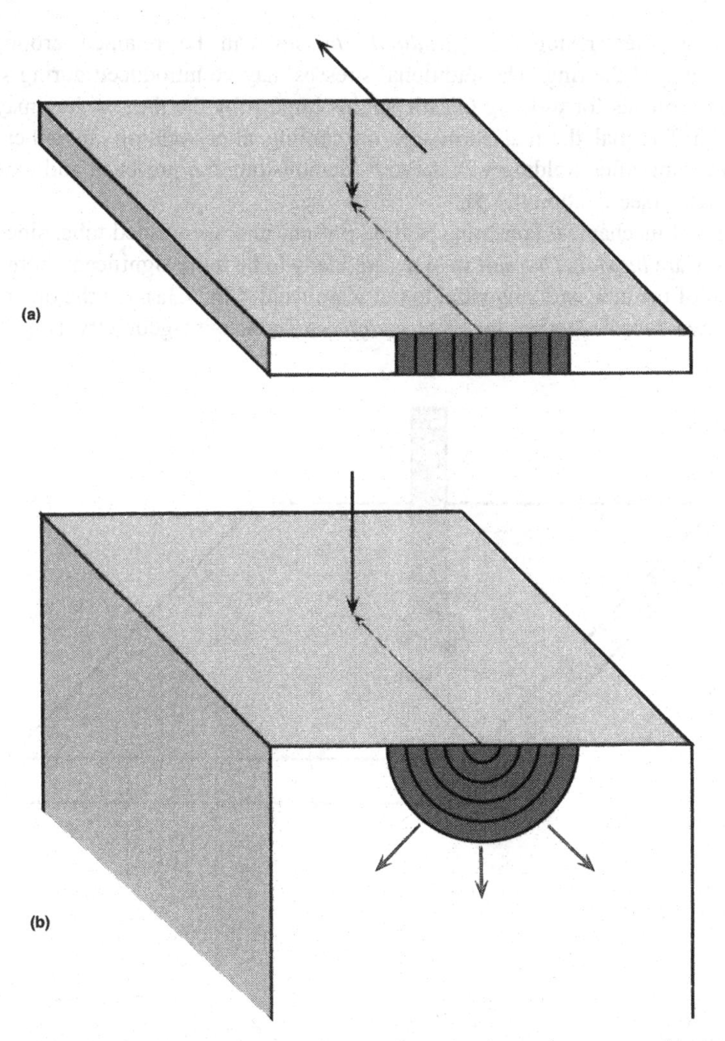

Fig. 5.1. Two limiting *heat transfer* conditions for a moving point source of heat: (a) *two-dimensional* heat transfer; (b) *three-dimensional* heat transfer to an infinite half-space.

welding of thick components, the final weld passes approximate this situation. Heat transfer is *three-dimensional* in the half-space beneath the weld pool and the stress distribution is *triaxial*. It follows that triaxial residual stresses are likely to be retained in the welded component.

As for *geometrical constraints*, spot welding a wire to a surface (typical of a microelectronic assembly, Fig. 5.2a) gives a join which is only loosely constrained along the axis of the wire, while butt welding a bar involves a more complex geometry with considerable *bending constraint* (Fig. 5.2b). If the bar is butt-welded into a ring, then residual *longitudinal stresses* can be retained around the circumference of the ring. The additional stresses may be introduced during setting up the components for welding (elastic strains imposd by the jig), or they may be a result of differential thermal shrinkage on cooling after welding. In either case, cutting the ring after welding will serve to demonstrate the presence and extent of these stresses (see Section 4.1.5).

Additional mechanical constraint will be present in seam-welded tube, since now the stresses are *biaxial*. *Thermal stresses* are likely to be more significant, both along the length of the tube and tangential to the seam weld (Fig. 5.3a). On the other hand, welding two tubes together involves a *different* constraint geometry (Fig. 5.3b).

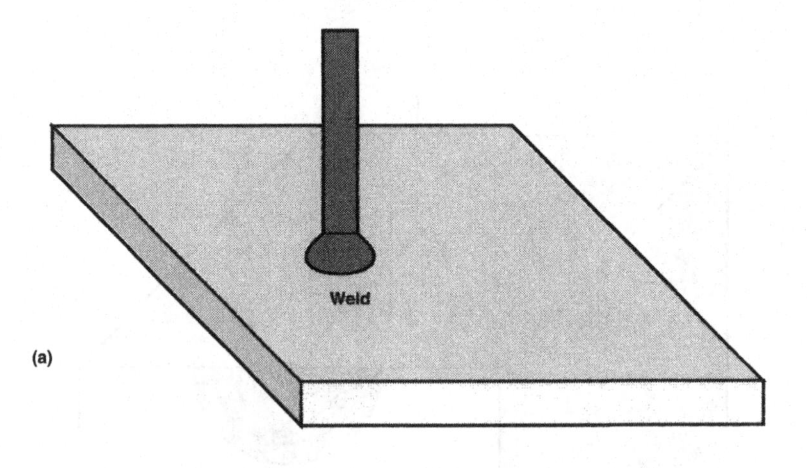

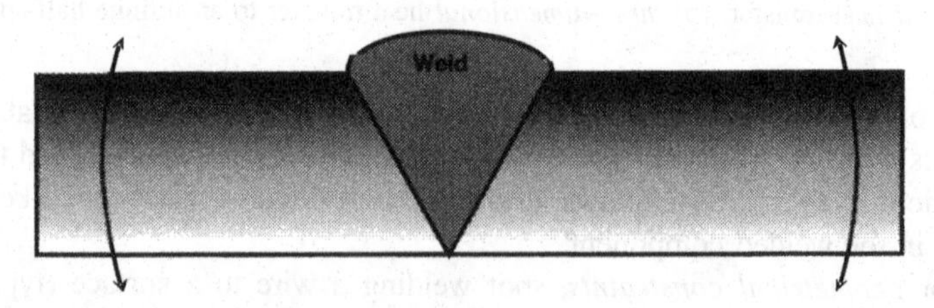

Fig. 5.2. *Mechanical constraints* depend on the weld *geometry*: (a) spot-welded wire experiences *minimal* constraint; (b) butt-welded bar is constrained in *bending* in two dimensions.

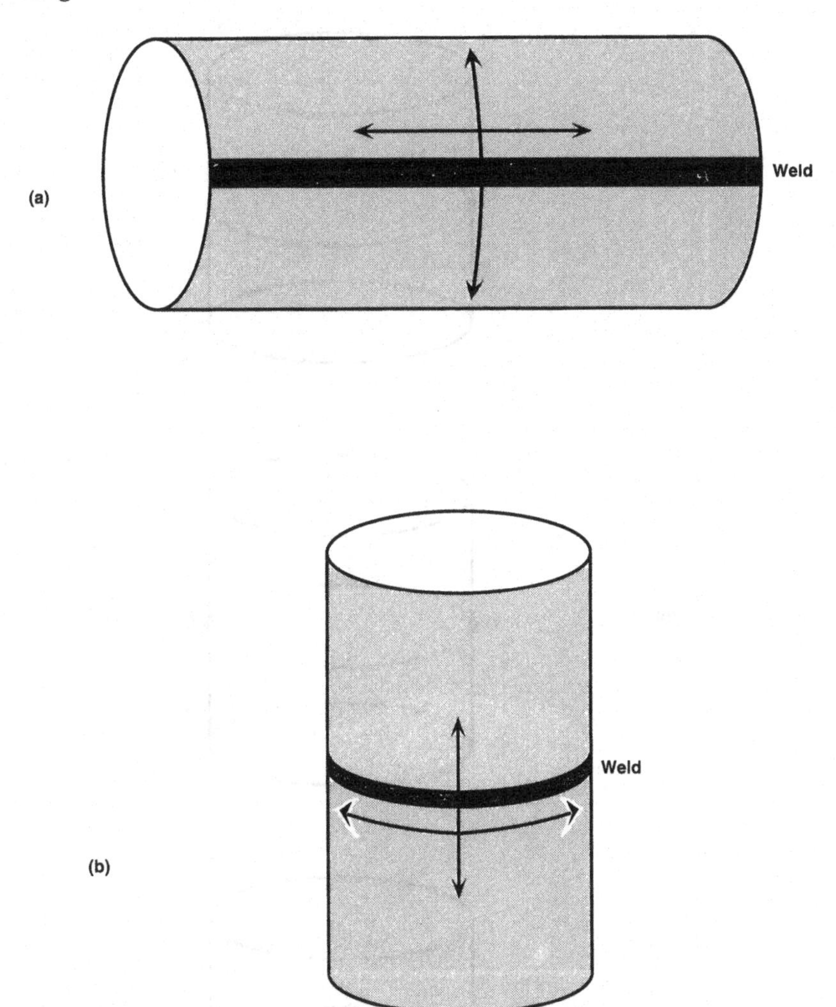

Fig. 5.3. *Residual stresses* in a thin-walled welded tube. (a) The *axial* and *radial* stresses are the principal stresses in a *longitudinally welded* tube. (b) Two tubes joined end to end by a *circumferential weld* will have a different pattern of residual stress.

Several strategies are possible for welding two large tubes together and these are illustrated in Fig. 5.4. All three start by tack-welding the two tubes in position. Thereafter we can use a continuous, single weld pass (Fig. 5.4a), two passes which start at the same spot but move in opposite directions around the perimeter (Fig. 5.4b), or two passes which start at the two ends of a diameter and move in the same direction around the tube (Fig. 5.4c). It is the last strategy, which is often preferred in welding large tubular structures, two welders working together to minimize the

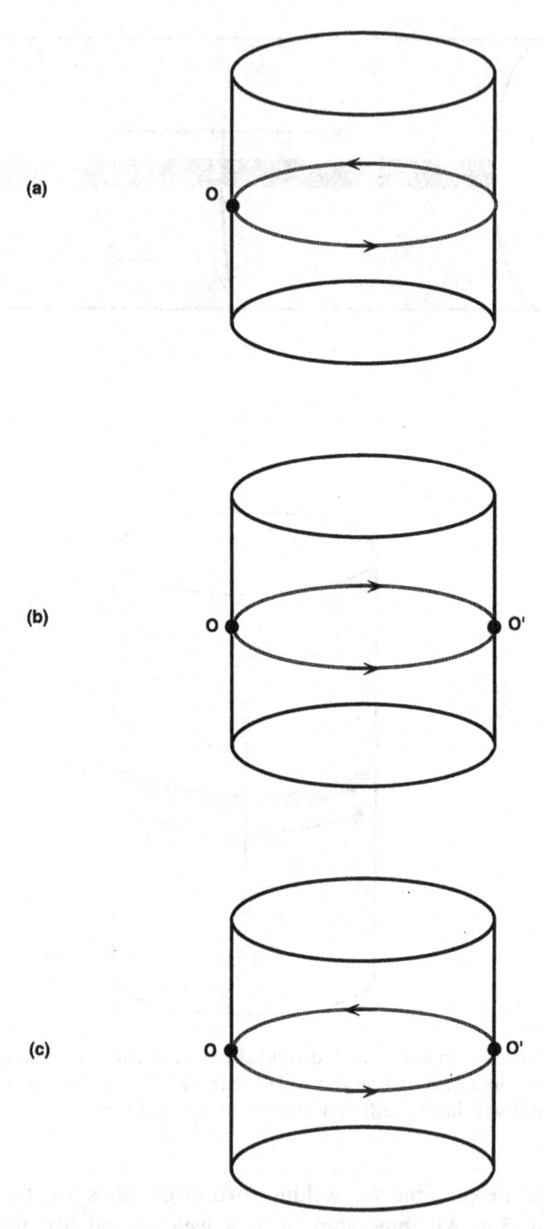

Fig. 5.4. The *residual stresses* in the circumferential welding of a tube will depend on the welding procedure: (a) single weld pass starts and finishes at O; (b) two weld passes are initiated at O, travelling in opposite directions to meet at O′; (c) two weld passes are initiated diagonally opposite each other, at O and O′, travelling in the same sense to finish at O′ and O respectively.

asymmetric thermal expansion about the tube axis. As in all welding of structures, minimizing the residual stresses developed *both* by the structural constraints *and* by the welding cycle is a major objective.

If a thick spherical shell is formed by welding together two hemispheres, then the mechanical constraint is still further increased, and will depend on the ratio of the shell thickness to the shell diameter. In this instance every precaution will be taken to allow residual stresses to relax, most commonly by a post-welding, *stress relief* heat treatment at a temperature which is sufficient to reduce the yield stress of the component, but not so high as to affect the microstructural integrity.

Weld geometries are frequently complex, and so far we have only mentioned simple spot-welds, butt-joints and seams. Edge welds require that the plates be *bevelled* to help control formation of the weld pool (Fig. 5.5a). Thin plates can be edge-welded in a single pass, but it is easier to avoid weld defects in thicker plate and reduce the width of the *heat-affected zone* (the region either side of the weld line with properties and microstructure modified by the welding cycle) if several weld passes are made (*multipass welding*, Fig. 5.5b). Although the yield strength of *weld filler metal* may *exceed* that of the bulk component, the strength in the heat-affected zone (HAZ) will most probably have been *reduced*. It is common practice to

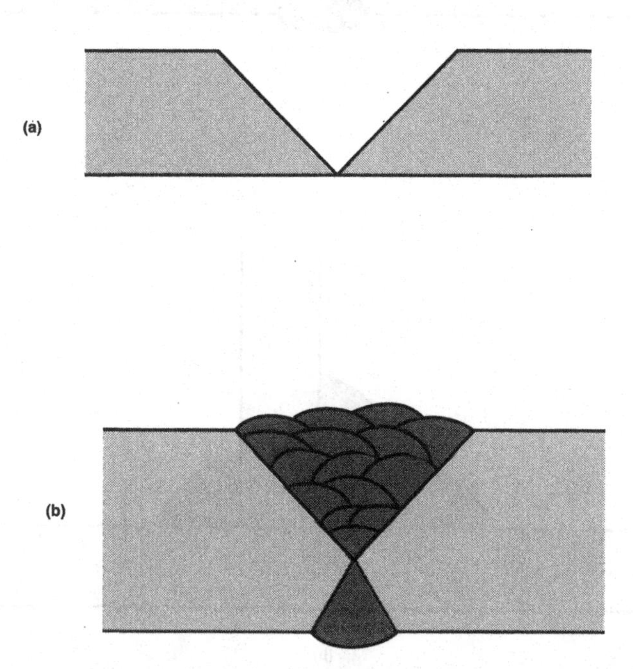

Fig. 5.5. (a) Two thick plates to be joined by butt-welding are *bevelled* to accommodate molten filler metal, improve the tolerances associated with setting up and ensure full penetration of weld metal. (b) Very thick plate may be bevelled from both sides and fixed in position by a low energy, locating *tack weld* before being *multipass welded*, as indicated.

reinforce a weld by doubling or tripling the thickness with additional plates which are *lap-welded* to the assembly (Fig. 5.6a). Similar principles of reinforcement are involved in the welding of a *reinforcing rib* to a flat bar or plate in a T-joint (Fig. 5.6b).

One last word should be reserved for *spot-welds*, as opposed to continuous welding (seam-welds). *Spot welding* is an important process in its own right, especially in the automobile industry, and we have already mentioned the spot welding of electrical leads in microelectronic assemblies (and will return to this in Section 11.4.1). Spot welding is also a frequently used method of *setting up* an assembly prior to structural welding. In particular, spot or tack welding can often be substituted for expensive jigging assemblies. However, spot-welds are sites of *stress concentration*, and hence potential sources of weakness.

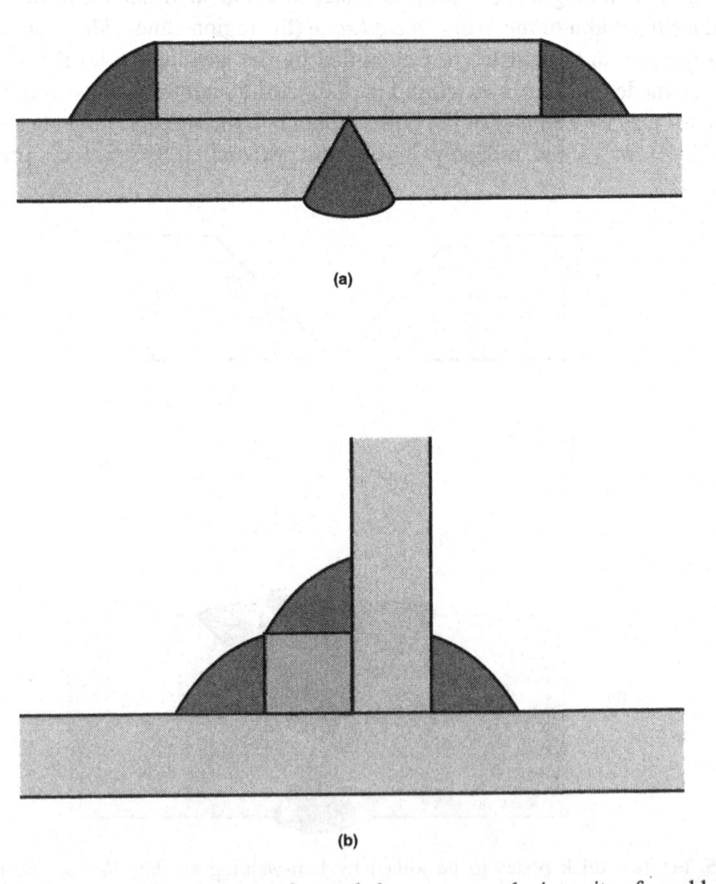

Fig. 5.6. Additional *reinforcement* may be needed to guarantee the integrity of a welded joint. (a) A reinforcing plate increases the load-bearing cross-section in the weld zone. (b) A reinforcing rib strengthens the weld at a T-junction.

5.1.2 Some Process Options

Apart from those few *solid state welding* processes, in which no liquid phase is involved, all other welding processes are more fully described as *fusion welding*. The various fusion-welding processes are classified according to the different sources of heat employed. *Electric arc welding* is the most common welding method for the assembly of large structures (buildings, ships, bridges, railways, pipelines, oil-rigs and electrical power plants). *Arc welding* has a long and respectable history dating back to the earliest days of industrialized mass production. The first patent was issued in England in 1885, and this was followed by a US patent in 1889. Since then arc welding processes have developed on three separate but related fronts: the properties of the *welding electrodes* and the metallurgy of the *filler metal*; flux compositions and the *protection of the weld pool* from oxidation and composition changes; *current* and *voltage control* in the power supply. We will return to these points in later sections.

Flame welding, generally based on the combustion of an oxygen–acetylene gas mixture, is probably the second commonest source of heat for the welding of metal components. Before the development of power sources for use in remote locations, the *oxyacetylene torch* was the preferred heat source for the *on-site welding* of steel structures. It is still commonly used for small tasks, where the transport of a power source for arc welding is neither practical nor economic. However, flame welding is less reliable than arc welding for controlling the welding variables (size and temperature of the weld pool and its rate of advance), and, as a result, the structure of the welded joint (that is, changes in weld composition and HAZ microstructure, the incidence of slag inclusions and porosity, and the residual thermal stresses).

Another important source of heat for fusion welding is the electrical resistance to an imposed current in the contact area between the components. *Electrical resistance welding* is most commonly used for the automated welding of metal sheet. The technique relies on the limited area of true contact at a loaded mechanical contact. The alloy sheets are compressed between two electrodes which are in *rolling contact* and the sheets are fed between the electrodes at a constant rate (Fig. 5.7a). It is almost impossible to maintain a constant rate of heat supply, since the area of true contact immediately increases as the temperature exceeds the melting point, lowering the resistance between the electrodes, so this is not even attempted. Instead the power source is *pulsed*, ensuring that power dissipation occurs primarily in the contact area between the sheets which lies immediately beneath the electrodes. As the electrodes advance, subsequent energy pulses repeat the process. The weld line is a 'seam' consisting of overlapping spot welds, and the process is referred to as *seam welding*.

Electron beam welding has developed rapidly with the growth of the aerospace industry. The process is expensive, requiring a massive investment in both vacuum equipment and power generation. The components to be welded are assembled in a large vacuum chamber using a suitable jig. The heat source is a high energy beam of

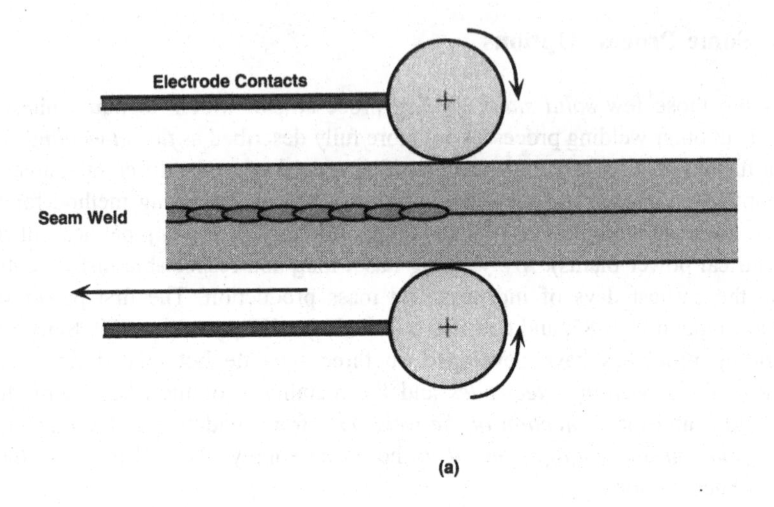

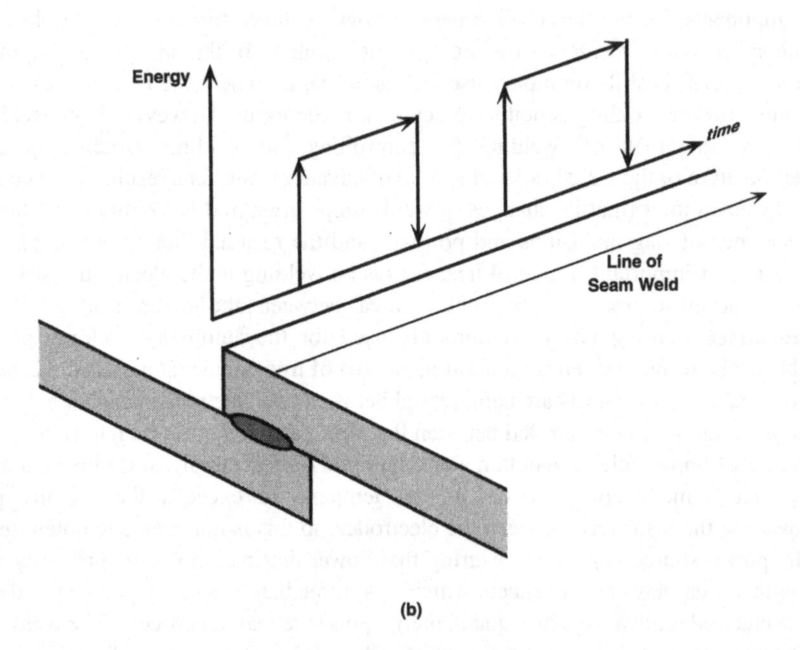

Fig. 5.7. *Seam welding* is frequently used in the aerospace industry in order to resistance-weld thin sheets. (a) The sheets to be welded are fed in contact at a controlled speed and pressure between rotating electrodes. (b) The welding current is *pulsed* along the line of the weld, forming the '*seam*' in a series of readily reproducible, overlapping spot-welds.

electrons, accelerated from a suitable source and focussed onto the workpiece by electromagnetic lenses. The workpiece is manipulated beneath the focus of the beam by *robotic control*, which determines both the site of the weld and the rate of advance of the weld pool. The welding of complex assemblies is achieved under clean (vacuum) conditions, reducing the incidence of welding defects. The only defects commonly encountered with this process are those associated with incomplete fusion or incorrect siting of the workpiece.

Laser beam welding has also found some areas of application over the two decades since it was first proposed. Unlike electron beam welding, no vacuum is required. The *focussed* electromagnetic radiation available from a pulsed power laser provides a localized energy density which may exceed that in an electric arc. However, a large and poorly controlled fraction of the electromagnetic energy is reflected from the workpiece, while fumes emitted during welding scatter and absorb the incident beam. Laser welding has developed rather slowly, and has been somewhat overshadowed by both *laser-cutting* and *laser-coating* technologies, both of which have proven industrial viability.

Finally, this brief account of the sources of heat for fusion welding would be incomplete without mentioning *exothermic chemical reactions*. The *thermite* process was originally developed for rapid welding repairs in remote locations, under conditions in which neither electric-arc nor flame welding were practicable. An example would be the welding of railway line in remote areas. The heat evolved in a strongly exothermic reaction melts the reaction product at the site of the join. Mixed powders of *iron oxide* and *aluminium metal* are tamped into the region of the join and ignited, when an exchange reaction occurs:

$$Fe_2O_3 + 2Al \rightarrow Al_2O_3 + 2Fe$$

The heat released is sufficient to form a weld pool of molten iron and allow time for the (much less dense) aluminium oxide to form a slag before the weld pool solidifies. The problems associated with this process are many: the large (greater than 50%) shrinkage associated with formation of the weld pool from the powder, the control of composition of the weld metal, the presence of heated gas trapped in the powder, and the limited time available for the escape of gas and oxide to the free surface. Nevertheless, the process is available and has its uses.

5.1.3 Weld Defects

Loss of structural performance associated with a welded joint often has little to do with either the metallurgy of the weld metal or the neighbouring HAZ, but is rather the presence of *processing defects* introduced during welding. The *prevention* and *detection* of welding defects are major factors which assure the success of a welding operation. In this section we describe some common welding defects.

Three classes of welding defect may be distinguished: (i) *geometrical* defects associated with weld and component geometry; (ii) defects within the *weld metal*;

and (iii) *cracking* occurring either during or after welding, and associated with both embrittlement and residual stress.

Examples of *geometrical defects* are shown in Fig. 5.8 and are associated with misalignment of either the components or the heat source, or badly chosen welding parameters which result in lack of weld *penetration* and *fusion*. Geometrical defects

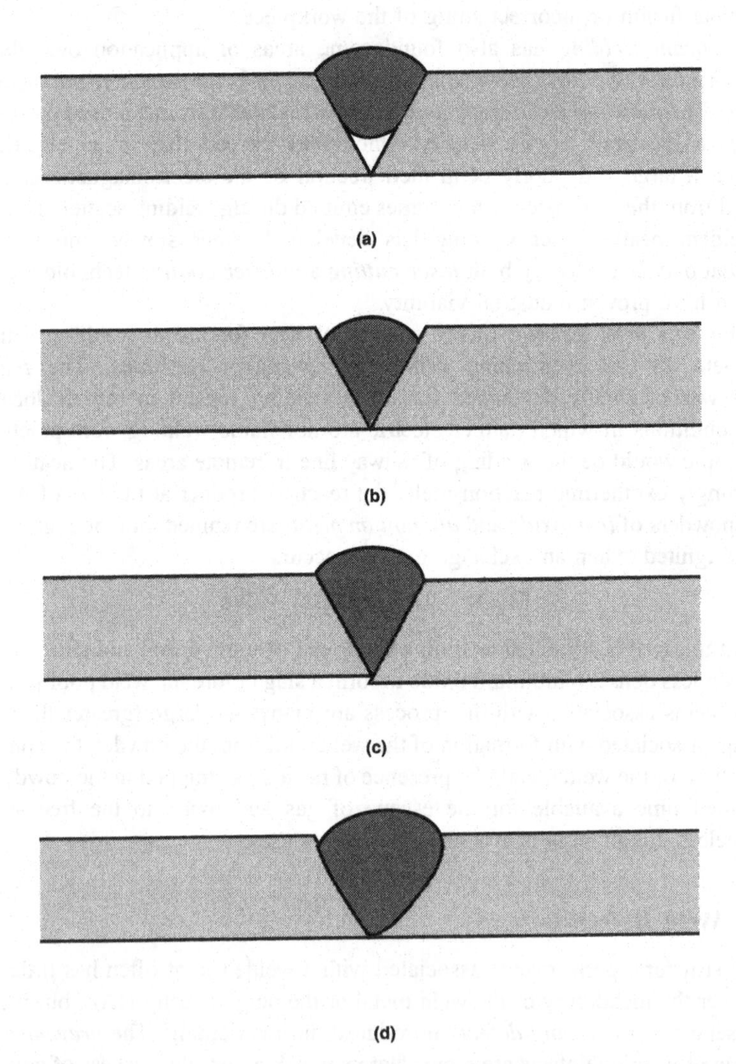

Fig. 5.8. *Geometrical defects* in welds may take several forms: (a) incomplete penetration of the weld bead; (b) insufficient filler metal in the bead; (c) inaccurate placement of the components before welding; (d) inaccurate placement of the heat source with respect to the components to be welded.

create *grooves* and *notches* which result in undesirable sources of *stress concentration* in the finished assembly.

Inclusions and *porosity* are common defects within the weld bead. Slag inclusions result from *incomplete separation* of the fluxed oxide from the weld pool, usually due to a combination of the limited time of fusion with a high melt viscosity. However, the real cause is commonly inadequate surface penetration and cleaning of the components prior to welding, and an excessive amount of slag due to an initially dirty surface. Grease, rust and foreign matter *have* to be removed by solvents, abrasion, or just a wire brush, prior to welding. There are no short cuts.

Porosity may result from gas evolution due to the presence of foreign matter on the components (especially *grease*), but can also originate from the electrodes and filler metal (*moisture* absorbed on flux-coated electrodes). Gas may also be *absorbed* by the molten weld pool from the surrounding atmosphere and released during solidification. Many gases are highly soluble in molten metal, but practically insoluble in the crystalline solid (Fig. 5.9). This is particularly true of nitrogen and oxygen, while dissolved oxygen will react with carbon in a steel to release *carbon monoxide* on solidification:

$$[C]_{Fe} + [O]_{Fe} \rightarrow (CO)$$

Hydrogen is soluble in both liquid and solid metals, and can diffuse rapidly at room temperature as an interstitial impurity, leading to *hydrogen embrittlement*.

Both weld porosity and slag inclusions act as *stress concentrators* and sources of failure initiation. The size and extent of the defects, and their distribution in the weld

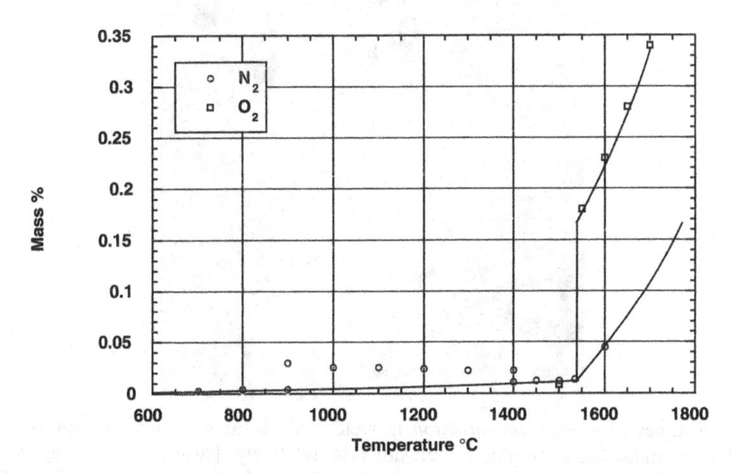

Fig. 5.9. Temperature dependence of the solubility of oxygen and nitrogen above and below the melting point of iron. Data from *Smithells Metals Reference Book*, Seventh Edition. Edited by E. A. Brandes & G. B. Brook, Butterworth-Heinemann Ltd., Oxford, 1992.

zone, determine how deleterious they are, and extensive *international standards* specify what is and what is not acceptable. These defects are commonly detected by *dye penetrant* tests, *X-ray radiography* or *ultrasonic inspection*, and graded according to their severity. Some examples are given in Fig. 5.10.

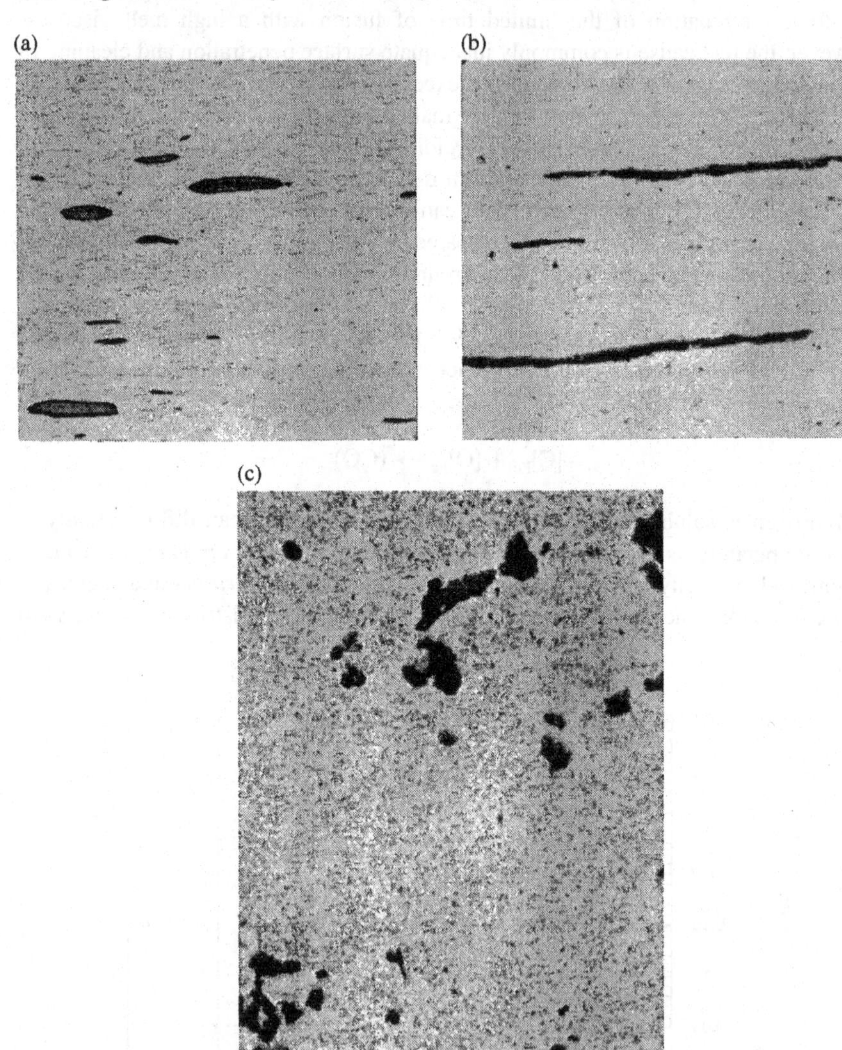

Fig. 5.10. Sources of *stress concentration* in welded steel. (a) A limited volume-fraction of small, ductile manganese sulphide inclusions is relatively harmless. (b) Slag stringers (primarily iron silicate) are brittle and seriously affect both the toughness and the fatigue strength. (c) Porosity reduces the load-bearing cross-section but *also* acts as a source of stress concentration. (a) and (b) are reproduced by permission of the Institute of Material. (c) From Greaves and Wrighton: *Practical Microscopical Metallography (1950)*.

The *third* class of weld defect comprises various forms of cracking and embrittlement and may have several contributing sources. *Shrinkage cracks* are associated with the volume changes which occur in the weld pool during solidification and subsequent cooling from the melting point. They are a result of *thermal stresses* and the cracks form during the welding cycle. They can usually be avoided by controlling the rate at which the welded zone is cooled, allowing sufficient time for residual thermal stresses to *relax* by plastic flow. Some cases of *hot-cracking* in steel are associated with the presence of embrittling *inclusions*, especially iron sulphide, FeS, which can nucleate cracking at relatively high temperatures. The presence of sufficient manganese leads to the preferential formation of *manganese sulphide* (MnS) inclusions, rather than iron sulphide, which are *ductile* in the critical temperature range, ensuring that plastic yielding precedes brittle cracking.

A common cause of fine *hairline* cracks is dissolved hydrogen, usually the result of *moisture* present during welding. Many structures have to be welded under less than ideal conditions, in wet or snowy weather for example, and it can be very difficult to prevent access of water vapour to the arc zone. If this occurs *atomic hydrogen* may be introduced into the weld pool and, in steels, will diffuse rapidly into the heat-affected zone. Embrittlement and cracking occurs under the influence of *residual stress* some time *after* the weld zone has reached room temperature. Welding under *dry* conditions is the best preventative, but allowing the hydrogen to diffuse out of the solidified metal at a temperature above that at which embrittlement can occur is a sufficient cure. The hydrogen ion is an isolated *proton*, and both its solubility and ionic mobility are high. At 100°C hydrogen diffuses rapidly in steel. *Hydrogen cracking* is avoidable with reasonable care, welding under dry conditions, using special electrodes and a post-welding low temperature treatment.

Stress corrosion cracking may be more difficult to prevent. The cracks propagate over a period of time as a result of the combined effect of *stress* and *environment*. Many alloys are susceptible, including several *stainless steels* and *aluminium* or *titanium* alloys. Cracking can occur in the bulk metal, and is not confined to welded structures, but the *microstructural changes* that occur in the HAZ of a weld, combined with residual *thermal stresses* associated with the welded zone, greatly enhance the probability of cracking. Stress corrosion cracking in the vicinity of a weld is frequently referred to as *weld decay*. A classic example occurs in the welding of stainless steels which contain small amounts of residual carbon. Chromium carbide (Cr_7C_3) is precipitated in a well-defined temperature zone in the HAZ, either side of the weld line, leading to *depletion* of chromium in solution and a more anodic electrode potential in this region (see Section 4.2.1). Chloride-containing solutions cause rapid attack in the presence of minimal tensile stresses, leading to through-thickness crack penetration in the space of a few weeks. The phenomenon is confined to an intermediate zone within the HAZ, since chromium carbide is *redissolved* in the high temperature region immediately adjacent to the weld, while the temperature at the outer edges of the HAZ is insufficient for carbide precipitation.

Examples of cracking diagnosed as due to *hot-cracking*, *hydrogen (hairline) cracking*, and *stress corrosion cracking* are given in Fig. 5.11.

5.2 TRANSPORT PROPERTIES

In this section we give an elementary account of those *heat* and *mass transport* processes which are important in welding. This introductory treatment is intended to provide some awareness of the factors which control the material history during a weld cycle, and hence determine the final structure of the weld and HAZ.

5.2.1 Heat Transfer in Welding Processes

The *thermal history* of a volume element of either the components being welded or the associated weld pool is determined by four factors: (i) the weld *geometry*, (ii) the *thermal properties* of the material, (iii) the *energy input* from the heat source and (iv) the *velocity* of the heat source with respect to the components. Both the two-dimensional and the three-dimensional limiting conditions relate to the *steady state* temperature distribution for a *constant*, *point* source of heat moving at a *constant* linear velocity across the workpiece. In the 2-D case, the through-thickness temperature distribution is ignored and heat propagation is by *conduction* confined to the *plane* of the workpiece.

The thermal conductivity κ and the thermal diffusivity α are *material constants* which are assumed independent of temperature but are related by the equation $\alpha = \kappa/(C\rho)$, where C is the heat capacity and ρ is the density. The *general solution* for the temperature distribution around a travelling heat source is of the form:

$$T = \frac{q}{2\pi\kappa r}\exp[-v(r-x)/(2\alpha)] \qquad (5.1)$$

where v is the linear velocity of the point heat source emitting at a rate q. The parameter r is the *radial* distance from the source *perpendicular* to the direction of travel, while x is the distance from the source *along* the direction of travel. These geometrical parameters are defined in Fig. 5.12.

The above solution of the heat equation is derived from the two standard equations which relate, firstly, the *heat flux* J to the temperature gradient, and, secondly, the *time dependence of temperature* to the heat flux gradient: $J = -\kappa \mathrm{d}T/\mathrm{d}x$ and $\mathrm{d}T/\mathrm{d}t = C\rho\ \mathrm{d}J/\mathrm{d}x$, respectively. Combining these two equations yields a second order differential equation which defines the time dependence of the temperature at a point: $\mathrm{d}T/\mathrm{d}t = \alpha\ \mathrm{d}^2T/\mathrm{d}x^2$. The solution of this equation for a travelling point heat source is the equation given above. This approximation ignores the properties of the *melt* in the weld pool and the effect of *convectional flow* in the melt, as well as the finite *size* of the heat source. It also assumes that thermal diffusivity and conductivity are *independent* of temperature.

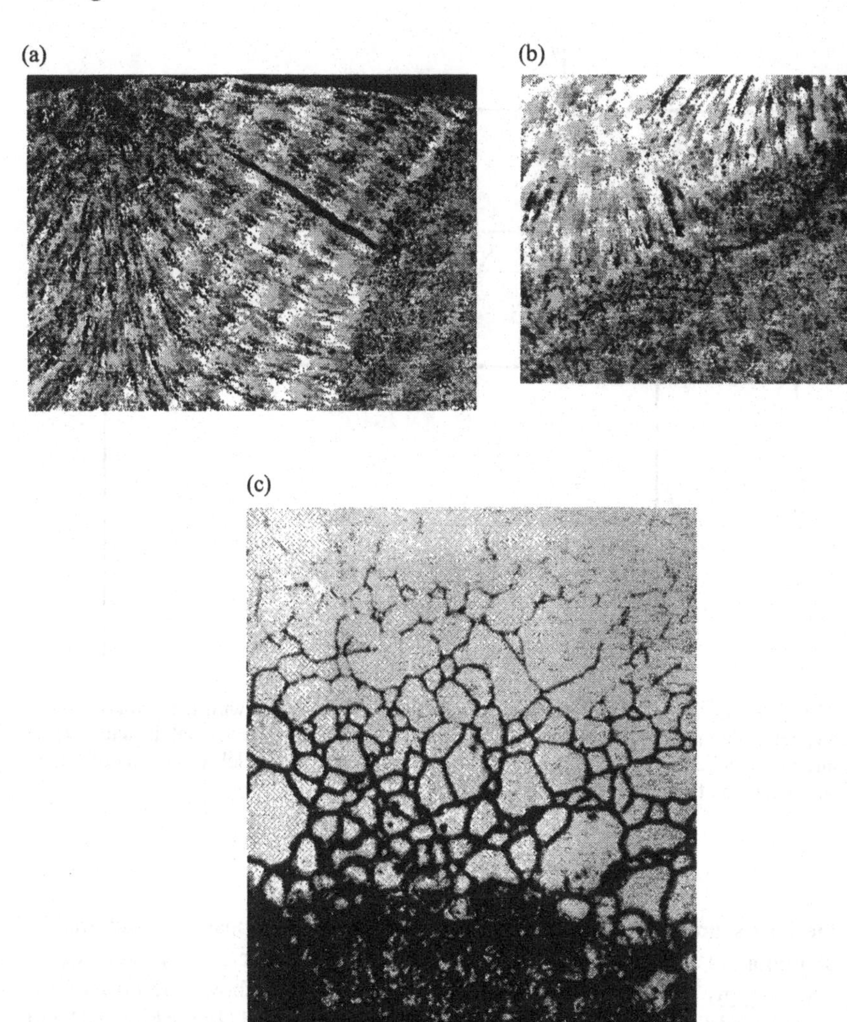

Fig. 5.11. *Cracks* in welded structures may have several causes. (a) *Hot-cracking* occurs during cooling when low melting point, non-metallic, grain boundary phases accommodate the solidification shrinkage in the weld pool by allowing grain separation. (b) *Hydrogen, hairline cracking* after welding is attributed to the reduction in surface energy due to surface adsorption of atomic hydrogen dissolved in the metal during the welding cycle. (c) *Stress corrosion cracking* is associated with the stress-assisted environmental attack of a susceptible microstructure. In welded structures the sources of stress are quite often *residual stresses*, unassisted by any appreciable operating stress. (a) and (b) reproduced from *Introduction to the Physical Metallurgy of Welding*, Easterling (1983) Butterworths. (c) From Metals Through the Microscope (1988) Tomer.

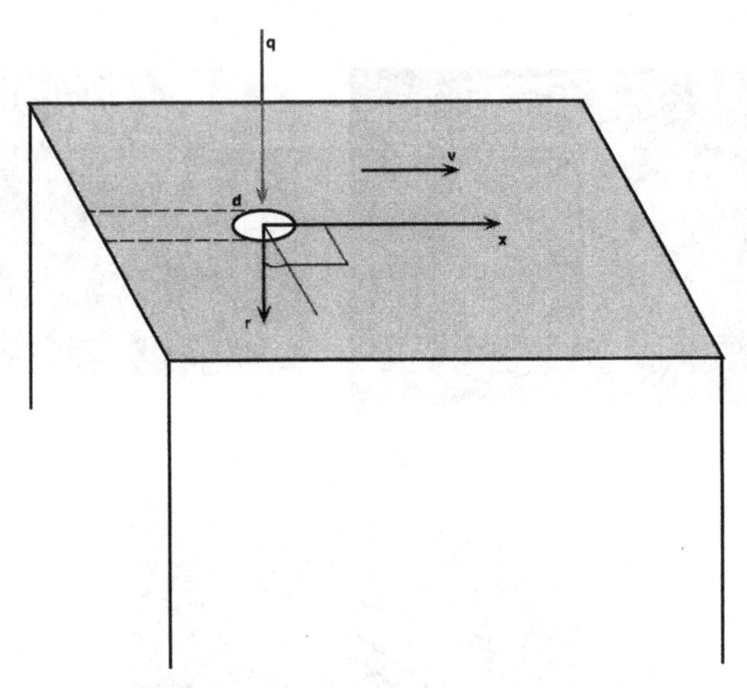

Fig. 5.12. The *rate of heat transfer* from a point source q, together with the *velocity* of the source v, determine the thermal history at a position (x, r) defined by the radial distance r from the source perpendicular to the axis of travel and the distance x parallel to the axis of travel. The parameter d is the diameter of the weld pool.

If the heat source creates a weld pool of diameter d, it is possible to normalize several important parameters for both the 2-D and the 3-D case (the latter being assumed to apply to an *infinite half space*). Thus Fig. 5.13 shows the *normalized* parameter $q/2\pi kT$ as a function of the *normalized* parameter $vd/4\alpha$ for both 2-D and 3-D heat transfer. For the 3-D case it is the temperature distribution *in the plane* of the surface which is plotted, although it is also possible to derive the temperature distribution *beneath* the surface. Figure 5.13 illustrates the approximately linear dependence between the normalized parameters over much of the range of interest. In particular, the *heat input* is linearly related to the *temperature* in the weld pool, while the *size* of the weld pool is inversely proportional to the *velocity* of the heat source. As must be the case, the temperature reached in the 3-D case at a given rate of heat input is lower than in the 2-D case, since heat is being lost in three, rather than two dimensions.

In addition to the *steady-state* distribution of the temperature it is important to know the *cooling rate* at a given location in the neighbourhood of the moving heat

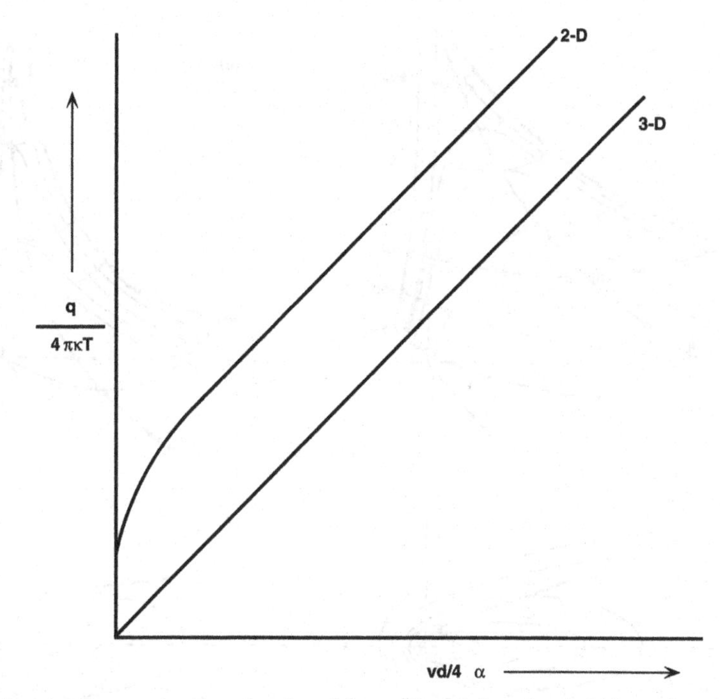

Fig. 5.13. To a good approximation, the *size of the weld pool d* can be plotted as a function of the *rate of heat transfer q* for both the 2-D and the 3-D cases, with both parameters normalized with respect to the thermal constants of the metal, α and κ, the temperature in the weld pool T, and the velocity of the heat source v.

source. In 3-D this is given approximately by the relation:

$$\frac{\mathrm{d}T}{\mathrm{d}t} = \frac{-vT}{r}\left[\frac{x}{r} - \frac{vr}{2\alpha}\left(1 - \frac{x}{r}\right)\right] \tag{5.2}$$

where r is now the radial distance from the weld line, rather than the radial distance from the heat source, and x is again the distance from the moving heat source along the axis of v, the x axis. In this equation the decay of the temperature is sensitively dependent on the velocity of the heat source. An example of the calculated temperature distribution is given in Fig. 5.14, both *parallel* and *perpendicular* to the line of the weld.

An alternative way of generalizing the thermal history is to restrict analysis to the cooling rate at the *solidification front* on the trailing edge of the weld pool. Ignoring all corrections due to the finite size of the weld pool and the heat of solidification, the appropriate relation is:

$$\frac{\mathrm{d}T}{\mathrm{d}t} = \frac{-2\pi\kappa T_{\mathrm{m}}^2}{(q/v)} \tag{5.3}$$

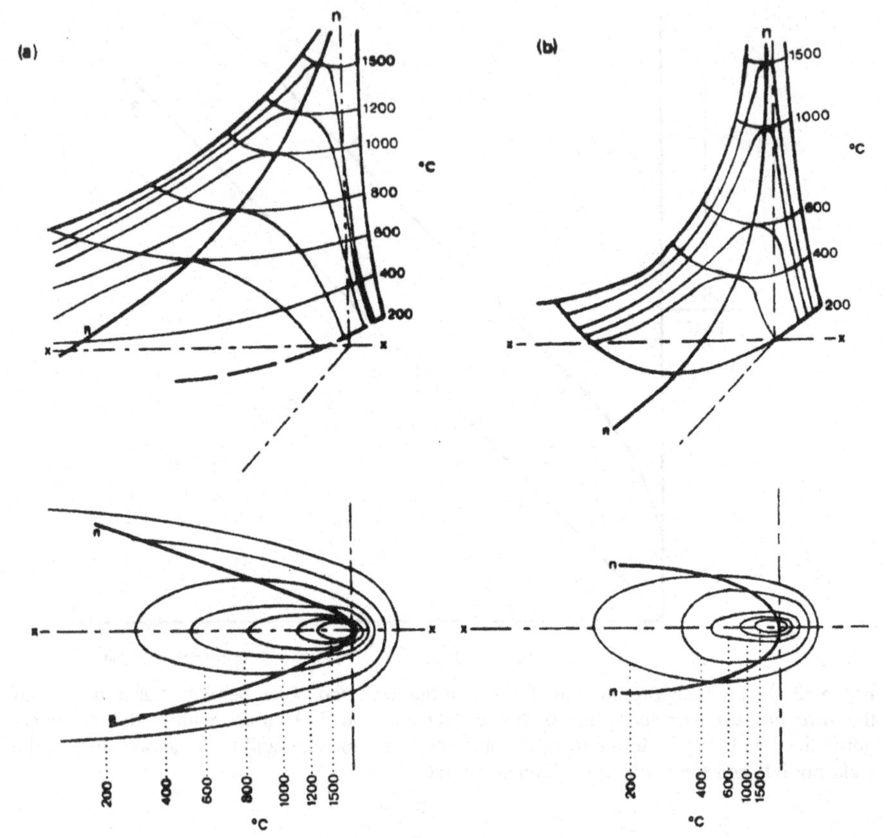

Fig. 5.14. The steady-state temperature distribution derived from the simple heat transfer model summarizes the *thermal history* of each volume element in the weld zone. (a) The *2-D* heat transfer case. (b) The *3-D* case. Reproduced from *Introduction to the Physical Metallurgy of Welding*, Easterling (1983) Butterworths.

where T_m is the melting temperature. This form brings out the reciprocal relation between the heat input q and the weld velocity v.

5.2.2 Diffusion Mass Transfer

We consider the mass transfer from a metal electrode or filler wire into the weld pool in a later section, but in the meantime we summarize some of the main features of *diffusion* and *mass transfer* in the *solid state*. As noted previously, changes in composition and in the distribution of the constituents may affect the final performance of the welded structure, especially the susceptibility to *brittle failure* associated with *hydrogen embrittlement* or *stress corrosion cracking*.

It comes as a surprise to many students that changes in composition can occur in the solid state, and that mass transport by solid state *diffusion* is a common phenomenon with major technological implications. Diffusion in crystalline solids usually occurs by the thermally activated transfer of an atom or ion from an *occupied* lattice site to a neighbouring, *unoccupied*, or vacant site. In a *thermodynamically stable* solid, which by definition has a *uniform* microstructure with phases of *homogeneous composition*, this transfer process occurs at *random*, leading to local concentration fluctuations of limited lifetime. However, if a *concentration gradient* is present, the diffusing species will acquire a drift velocity, and the equations governing the rate of mass transfer in a concentration gradient, and the resultant *changes* in concentration, are directly analogous to those quoted earlier for *heat transfer* in a temperature gradient, and the resultant changes in temperature. *Fick's* first law of diffusion gives the *mass flux* as a function of the *concentration gradient*: $J = -D\,\mathrm{d}c/\mathrm{d}x$, while *Fick's second law* gives the *time dependence* of the concentration on the *flux gradient*: $\mathrm{d}c/\mathrm{d}t = \mathrm{d}J/\mathrm{d}x$. As with the corresponding heat transfer equations, combining Fick's two relations yields a second order differential equation which can be solved for specific boundary conditions: $\mathrm{d}c/\mathrm{d}t = -\mathrm{d}(D\,\mathrm{d}c/\mathrm{d}x)/\mathrm{d}x$. The constant D is termed the *diffusion coefficient* and is a characteristic of the diffusing species. Provided the diffusion coefficient is *independent* of composition, the combined expression for Fick's law of diffusion can be written: $\mathrm{d}c/\mathrm{d}t = -D\mathrm{d}^2c/\mathrm{d}x^2$. A common case in joining technology is *planar diffusion* perpendicular to a planar interface maintained at *fixed* composition c_0 by an essentially infinite source of the component. The solution of the diffusion equation is then a *Gaussian error function*:

$$(C_x - C_0)/(C_s - C_0) = 1 - \mathrm{erf}(x/[2\sqrt{Dt}]) \quad (5.4)$$

The values of the error function for any given value of x can be read from standard tables, using the distance from the planar source x, normalized by the parameter $2\sqrt{(Dt)}$, often termed the *diffusion distance*. Some calculated curves are given in Fig. 5.15, and illustrate how the diffusion gradient decreases as the diffusing species migrates into the bulk from the interface.

Since diffusion is a *thermally activated* process it is a strong function of temperature. This dependence on temperature is usually described very adequately by the *Arrhenius* equation: $D = D_0 \exp(-Q/RT)$, where R is the gas constant and Q is an effective *activation energy for diffusion*. This is the energy required to raise the internal energy from the ground state for an atom or ion to a mobile, activated energy state. The constant D_0 is called the *pre-exponential factor* or *entropy factor*, and is related to the changes in configurational and vibrational entropy which accompany the transition of the diffusing species to the activated state prior to its migration from one lattice site to another. The pre-exponential also includes the probability that the migrating atom may *return* to its original position within the time frame of the same excitation event.

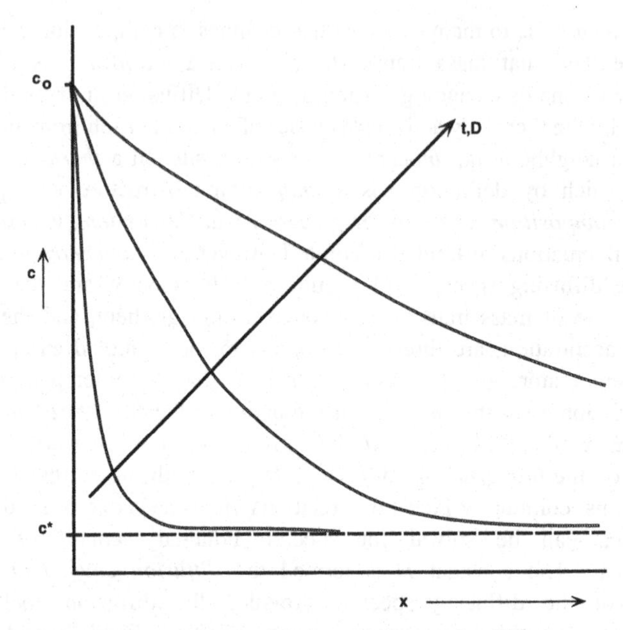

Fig. 5.15. The one-dimensional error function solution to the diffusion equations, assuming constant surface and bulk concentrations, c_0 and c^*, respectively, as a function of time t and diffusion coefficient D.

The effective activation energy Q may be related to the diffusion *mechanism*. For example, *interstitial* species, such as carbon in steel, are present in low atomic concentrations, so that most interstitial sites are unoccupied. Diffusion then only requires that sufficient *internal energy* should be available to enable the interstitial to make the jump to an empty site, while the *number* of empty sites remains a constant. Under these circumstances, the activation energy can be associated directly with the energy required to overcome a *potential barrier* between *occupied* and *unoccupied* sites.

Substitutional diffusion usually occurs by atom transfer to a neighbouring vacant site. Vacant sites (*vacancies*) are present in metals in *thermal equilibrium*. Their *concentration* depends on the increase in internal energy of the lattice which is associated with their formation. The probability of a site being vacant is then given by: $c_v = c_0 \exp(-E_f/RT)$, where E_f is the *molar energy* of vacancy formation. If the excess internal energy required to make the jump from an *occupied* lattice site to a neighbouring *unoccupied* site is E_m, while the *probability* of the site being unoccupied is c_v, then the *effective* activation energy for *substitutional* diffusion will be: $Q = E_f + E_m$, that is, the *sum* of the excess internal energies required for vacancy *formation* and for vacancy *migration*.

In many practical cases more than one species migrates and the atomic species *interdiffuse*. For example, in a *diffusion bond* between nickel and copper, *both* the nickel and the copper will migrate. The *diffusion coefficient* determined by measuring the composition of one or other of the species as a function of distance from the original interface is then termed the *chemical diffusion coefficient*, and it will have an exponential dependence on temperature characterized by an *effective activation energy*. However, this activation energy bears no direct relation to the activation energies associated with migration of the individual species.

Impurity diffusion in ionic crystals, important in the bonding of *ceramic* components, can lead to comlex situations for which the effective activation energies for diffusion, determined by experiment, defy any simple atomic description. This is especially the case when a diffusing substitutional *anion* or *cation* differs in *valency* from the parent ion, leading to the presence of additional, charge-balancing *interstitial* or *vacancy* defects whose concentration depends on the concentration of the substitutional impurity.

5.3 ENERGY SOURCES

In this section we return briefly to the subject of heat sources before describing the physical characteristics of the *electric arc* and the *liquid metal transfer* processes which accompany the formation of a weld pool in metal arc welding.

The *energy* required to bond two components by welding is usually provided by direct heat, the exception being *explosive bonding*, which uses the *kinetic energy* released during impact of a moving component and a stationary target, and *friction welding*, which combines *frictional heating* at the interface with *localized plastic shear*. The two commonest sources of direct heat are those derived from either a *chemical reaction* or *electrical energy*. *Chemical energy* is released as heat in an *oxyacetylene flame*, as well as in the *exothermic reaction* between aluminium powder and iron oxide; the 'thermite' reaction.

Electrical energy is released as heat when an electric current is passed down a potential gradient. For materials that obey *Ohm's law*, the energy released per unit time is I^2R, where I is the current passed and R is the resistance across the contact interface. This is the basis of *resistance welding*, in which the heat is released at the contact interface between the two components. When an *electron beam* is accelerated through a potential drop V and absorbed in a solid target the energy of the beam is also dissipated as heat. If the total current in the beam is I, then the *rate of energy release* is just IV, and this is the heat source available for *electron beam welding*.

The situation in an *electric arc* is somewhat more complex. The heat required for welding is derived from the *recombination* which occurs when excited species in the arc reach the target surface. The source of heat is now a *plasma*, which consists of atoms in the gas phase which have been stripped of their outer electrons. Heat is generated from the energy dissipated in the arc.

5.3.1 Arc Characteristics

In *tungsten inert gas* (TIG) welding the arc is formed in an inert gas (argon) between a tungsten *cathode* and the workpiece, which is made the *anode*. In TIG welding a *stable electric arc discharge* can be maintained at a potential or between 10 and 50 volts. The *current* passing through the plasma may be anything from 10 to 2000 amps, and corresponds to rates of energy dissipation of several kilowatts. The *structure of the arc* is shown schematically in Fig. 5.16a, while the approximate *potential drop* between the cathode and anode is illustrated graphically in Fig. 5.16b.

Bright spots are observed at both the cathode and the anode and are separated from the *plasma column* by narrow potential drop zones. The *cathode drop* is typically about 8 volts over a distance of between 10 and 100 μm, while the *anode drop* is of the order of 3 volts over a distance of about 10 μm. These values correspond to the *electron mean free path* in the gas phase. The *plasma temperature* in the TIG welding process may be anything between 6000 and 20 000°C, and when a wire of iron-based filler metal is introduced into the plasma, the optical emission spectrum is that of *iron vapour* at the appropriate temperature.

If the potential is *reversed* and the *filler* metal is made the *anode* while the workpiece is made the *cathode*, then the configuration corresponds to the *metal inert gas (MIG)* welding process. The filler metal electrode must be fed continuously into the arc to maintain the plasma gap as molten metal is transferred from the anode to the weld pool. The plasma can be held stable if the plasma voltage drop is kept between 0.5 and 5 volts. In addition to the *inert gas (argon) plasma* it is possible to work with other gases, most notably CO_2. The temperature at which the gas forms a *plasma* (when thermal excitation strips away the outer electrons) varies from 3800 to 8300 K, as summarized in Table 5.1.

Since the electrons have a much lower mass than the ions, they have a much higher *mobility* at any given temperature. It follows that most of the current is carried by the *electrons*. The velocity of the stream of metal droplets, the *metal jet*, is of the order of 0.1 km s^{-1}, and this jet is subject to strong *electromagnetic forces*, associated with the high electron current in the plasma.

The voltage–current characteristics of the arc vary appreciably with the type of process. Figure 5.17a compares schematically the voltage–current relations for stable arcs in the *MIG* and *TIG* processes, as well as those of a *submerged alternating current (AC) arc*, and a *flux-coated electrode* (used without a protective atmosphere). These arc characteristics assume that the *arc length* is continuously adjusted to maintain a stable arc as the current is increased. In Fig. 5.17b the voltage required to maintain a *specific arc length* is plotted for the *MIG*, *TIG* and *coated electrode* processes. The coated electrode, which has to form the plasma in air,

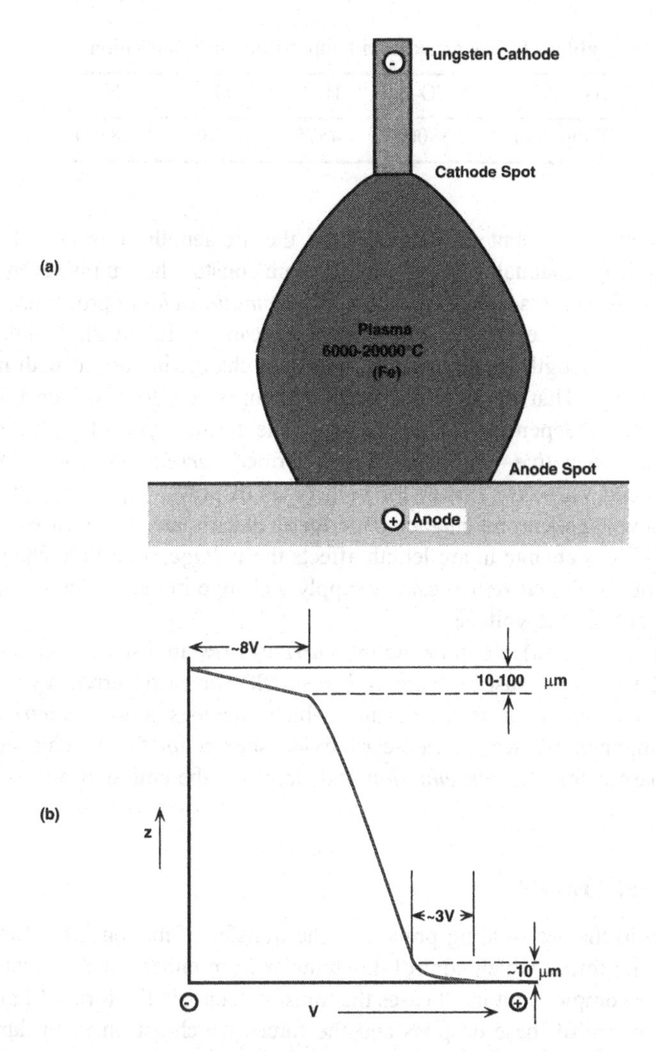

Fig. 5.16. The structure of an electric arc formed during welding. (a) The *tungsten inert gas* arc (TIG) showing the cathode and anode spots and the expected plasma temperature for steel welding. (b) The potential distribution in the arc, showing the anode and cathode drops and their approximate dimensions.

requires higher voltages, consistent with the high plasma formation temperatures of oxygen and nitrogen (Table 5.1).

Power supplies for arc welding are of two general types, depending on whether they maintain the *current* or the *voltage* at a specific value. In the former type a fluctuation in the *voltage* leads to only a small change in the current. This is desirable in hand-held, *manual welding*, where it is important to minimize variations

Table 5.1. Gas breakdown temperature (90% ionization)

Gas	CO_2	H_2	O_2	N_2
Temp. (K)	3800	4575	5100	8300

in the welding current but difficult to keep the arc length constant. That is, the primary concern in manual welding is to maintain constant heat input, even if the arc length varies. The reverse situation is true of *automatic welding* processes, in which the geometry of the electrode and workpiece can be robotically controlled. A fluctuation in arc length should be corrected by a change in current with minimum change in voltage. That is, the primary requirement is now for a *self-correcting arc length*, sensibly independent of the voltage. The former type of power supply is termed *voltage sensitive*, while the latter is termed *current sensitive*. Figure 5.18 shows schematic *response curves* for both types of power supply, together with a hypothetical voltage–current characteristic for an electric arc. In the case of *voltage-sensitive* supply, a change in arc length affects the *voltage*, with little change in the current, while for the *current-sensitive* supply a change in the arc length affects the *current* rather than the voltage.

As noted (Fig. 5.17a), *AC (alternating current) arcs* are used for submerged arc welding. The AC arc must be *reignited* (restruck) for each current cycle. This is made possible by an electrode coating which includes a low *electronic work function* component (typically a rare-earth oxide, see Section 5.3.2). This reduces the activation barrier for *electron emission* and stabilizes the emission process.

5.3.2 Metal Transfer

A key stage in the arc welding process is the transfer of molten filler metal to the *weld pool*. The forces which control this transfer from either a metal electrode or a filler wire are complex, but in all cases the transfer occurs in the form of liquid metal droplets. The *size* of these droplets and the forces which act on them during their formation determine the mechanisms of metal transfer.

There are *four* main forces. The *first* is due to the *surface tension* of the molten metal drop and is given by $F_s = 2\pi\gamma a$, where γ is the surface tension and a is the radius of the metal electrode or filler wire. The *second* is the force of *gravity*, which acts to pull the drop off the electrode *against* the surface tension: $F_g = \rho g V$, where ρ is the density of the molten metal, g is the gravitational constant and V is the volume of the drop. The *third* force is the *electromotive force* (EMF) acting on the drop once it has lost contact with the electrode. The EMF is difficult to model. It depends on the effective charge on the drop, the sign of the voltage drop, the electron current density and the ratio of the size of the drop to the diameter of the filler wire. The *fourth* force is a *hydrodynamic* force acting on the falling drop, and results from the

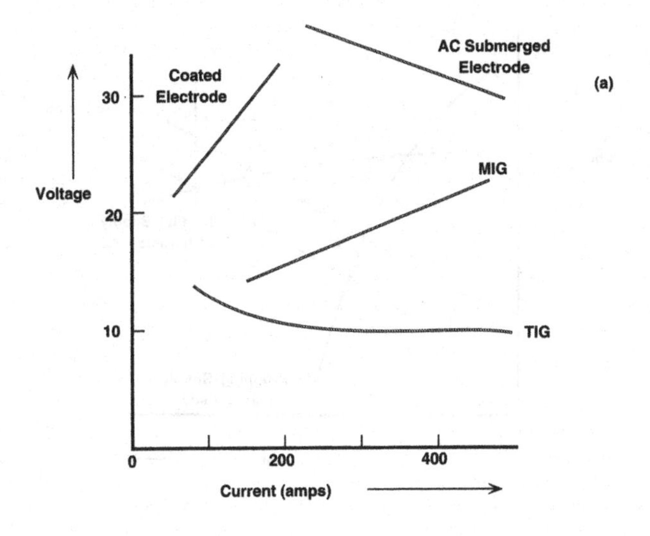

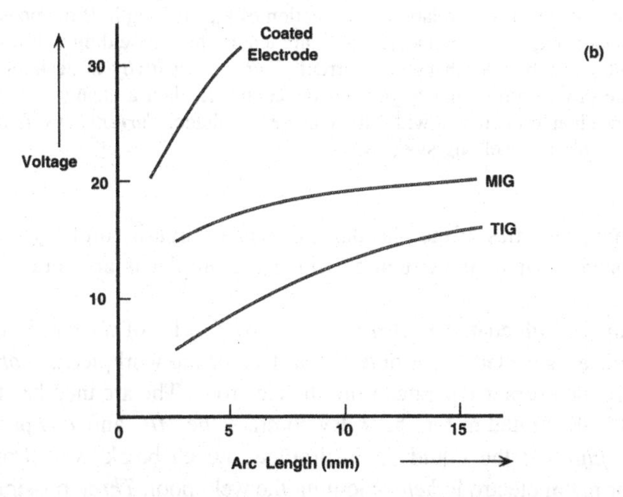

Fig. 5.17. (a) Approximate current–voltage relations for four different arc welding processes: *metal inert gas* (MIG), *tungsten inert gas* (TIG), *coated electrode* CO_2 gas, and AC (alternating current) *submerged arc*. (b) The relation between arc voltage and arc length for three different arc welding processes: MIG, TIG and the coated electrode process.

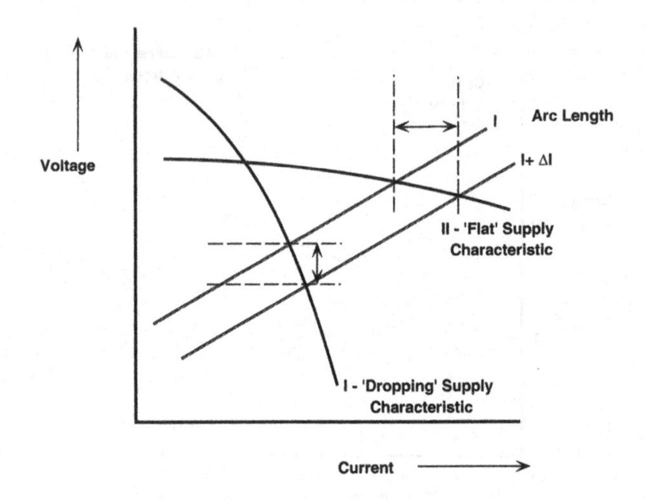

I - Voltage Sensitive: Manual Welding

II - Current Sensitive: Automated Welding

Fig. 5.18. The voltage–current relation is a function of the arc length. If the power supply has a '*dropping*' characteristic, I, then a change in arc length during welding will result in a large change in voltage with little change in current. *Voltage sensitivity* is desirable for *manual* welding. If the power supply has a '*flat*' characteristic, II, then a change in arc length will result in a large change in current with little change in voltage. *Current sensitivity* is desirable for *automated* (robotic) welding systems.

hydrodynamic properties of the plasma. It depends primarily on the jet velocity and the droplet size. That is, the extent to which the droplet is accelerated by the gas stream.

Welding in *air* with coated electrodes occurs by a series of *short circuits*, in which the molten metal is pulled down onto the surface of the workpiece, *short-circuiting* the arc before the droplet separates from the electrode. The arc then has to *re-ignite*. The process is illustrated in Fig. 5.19. By contrast, the *MIG* and *TIG* processes both involve *free flight* of the liquid metal droplets, which break away from the filler metal wire or metal electrode *before* joining the weld pool. *Three* possible situations can be distinguished (Fig. 5.20). In the first *small* droplets are pulled off the filler wire by the *EMF* and propelled into the weld pool by *hydrodynamic* forces. In the second case (typical of CO_2 gas welding) the droplet is *repelled* by the *EMF*. Only the larger droplets, travelling under *gravitational* forces, contribute to the weld pool, while the smaller droplets are lost from the weld pool by 'splashing' or splatter. In the third case, the metal at the end of the filler wire forms a *large* droplet which only detaches when the *gravitational* force overcomes the *surface tension*. This last situation is typical of the welding of aluminium alloys.

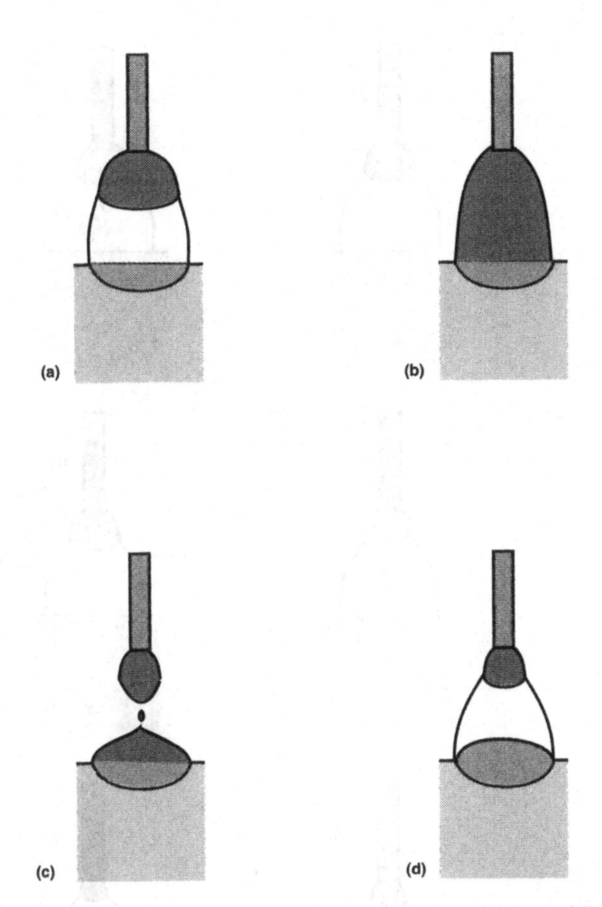

Fig. 5.19. *Coated electrodes* repeatedly make and break the arc during welding. (a) The arc melts forming a liquid pendulous drop on the electrode. (b) The drop makes contact with the weld pool and short-circuits the arc. (c) The drop breaks away from the electrode to join the weld pool. (d) The arc is then restruck as the tip of the electrode approaches the pool.

We can therefore summarize the four dominant modes of liquid metal transfer:

1. *Direct transfer* of molten metal occurs from a coated electrode to the weld pool with periodic *short-circuiting* of the arc.
2. Small *droplets* are pulled off the filler wire or electrode by the *EMF* and propelled into the weld pool by *hydrodynamic* forces.
3. Droplets are subjected to a *repulsive* EMF in the arc and arrive at low velocities, with the smaller droplets being lost to the weld pool.
4. *Large* droplets are detached from the filler wire or rod by *gravitational* forces and fall into the weld pool.

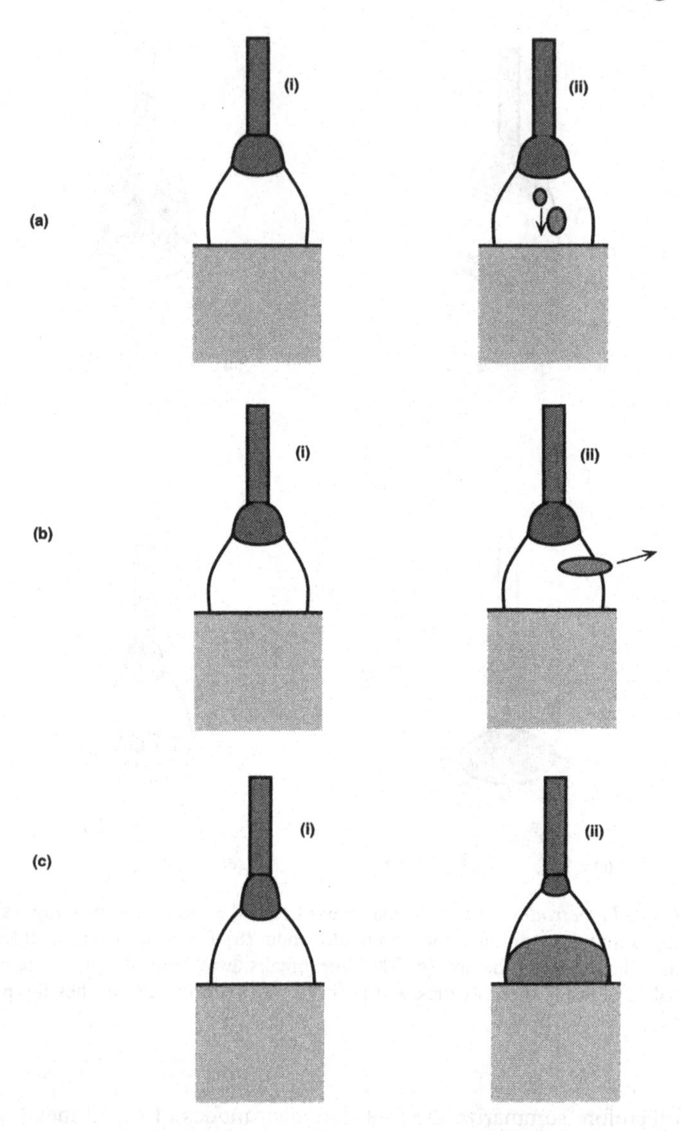

Fig. 5.20. In the *TIG* and *MIG* processes three separate metal transfer mechanisms can be distinguished, all involving free- flight of metal droplets through the arc. (a) The arc melts the tip of the electrode (MIG) or filler wire and the electromotive force (EMF) pulls away small droplets and accelerates them into the weld pool. (b) In reverse potential the reverse sign of the EMF repells small droplets and, under the force of gravity, only the largest can reach the weld pool. (c) At low welding voltages the EMF is insufficient to pull away small droplets and mass transfer is then dominated by large droplets which fall into the weld pool under gravity alone.

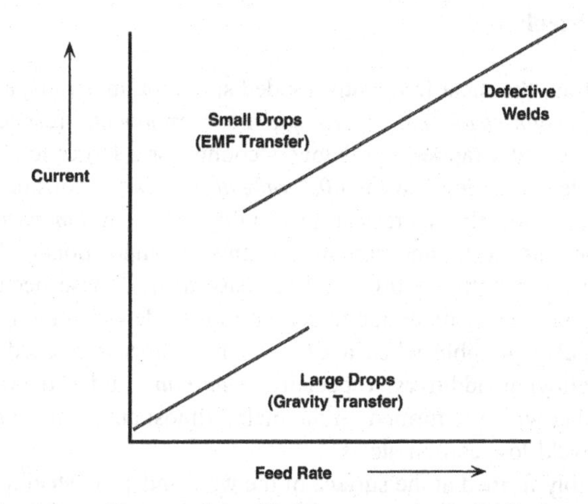

Fig. 5.21. *Aluminium alloys* can be welded in two regimes. At *low* currents and feed rates metal transfer is dominated by large molten drops detached from the feed wire by gravitational forces. At *higher* currents and feed rates the electromotive forces (EMFs) start to dominate mass transfer and there is an abrupt increase in the current, together with a decrease in the drop size. At *very high* feed rates instability sets in and the final weld becomes defective.

It is sometimes possible to observe a *transition* in these metal transfer processes as the *feed rate* for the filler is changed. This is illustrated in Fig. 5.21. At low feed rates (and low currents) the liquid drop grows until it is detached under gravity, but at a rapid rate of feed the droplets are pulled off by electromotive forces. There is a sharp increase in current, and correspondingly a sharp decrease in droplet size, when the mechanism changes in the transition region. If the current (and corresponding feed rate) are raised too far, then the high droplet velocity leads to defective welds associated with extensive splatter.

The complexity of molten metal transfer should not obscure the need to control the process. This is achieved by seeking an appropriate *working window* in which the rate of filler metal supply is adapted to both the current–voltage characteristics of the arc and the rate of welding (the velocity of the heat source).

5.4 ALLOY WELDING

In the next chapter we consider the metallurgical factors which determine the *microstructure* of weld metal and the *microstructural changes* which occur in the heat-affected zone (HAZ). In the present section we are primarily concerned with the effect that the welding parameters have on the metal alloy which is to be welded. We therefore restrict ourselves to the practical factors which determine the *weldability* of an alloy, especially those which are characteristic of specific *alloy systems*.

5.4.1 Carbon Steels

Low carbon steels are the most frequently welded structural materials, although the welding of cheap, *non-structural*, *thermoplastic* components (especially heat-bonded, plastic bags and wrappers) probably accounts for a larger total volume of material. Only steels containing less than *0.3 wt% of carbon* are considered suitable for welding, because of the increased probability of both *embrittlement* and *compositional changes* at higher carbon contents. Compositional changes are nevertheless common in both the filler and the base metal. These occur primarily through the loss of carbon by oxidation to carbon monoxide gas, although a *gain* in carbon content is also possible when a CO_2 welding process is used. The more readily oxidized alloying additives, especially *aluminium* and *titanium*, are also easily lost to the slag which is formed on the melt. Almost any heat source can and has been used to weld low carbon steels.

A *slag* is invariably formed at the surface of the weld and, as in steel making, may be either *acid* or *basic*. Acid slags are high in *silica* and have low viscosities. Basic slags have a high *manganese oxide* content and much higher viscosities. The slag serves a useful function by protecting the metal in the weld pool against excessive oxidation.

The *mechanical properties* of plain carbon steel welded joints are of direct concern in a wide range of engineering structures. Four factors should be considered: (i) the influence of *weld geometry*, (ii) the effect of *inclusions*, (iii) the presence of *embrittling contaminants* and (iv) the influence of *porosity*. These factors parallel the previous discussion on weld defects previously given in Section 5.1.3.

Fatigue cracks in a welded joint typically initiate from a *reentrant notch*, such as that formed on the free surface at the edge of the weld pool (Fig. 5.22a). The effect of the notch angle is directly attributable to a *stress concentration factor* which reduces the fatigue limit (Fig. 5.22b). *Inclusions* also act to reduce the fatigue limit, the *reduction* in fatigue limit roughly following the $\sqrt{c}$ dependence on defect size predicted by fracture mechanics (Fig. 5.23a). The embrittling effect of *hydrogen* contamination on the fatigue life is also indicated. *Sulphur* is also an embrittling impurity, but the deleterious effects of sulphur can be mitigated by the presence of manganese, which precipitates ductile *manganese sulphide*, MnS, in place of the brittle FeS. Figure 5.23b shows schematically how the susceptibility to brittle cracking depends on both the carbon content and the Mn/S ratio.

Finally, *porosity* is associated both with the evolution of gas in the solidifying weld metal and the *shrinkage* which accompanies the solidification process. The common gases responsible for porosity are H_2O, CO, H_2 and N_2, all of which dissolve in the melt and are then evolved when the melt solidifies. The *distribution*, as well as the *size* and *volume fraction* of the porosity, determines the extent of any degradation in mechanical properties, and we have already noted the importance of industrial *international standards* in *non-destructive evaluation* for grading the severity of welding defects.

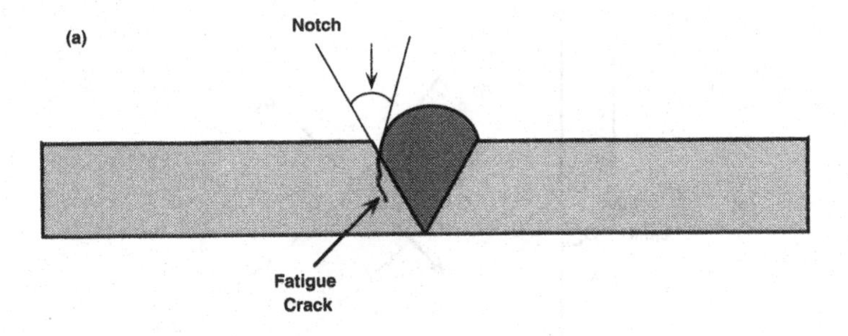

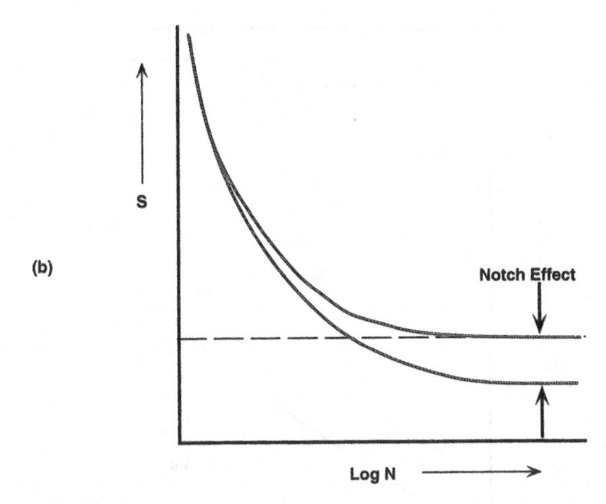

Fig. 5.22. Any *reentrant notch* in the weld geometry can act as a *stress raiser*. (a) This initiates a fatigue crack under cyclic loading, and (b) reduces the fatigue limit with respect to that of a notch- free weld.

5.4.2 Alloy Steels

We define an *alloy steel* as one that contains at least 80 wt% iron (that is, *less* than 20 wt% alloying additions). Fusion welding of high strength, high toughness structural steels requires careful control of the welding parameters to avoid brittle failure, which may occur either during the cooling stage of the weld cycle, *hot-cracking*, or subsequently, either *before* or *after* the weld is put into service.

As noted, *hot-cracking* is commonly associated with the presence of a low melting point, intergranular liquid phase. An example is *boron* residues which form a low melting point borate flux. The boundary failure is due to a combination of the presence of a liquid phase at the boundary with *thermal shrinkage*. Hot-cracking is readily recognized from the wide crack opening of the cracks (Fig. 5.11). The

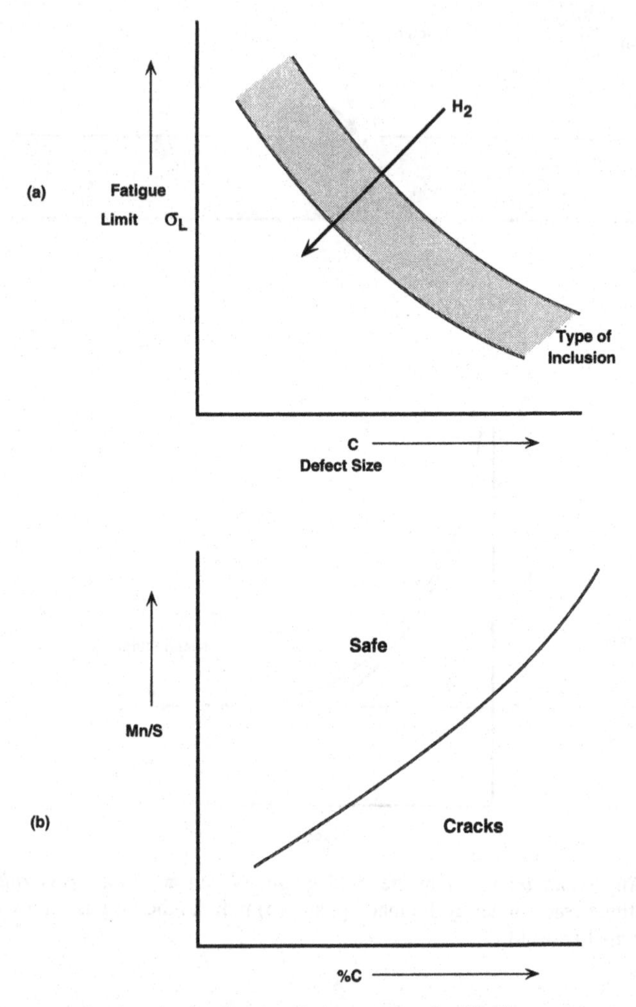

Fig. 5.23. (a) The reduction in the fatigue limit associated with the presence of *inclusions* depends on both the type and size of the inclusions, and the fatigue limit will be still further reduced if *hydrogen* is occluded during the welding process. (b) Increased *carbon* and *sulphur* contents in a weld both increase the probability of cracking, but *manganese* traps sulphur as ductile MnS inclusions, leading to a dependence of the *crack susceptibility* on the concentrations of all three elements.

presence of *sulphur* is a comon cause of hot-cracking in welded low manganese steels, as discussed in the previous section. For reasons which are unclear, high nickel alloy steels seem to be *more* susceptible to hot-cracking than other compositions.

Cold-cracking may be observed in either the HAZ or the weld itself. Four contributing causes have been identified: (i) the presence of atomic *hydrogen* introduced as *moisture* during the welding process, (ii) the formation of *martensite* during the weld cycle, (iii) unrelaxed *residual tensile stresses* in the completed weld, and (iv) a welding cycle which results in any other unsatisfactory metallurgical outcome!

The effect of small amounts of *hydrogen* on the Charpy notch energy has been clearly demonstrated, and both the notch toughness and the ductile–brittle transition temperature are affected (Fig. 5.24a, b). A fuller discussion of residual stress in the

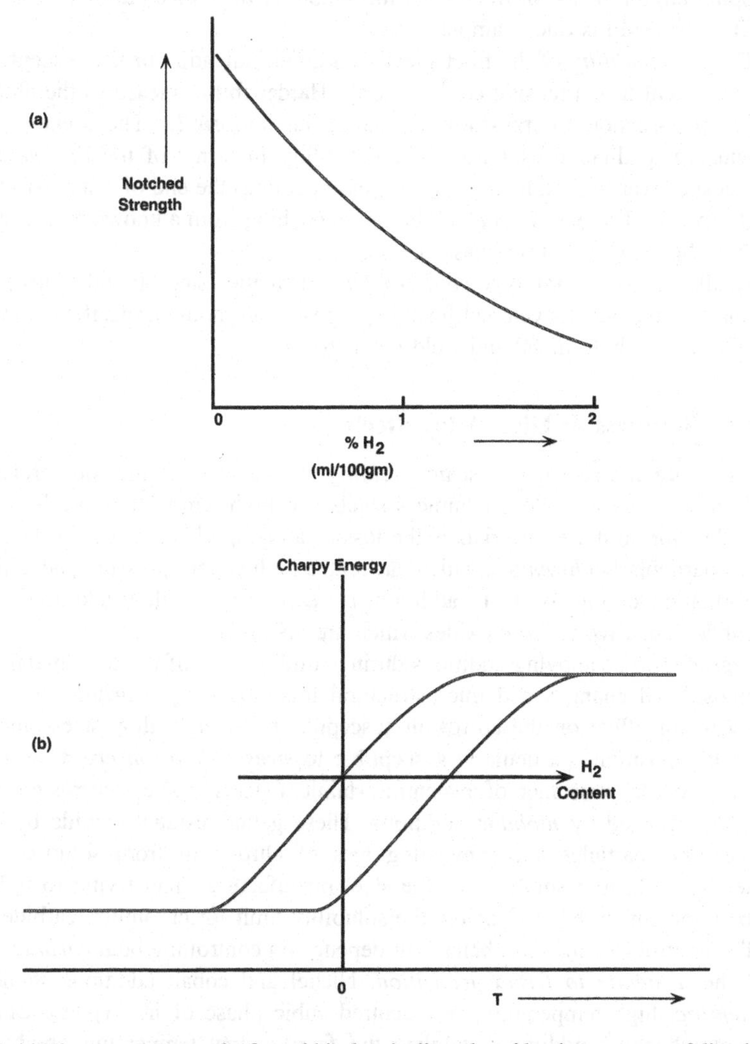

Fig. 5.24. (a) The *impact strength* of a weld is drastically reduced by the presence of dissolved hydrogen. (b) The ductile–brittle *notch transition temperature* is also raised.

weld zone and metallurgical changes in the heat-affected zone (HAZ) will be deferred until the next chapter.

Because of the major metallurgical changes that occur in the HAZ, and the very different microstructure of the cast weld metal from that of either the bulk components or the HAZ, filler metal compositions do not approximate the bulk composition. A major consideration is the danger of *embrittlement* associated with the formation of *martensite* during the cooling of the weldment to ambient temperature. One solution is to use an *austenitic steel* for the filler metal, thus side-stepping any problems of martensite formation. A filler metal alloy of composition 18Cr–10Ni–3Mo is one example.

The *hardenability* of the filler metal is a good indication of the susceptibility of the weldment to martensitic cold-cracking. Hardenability measures the ability of a large cross-section to transform to martensite *uniformily*. The *Jominy bar* test provides a qualitative estimate of hardenability in terms of the *thickness* of the martensite layer formed by *quenching* (rapid cooling) the end of a standard bar with a jet of water. The bar is *annealed* before quenching from a known temperature and time in the fully-austenitic phase region.

Finally, the susceptibility to *cold-cracking* of an alloy steel should be tested. Some standards have been developed for this purpose, and monitor specific combinations of filler metal, bulk metal and weld geometry.

5.4.3 Stainless & High Alloy Steels

Arc welding and *resistance* (seam) *welding* are the welding methods preferred for high alloy, iron-base alloys (stainless steels and high temperature steels). A major consideration in these materials is the loss of alloying elements during the welding cycle, particularly *chromium*, which has both a high vapour pressure and a high heat of formation of the oxide. In addition, *refractory metal* alloy additions, such as molybdenum, have *volatile* oxides which are lost as *fume*.

Segregation of alloying additions during solidification of the weld metal leads to compositional changes and microstructural inhomogeneity. *Carbide precipitation* has a major effect on the corrosion susceptibility of high alloy steels and makes some compositions particularly susceptible to *stress corrosion cracking*, as noted previously. The resistance of austentitic stainless steels to stress corrosion cracking can be improved by *niobium* additions. These getter residual carbide by forming stable NbC particles and preventing loss of chromium from solution. This is generally a cheaper solution for the steel manufacturer than trying to reduce the carbon content of the steel below the solubility limit for chromium carbide.

The control of alloy steel behaviour depends on controlling both *carbide stability* and the *austenite to ferrite transition*. Nickel and cobalt additions stabilize the *austentitic*, high temperature, face-centred cubic phase of iron (γ-Fe). Chromium and molybdenum additions stabilize the *ferritic*, low temperature, body-centred cubic iron phase (α-Fe). Neither nickel nor manganese form carbides in iron alloys.

Chromium carbide, Cr_7C_3 (and mixed chromium–ion carbide, $M_{27}C_6$), is more stable than iron carbide, Fe_3C. *Molybdenum* and *tungsten carbides* are even more stable, while additions of *vanadium*, *titanium* and *niobium* lead to the formation of carbides which are only dissolved in iron at very high temperatures and long times (see Section 6.3.3).

Some high-performance, specialty steels depend on the precipitation of *intermetallic phases* to attain their high strength and toughness, and several of these compositions can be successfully *heat treated* after welding. The *maraging* series and the *PH (precipitation hardening)* stainless steels are the most important examples. Heat-treatable, welded steels do *not* contain carbon, which would be deleterious both to the *weldability* and to the *heat treatment* response.

5.4.4 Non-ferrous Alloys

The classification of engineering alloys into just two groups, *ferrous* alloys and *non-ferrous* alloys, is justified by the remarkable cost-effectiveness, strength and toughness of the ferrous alloys. Similarly, the large amount of space we have allotted to the welding of ferrous alloys reflects the need to preserve their excellent mechanical properties in welded assemblies. Steels melt at temperatures approaching 1600°C, and the need to attain and control temperatures in this range over the limited dimensions of a welded joint places very stringent requirements on the welding process. The *non-ferrous* alloys melt at temperatures that range from below 600°C (for magnesium and aluminium alloys) to 1200°C (for copper and nickel alloys) and above (for titanium and specialty alloys). The heat-source requirements are generally *less* stringent for these lower melting point alloys. However, other problems are not lacking, and we summarize briefly some special requirements for each class of non-ferrous alloy.

5.4.4.1 Aluminium Alloys

Stoichiometric *aluminium oxide* forms spontaneously on aluminium as an adherent, stable, non-conducting and protective film. This oxide film must be removed before welding can take place. *Friction welding* and *ultrasonic welding* remove the oxide by self-abrasion at the contact surface and solid state bond without exceeding the melting point of the alloy. The application of these processes is limited to simple geometries. In *fusion welding*, molten halide fluxes are used to dissolve the oxide at temperatures below the melting point of the alloy. An *AC arc* is commonly used as the heat source for welding aluminium alloys, and the frequency of the pulsed arc is dictated by the frequency of the AC supply.

Porosity is a major defect in aluminium alloy welds, and is associated with *hydrogen* evolution, originating from hydrated surface films and moisture, together with excessive heat input. Aluminium has a high coefficient of thermal expansion (twice that of steel), so that, although the temperatures involved are 1000°C lower than in the welding of steel, the residual thermal stresses can still be high. *Constrained* welds, especially in Al–Si alloys which have limited ductility, are susceptible to cracking associated with thermal contraction. The susceptibility depends on the ratio of the thermal stress to the failure strength.

As in the case of specialty steels, some high-performance (wrought) aluminium alloys can be *heat treated* after welding to optimize their mechanical properties. These heat-treatable welded alloys all require a *solution treatment* followed by *age-hardening* in order to develop a uniform precipitate distribution.

5.4.4.2 Magnesium Alloys

Magnesium alloys are no longer regarded as difficult (or dangerous) to weld, although some fire hazard undoubtedly exists. However, magnesium is a very *volatile* metal and care must be taken to avoid overheating.

5.4.4.3 Copper Alloys

Porosity is a common defect in welded components of copper alloys, and is usually associated with the evolution of *steam* due to the reaction of dissolved hydrogen and oxygen. Copper alloys frequently contain *zinc*, which is even more volatile than magnesium, so that loss of zinc is to be expected during welding (and may be a further source of porosity in the weld metal). In general, copper alloys are more frequently *brazed* than welded, and the lower brazing temperature helps to avoid zinc loss.

5.4.4.4 Nickel Alloys

Welding of nickel alloys requires temperatures approaching 1200°C. Either *TIG* or *MIG* processes can be used. Contamination by *trace elements* is the main source of weld defects. The presence of *sulphur*, *lead* or *phosphorus* can lead to *hot-cracking* through the formation of low melting point non-metallics which wet the grain boundaries. Typical sources of contamination are, as in the case of steel, *grease* or *oil* which is present on components which have been inadequately cleaned.

Porosity in welded nickel alloys has been traced to the presence of oxide or nitride contamination, leading to the evolution of CO or N_2 during welding.

Summary

Welding commonly refers to the joining of two metallic components by heating the region at the interface above the melting point of one of the components, although explosive bonding and friction bonding are also referred to as solid state welding processes. Welding is thus distinguished from brazing and soldering, in which a low melting point filler metal is used to make the join. The technological importance of welding derives from the high strength to weight ratio of welded joints and their generally good resistance to environmental attack.

Welding is an extremely versatile process, encompassing a wide range of joint geometries and dimensions, but in all cases a major concern is the extent of elastic constraint at the joint, which to a large extent determines the level of residual stress in the final assembly. A variety of heat sources are used for welding, of which by far the most common is the electric arc. Electrical resistance and electron beam welding also make use of electrical power, but other alternatives are gas welding, laser welding and the use of the chemical energy in an exothermic reaction.

The strength and ductility of a welded joint are often limited by the presence of defects introduced by the welding process. These include geometrical defects, due to either misalignment of the components or poor control of the melted zone, as well as non-metallic inclusions and porosity. They also include microcracking associated with thermal shrinkage and embrittlement, often due to the presence of atomic hydrogen. Stress-corrosion cracking may be due to a combination of embrittlement of the metal, unrelieved thermal stresses and the presence of a reactive environment.

The welding process is controlled by heat and mass transfer, and hence depends on the transport properties of the components. It is convenient to distinguish between the two-dimensional transport characteristic of thin sheet and three-dimensional transport in which the assembled components constitute a half-space. In both cases approximate analytical solutions are available to determine the temperature profile in the region of the weld pool and the thermal history of any volume element of the components. Similarly, solutions of the diffusion equations permit reasonable estimates to be made of the extent of mass transport in the diffusion zone during the welding cycle.

The process of metal transfer in the electric arc may have a large effect on weld quality, and is not easy to control. Several transfer processes have been distinguished: direct transfer with periodic short-circuiting of the arc, small droplets detached by electrostatic forces and large droplets detached from the feed source by gravitational forces. The current–voltage characteristics of the arc and the rate of advance of the heat source determine the extent of the working window available for successful welding.

Many metals and engineering alloys cannot be successfully welded, but the list of those that can is impressive: low carbon and low alloy steels, stainless steels and high temperature alloys, nickel and copper alloys, aluminium and magnesium alloys, titanium alloys and many others.

Further Reading

1. K. Easterling, *Introduction to The Physical Metallurgy of Welding*, Butterworth, London, 1985.
2. J. E. Lancaster, *Metallurgy of Welding*, Fourth Edition, Allen & Unwin, London, 1987.

3. R. W. Messler, Jr., *Joining of Advanced Materials*, Butterworth–Heinemann, London, 1993.
4. Metals Handbook, Eighth Edition, Volume 6: *Welding and Brazing*, American Society For Metals, Metals Park, Ohio, 1971.

Problems

5.1 List three main types of heat source for welding. Explain the basic processes of heat generation and list the important differences between the classes of heat source. For each type of heat source, give one specific example of a welding technique and discuss the temperature distribution in the components during the welding process.

5.2 Describe the process of electron beam welding. Discuss the advantages and disadvantages of this process. Give two classes of alloy which are often electron beam welded and explain why.

5.3 An oxyacetylene torch can be operated with either a *rich* or *poor* oxygen flow. How are these operating conditions likely to influence the weld morphology and weld strength?

5.4 Define *hot-cracking*. Describe the influence of the Mn/S ratio in a low alloy steel on its susceptibility to hot-cracking. How is the susceptibility affected by the carbon content?

5.5 Define *cold-cracking*. List and explain the principal reasons for cold-cracking, and the precautions which can be taken to prevent this phenomenon.

5.6 Give three possible sources for the evolution of hydrogen during welding. What mechanisms have been suggested for the hydrogen embrittlement of a weld?

5.7 Check a handbook to find the reasons for recommending electrodes of type ASTM 7015 and ASTM 7016.

5.8 Porosity in the weld region can seriously lower the strength of a weld, but may also (in some cases) prevent crack propagation. List some common causes of weld porosity.

5.9 A welded stainless steel assembly failed during service by stress corrosion at the weld. Explain the probable causes, and suggest what might have been done to prevent failure.

5.10 What is the usual reason for the appearance of martensite after welding a carbon steel? How would the presence of martensite be expected to affect the mechanical properties of the joint?

5.11 List the main difficulties in welding wrought aluminium alloys.

5.12 Explain how steam may be generated during the welding of copper, and discuss how this would be expected to influence the mechanical properties of a welded joint.

6

Weld Metallurgy

A wide range of mechanical and physical properties can be developed in a single engineering material of given composition by altering the *microstructure*. Such properties are said to be *structure sensitive*. The control of microstructure, through appropriate *processing*, allows the engineer to obtain the required material *performance*. However, there are usually several processing options available for achieving a given value of an engineering property. For example, the *same* hardness can be achieved in a given low carbon steel having several *different* microstructures. These are developed by a combination of *cold work* and *heat treatment* to achieve a range of *grain size* in *tempered martensites*, or a variety of *pearlitic* or *bainitic* structures.

In the welding of *like materials*, a primary aim is to approximate selected engineering properties of the *bulk* components in the region of the *weldment*. This *cannot* be done by preventing microstructural changes, but it can be achieved by careful attention to the *weld* and *HAZ metallurgy*, although we must recognize that only *some* of the properties will be retained.

We first consider the local effects of geometry, residual stress and thermal history in the *weld bead*. Previous discussion of these factors covered the *macroscopic* effects, on the scale of the *component* dimensions, while the present analysis is on the *mesostructural* scale, defined by the size of the weld pool and the width of the HAZ. This is first followed by a description of the microstructural morphology in the *weld bead* (the region which underwent solidification), and then an analysis of the complex microstructural changes which occur in the *HAZ*. Once more, the emphasis is on *steel*, although much of the discussion applies to welds formed in *other* metals and alloys. The weld is assumed to be between two *like* components of similar bulk composition and microstructure.

6.1 GENERAL CONSIDERATIONS

The local *weld topology* is dictated by the processing history, which may involve several heating cycles (*multipass welding*) and post-welding *heat treatment*. This is illustrated in Fig. 6.1. A simple weld between two bevelled plates is first *tack-welded* in position by a series of spaced *spot-welds*. The structural weld is then made in a *single pass*, to give a structure with a mirror symmetry plane passing through the *centre* of the weld and *perpendicular* to the plane of the welded plates.

The *second* example is a *multipass weldment* between thick bevelled plates. The *smaller* bevel is welded first, in either one or two passes, and, welding is then continued on the *reverse* side by multipass welding along the *larger* bevel. Each weld pass modifies the microstructure of both the previously-cast weld metal and the neighbouring bulk metal (in the HAZ). The *residual stress pattern* is also modified with each weld pass, and *interdiffusion* between the filler metal and the HAZ occurs during the repeated heating and cooling cycles.

6.1.1 Localized Residual Stress & Distortion

The *residual stresses* in the vicinity of a weld develop as a function of the *thermal history*, starting at the trailing edge of the weld pool (Fig. 6.2a). We confine ourselves to a *qualitative* discussion of the development of the *axial* stress distribution *parallel* to the weld line and *in the plane of the surface*, as the weld pool moves away. The temperature distribution across the weld line is illustrated in Fig. 6.2b. At the *edge* of the melt (*section I*) the axial stress must be zero (Fig. 6.2c). A *compressive stress* is developed in the hot zone near the solidification front, as a result of local *thermal expansion*. This is balanced by a *tensile stress* in the adjacent, comparatively cooler, bulk metal. At a later stage, *section II*, thermal *shrinkage* of the solidified weld metal imposes a *tensile stress* in the weld bead, while the tensile stress in the developing HAZ is *decreased*, relieved by thermal *expansion*. When cooling is nearly complete (*section III*) the axial stress in the weld zone is characterized by residual *tension*, while that in the HAZ is dominated by residual *compression*. *Shear stresses* are a result of the tensile stress gradient, and are a *maximum* in the transition zone between the cast weld bead and the HAZ. It is not difficult to predict that embrittlement of the *weld metal* will lead to cracking *across* the line of the weld bead, while loss of ductility in the *HAZ* will lead to cracking *parallel* to the line of the weld and close to the region of maximum shear stress.

The major *volume changes* are associated with the liquid to solid transformation in the weld pool and the thermal contraction which occurs on cooling from the melting point, but *phase changes* in the solid state are additional sources of residual stress. The transition from the face-centred cubic, *austentitic* structure of steel, to the body-centred cubic, *ferritic* structure is accompanied by a small increase in volume (Fig. 6.3). It follows that, in a ferritic steel, we can expect two regions *within* the HAZ, an *inner* region, near the weld bead, which has been heated sufficiently during the welding cycle to transform to *austenite*, and an *outer* region, adjacent to unaffected bulk material, which has remained *ferritic* throughout.

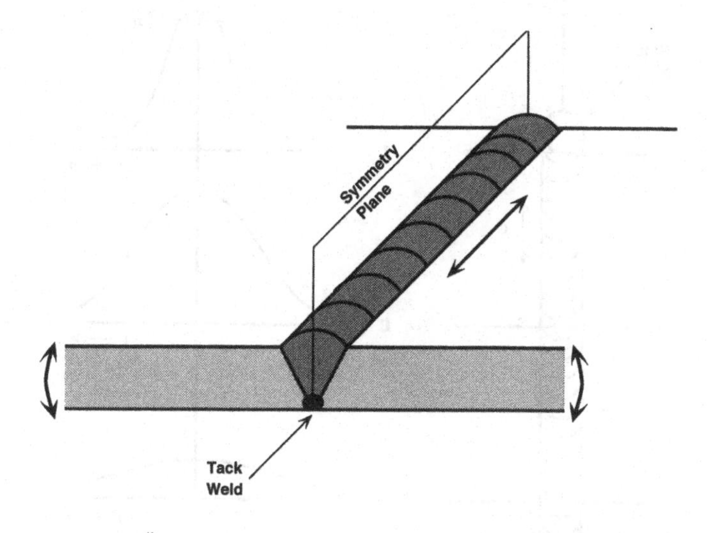

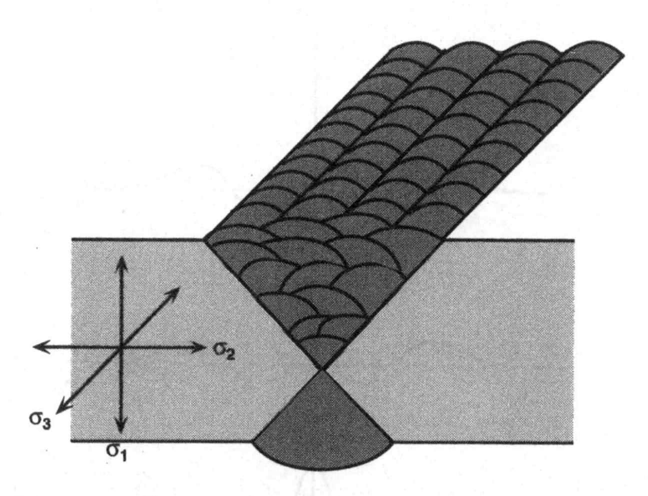

Fig. 6.1. In a single pass weld the *longitudinal shrinkage stresses* and the *transverse bending moment* dominate the residual stress pattern. In a multipass weld the *triaxial stress* pattern is complex and varies throughout the welded zone.

Distortion of the welded assembly is a side effect of residual stress and is dependent on the constraints imposed by the *geometry* of the welded structure. It is useful to distinguish between *bending moments*, which reflect the through-thickness weld asymmetry and give rise to *bowing* (Fig. 6.4a), and *tensile stresses*, confined to the plane of the welded components, which give rise to both radial and longitudinal shrinkage (Fig. 6.4b).

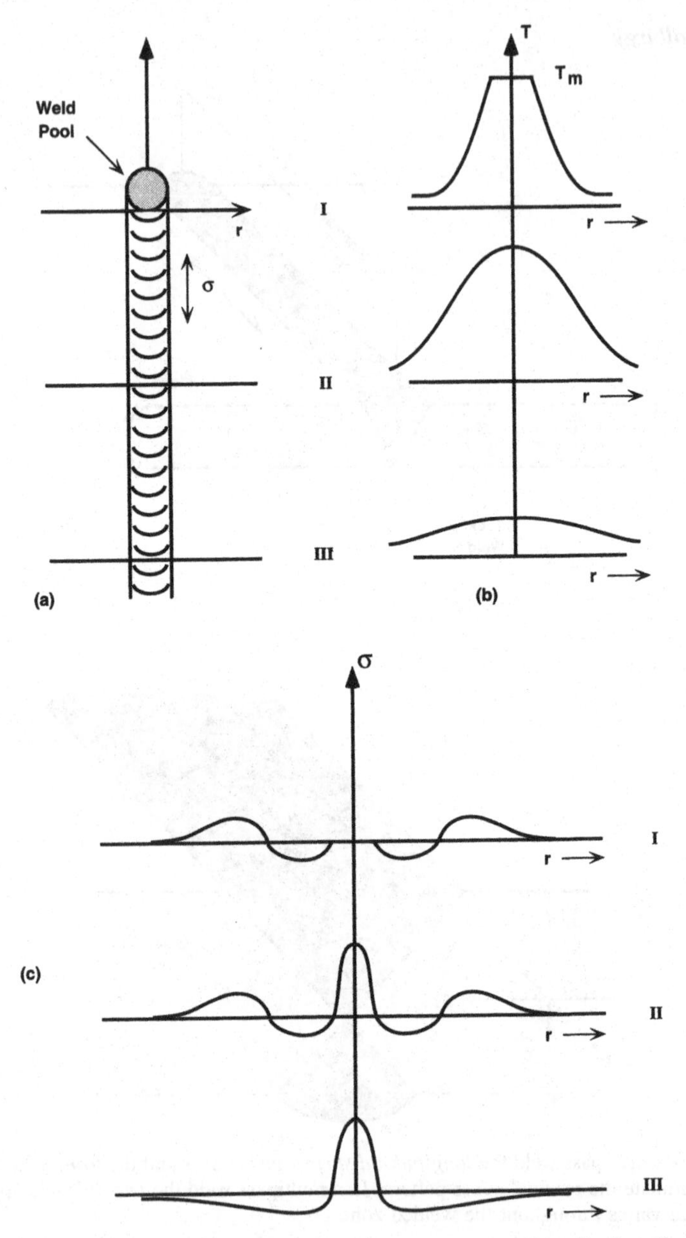

Fig. 6.2. (a) The *longitudinal* residual stress distribution varies along the *length* of the weld pass. (b) Three residual stress zones can be distinguished by their *temperature distribution*: I. The edge of the weld pool, II. The fully-solidified zone behind the weld pool, III. The final, fully-cooled weld. (c) In zone I *thermal expansion* puts the solid matrix into *compression*, and this is balanced by *tensile stresses* far from the weld line. In zone II solidification of weld metal leads to *tensile stresses* in the weld bead which are balanced by *compressive stresses* in the surrounding matrix, while the thermal gradient continues to generate a region of *tensile stress* far from the centre line. In zone III the maximum tensile stresses in the weld bead have relaxed somewhat and are balanced by a *longitudinal* compressive stress in the HAZ.

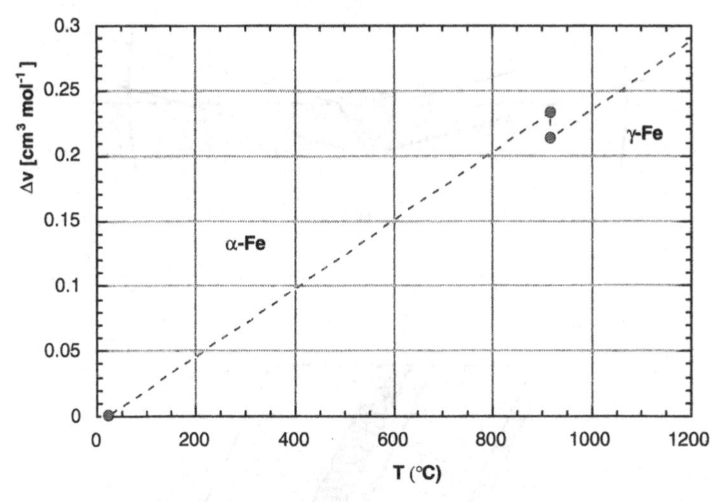

Fig. 6.3. The density of γ-Fe is *higher* than that of α-Fe, leading to an abrupt increase in specific volume as the metal is cooled through the transformation temperature (910°C).

6.1.2 Weld Parameters & Temperature Distribution

The equations of *heat transfer* which control the distribution of temperature in the bulk components surrounding a heat source have already been discussed, but now we are concerned with the development of the microstructure in both the weld metal and the HAZ, and it is the *thermal history* and *temperature profile* in these local regions which will now be considered.

We discuss *one* geometry, the *butt-welding* of two plates of identical composition and thickness, and consider *three* factors: (i) the *thermal conductivity* of the bulk metal, (ii) the *velocity* of the heat source, and (iii) the *thickness* of the plate. Figure 6.5 shows schematic temperature profiles along the line of the weld at the *surface* of the plate. *Three* materials are compared: an *aluminium alloy*, which has the *highest* thermal conductivity, a *carbon steel* with intermediate thermal properties, and an *austenitic stainless steel*, which has the *poorest* thermal conductivity. As the thermal conductivity *decreases* the temperature profile around the moving point heat source becomes increasingly *asymmetric.* The same effect is produced by *increasing* the velocity of the heat source—the hot zone *trails* further behind the position of the source as the velocity *increases. Increasing* the plate thickness, on the other hand, *reduces* the axial asymmetry of the temperature profile, as more heat is then lost in the through-thickness direction and the governing heat transfer equations become increasingly *3-D* rather than *2-D*.

In summary, the final distribution of *residual stress*, the dimensions of the *HAZ* and the microstructural changes that occur in the HAZ and the weld bead depend on the weld parameters *and* the thermal properties of the material *and* the geometry of the components.

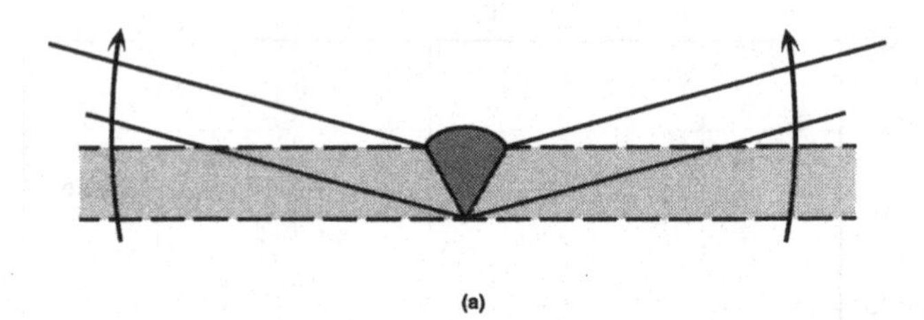

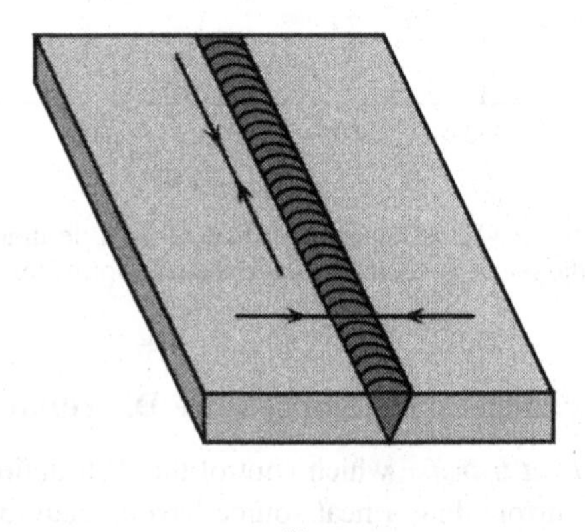

Fig. 6.4. (a) *Asymmetry* perpendicular to the weld line leads to a residual bending moment about the weld axis. (b) The principal *residual stresses* in the plane of the surface are *shrinkage* stresses parallel and perpendicular to the weld line.

6.2 WELD METAL MORPHOLOGY

Solidification of the filler metal in a weld pool is a *casting* process, similar to casting in a mould. In this case the mould has limited dimensions and casting during welding is a *continuous* process which occurs behind the moving heat source. Just as the microstructure of cast components may exhibit *dendrite formation*, *columnar growth* or *equiaxed grains*, so the microstructure in a weld bead may show the same features (Fig. 6.6). *Segregation* effects are also observed. The *partition coefficient* is defined as the equilibrium ratio of the concentrations of solute within the solid and the melt respectively, C_S/C_L. Segregation may result from the *rejection* of solute into the melt if the partition coefficient is *less* than unity, or *dilution* in the melt in the less common event that the partition coefficient is *greater* than unity (Fig. 6.7). *Dendrites* are branched single crystals which develop as the result of the more rapid heat transfer at the corners of partially-facetted growing grains. Dendrite formation

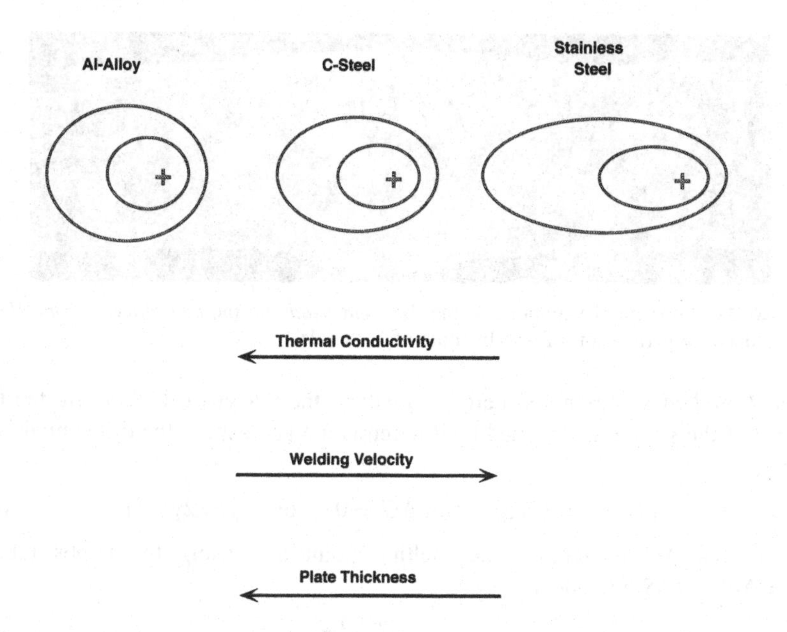

Fig. 6.5. The approximate shape of the free-surface *isotherms* as a function of thermal conductivity, welding velocity and plate thickness. The isotherms approximate the expected behaviour for an *aluminium alloy*, a *plane carbon steel* and a *stainless steel*.

is frequently revealed by *chemical etching*, since the rejection of solute between the branches of the dendrite commonly leads to a concentration-dependent chemical response. *Columnar growth* is observed as a consequence of the large temperature gradients developed during welding, and associated with the directionality of the *heat flux* away from the heat source. That is, crystals nucleated at the *edge* of the weld pool grow *in* towards the centre of the weld bead. *Equiaxed grains* are observed at the centre of the wild bead, when grains nucleated in the melt dominate the solidification process.

6.2.1 Grain Size & Cooling Rate

The *grain size* in the weld metal is primarily a function of the *cooling rate*, and can be described by classical nucleation theory. The total change in *free energy* of the system must be *negative* if crystal growth is to occur from the melt. Assuming *spherical* nuclei of the solid with *interfacial energy* γ, then the change of total *free energy* on introducing a *single* spherical nucleus of radius r is given by: $U = 4\pi r^2\gamma - 4/3\pi r^3\Delta G$, where ΔG is the *molar free energy change* accompanying the transition from liquid to solid (Fig. 6.8). *Small* nuclei are therefore unstable, because they can only grow by *increasing* the total energy of the system. The

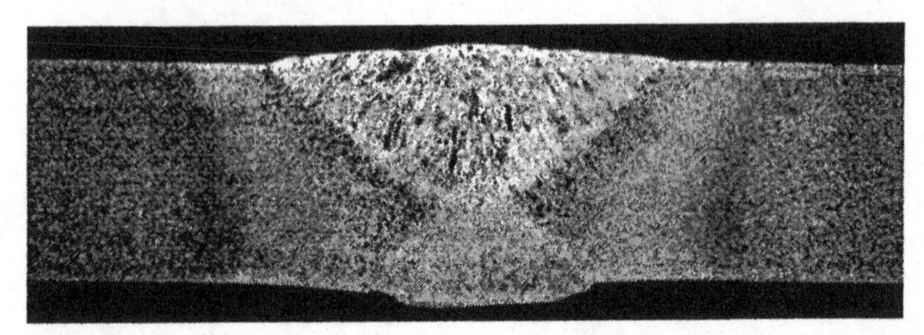

Fig. 6.6. Microstructural variations within the *weld bead* and the *heat-affected zone (HAZ)*. Reproduced by permission of The Institute of Materials.

critical nucleus size required before growth of the nucleus can *decrease* the total energy of the system is obtained by differentiating and setting the differential equal to zero:

$$\mathrm{d}U/\mathrm{d}r = 8\pi r_{\mathrm{c}}\gamma - 4\pi r_{\mathrm{c}}^2 \Delta G = 0, \quad \text{or} \quad r_{\mathrm{c}} = 2\gamma/\Delta G \tag{6.1}$$

Noting that ΔG is zero at the melting point and using the Gibbs relation $\Delta G = \Delta H - T\Delta S$, we obtain:

$$r_{\mathrm{c}} = \frac{2\gamma}{\Delta H} \cdot \frac{T_{\mathrm{m}}}{\Delta T} \tag{6.2}$$

where ΔT is the *degree of supercooling* below the melting point. Substituting this value for r_{c} in the previous relation for U gives the energy associated with the formation of a *critical nucleus*:

$$U_{\mathrm{c}} = \frac{16\pi}{3} \cdot \frac{\gamma^3}{\Delta H^2} \left(\frac{T_{\mathrm{m}}}{\Delta T}\right)^2 \tag{6.3}$$

Assuming *Bolzmann statistics* gives the probability of forming a critical nucleus at any temperature T *below* the melting point T_{m} as: $P_{\mathrm{c}} = \exp(-U_{\mathrm{c}}/RT)$. The number of nuclei per unit volume is therefore expected to *increase* rapidly as the super-cooling increases, leading to the formation of fine, equiaxed structures in rapidly cooled welds.

6.2.2 Phase Stability

The *stability* of the phases formed in the weld during solidification is a function of the *equilibrium phase relations*, which are given by the *phase diagram*. These equilibrium structures will be modified by the *cooling rate* and the presence of *impurities*. In the welding of low carbon steels it is the *iron–carbon* binary phase diagram that dominates the development of the microstructure (Fig. 6.9). In plane carbon steels containing less than 0.51 wt% of carbon equilibrium solidification initiates with the formation of BCC δ-ferrite, which then undergoes a *peritectic reaction* to form the γ-austenite, FCC phase. Figure 6.10 compares the diffusion

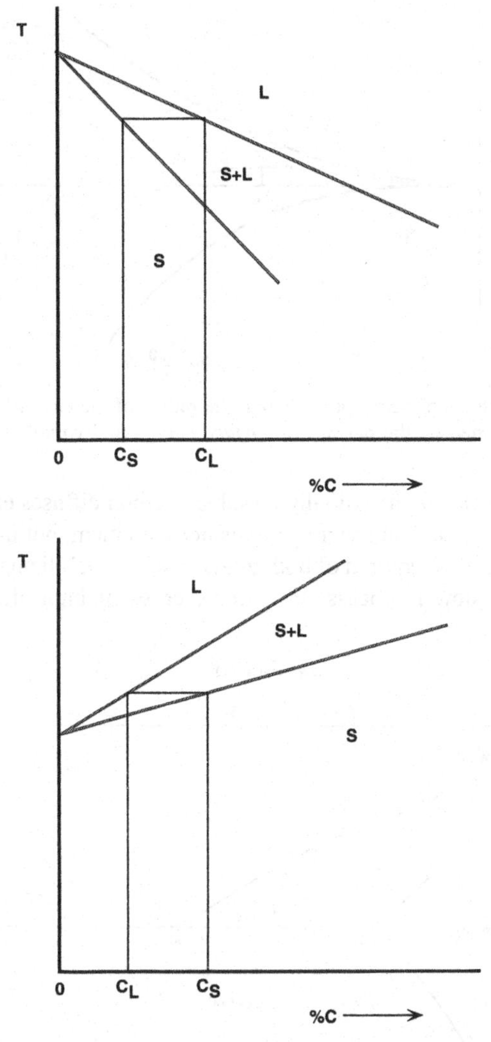

Fig. 6.7. The *partition coefficient*, defined as C_S/C_L, may be either *less* than or *greater* than unity, and determines whether *segregation* or *desegregation* (rejection) of solute is to be expected.

paths in a *peritectic* reaction with those in a *eutectic* reaction. In the eutectic reaction the constituents migrate easily through the *liquid phase* to the advancing front of the two solid phases, which constitute the eutectic product, while in the peritectic reaction diffusion has to occur through the single phase, solid reaction product, and is much slower. *Segregation* and incomplete reaction are common in peritectic transformations.

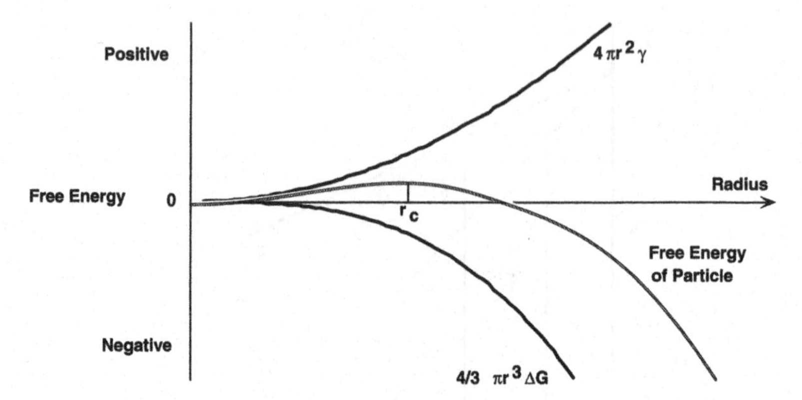

Fig. 6.8. Classical nucleation theory predicts that the radius of the *critical nucleus* for stable growth will be determined by the balance of *surface energy* and *volume free energy*.

In plain carbon steels the interstitially dissolved carbon diffuses extremely rapidly at the peritectic temperature and segregation is not a problem, but in low alloy steels *substitutional* diffusion is required of both α- and γ-phase-stabilizing alloy additions, and this is a much slower process. In extreme cases of high alloy contents and

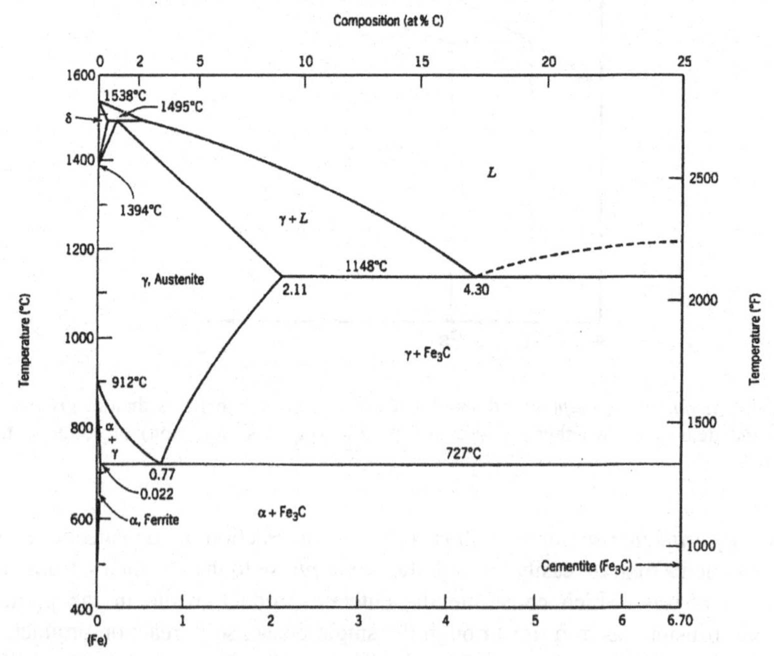

Fig. 6.9. The binary *iron–carbon phase diagram* is probably the most important figure in this text! Reproduced by permission of John Wiley & Sons Inc.

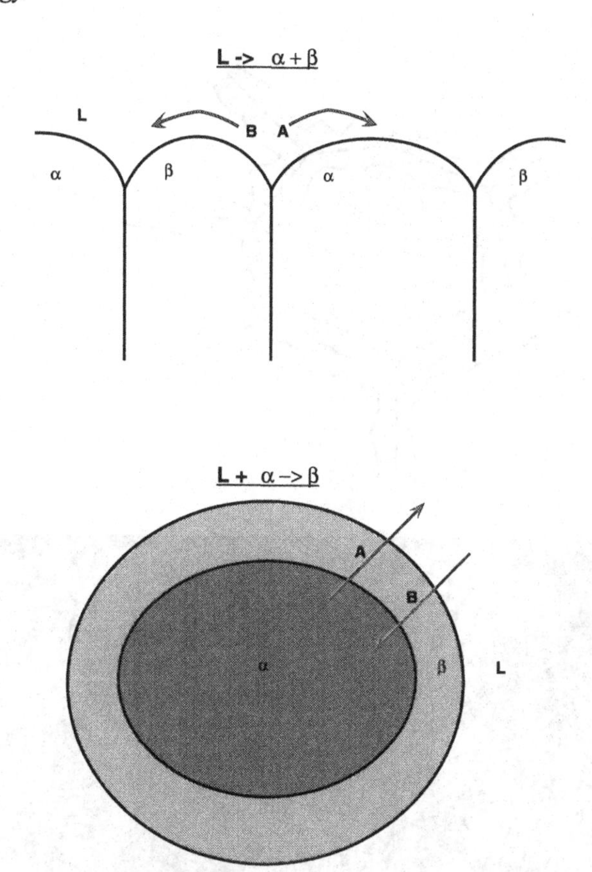

Fig. 6.10. Eutectic growth, $L \rightarrow \alpha + \beta$, occurs by mass transport of the diffusing species in the *liquid phase* and equilibrium concentrations are achieved fairly readily. *Peritectic transformations*, $L + \alpha \rightarrow \beta$, require diffusion of the constituents through the *product phase*, and phase equilibrium is difficult to achieve.

rapidly solidified weld metal, some δ-ferrite may persist throughout the γ-phase region. It will then be retained to room temperature, since the crystal structures of δ- and α-ferrite are identical.

As the temperature of a plane carbon steel approaches the *eutectoid temperature* (723°C) ductile, pro-eutectoid ferrite forms at the austenite grain boundaries, and may either form a continuous layer or grow into the austenite grains in an *acicular* morphology (a *Widmanstätten* structure) (Fig. 6.11a, b). The remaining austenite can transform to *pearlite* below the eutectoid temperature, an intimate mixture of α-ferrite and cementite (Fe_3C), with a lamellar morphology (Fig. 6.12). The kinetics of pearlite formation and the resultant pearlite morphology depend on the *cooling rate*

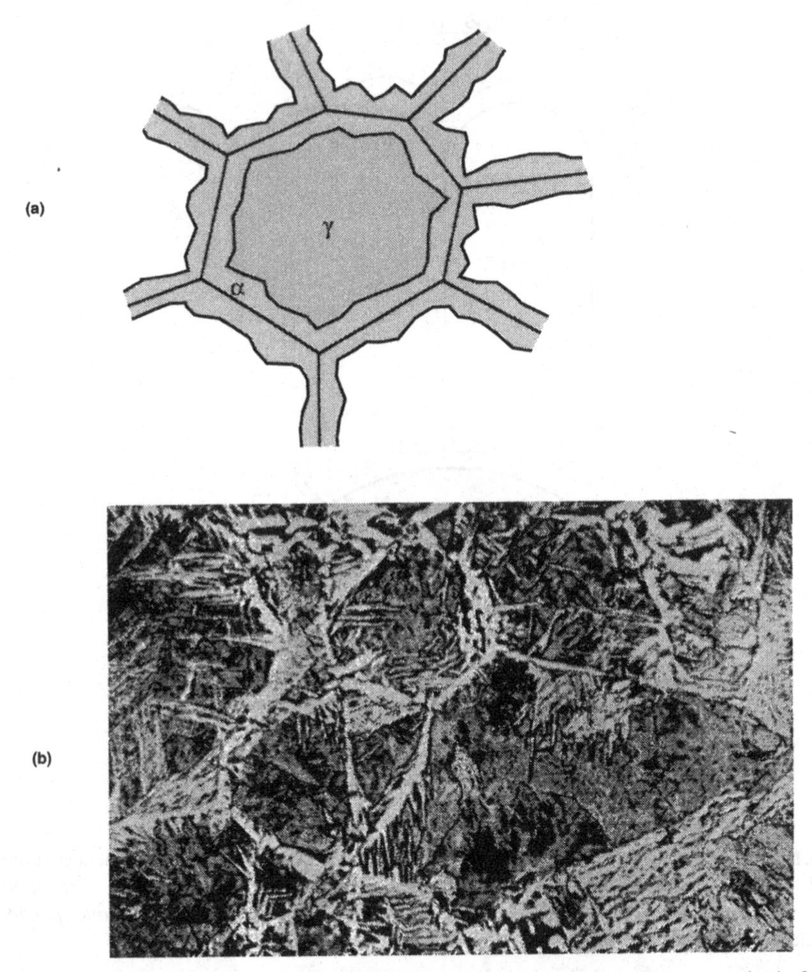

Fig. 6.11. During slow cooling of steels the first phase to precipitate from austenite is ferrite: (a) grains of *soft ferrite* typically coat the austenite grain boundaries. (b) Alternatively, ferrite nucleates at the austenite grain boundaries, but grows into the austenite grains in an acicular (*Widmanstätten)* morphology. (b) reproduced by permission of The Institute of Materials.

of the weld metal. The separation of the carbide lamellae *decreases* as the cooling rate is *increased*, to yield finer pearlitic structures at higher cooling rates.

Minor alloy additions interact to have major effects on the weld properties. Consider the combined effect of *manganese* and *sulphur* on the phase stability in steels. The iron–sulphur binary phase diagram is shown in Fig. 6.13. The liquid phase solution of sulphur in iron is retained at temperatures down to the *eutectic temperature* (988°C), when brittle, stoichiometric iron sulphide forms during the

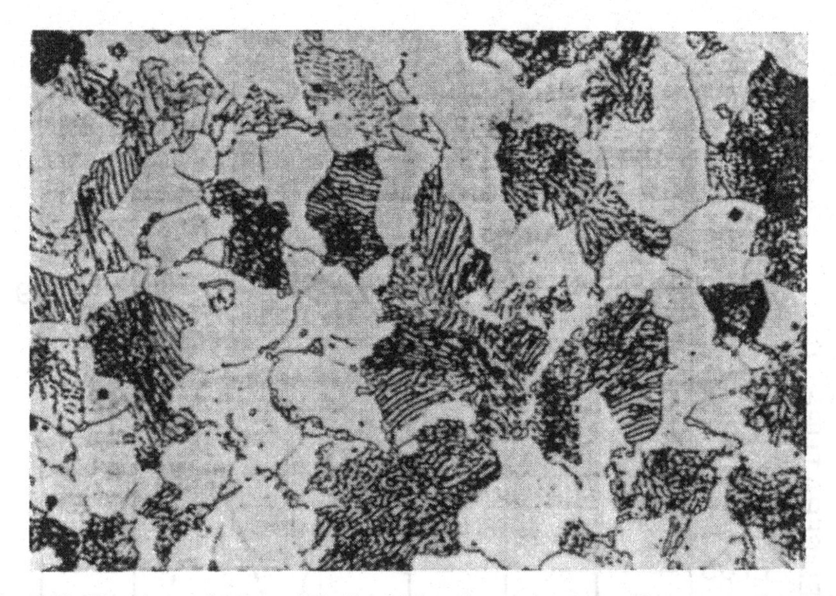

Fig. 6.12. The plates of α-Fe and Fe_3C, which make up the eutectoid structure of *pearlite*, nucleate and grow into the austenite grains in '*colonies*'. Within each colony all the plates are approximately parallel. Reproduced by permission of John Wiley & Sons Inc.

eutectic reaction. The liquid phase is present at the austenite grain boundaries and leads to *hot-cracking* associated with the relief of the thermal shrinkage. The brittle FeS inclusions, on the other hand, can lead to *cold-cracking* after welding. The presence of *manganese* is desirable, since MnS is thermodynamically more stable than FeS and is not brittle but ductile. The sulphur is *gettered* by the manganese to form isolated, ductile *manganese sulphide* particles which prevent the formation of a continuous liquid film at the austenite grain boundaries. Figure 6.14 shows the appearance of these MnS inclusions, usually described as *dove-grey* in appearance, a description which gives some idea of the affection with which steel-makers regard this particular non-metallic inclusion!

6.2.3 Planar & Cellular Growth Fronts

The presence of a solute element which is *partitioned* between the liquid and the solid in the two-phase region (Fig. 6.15) gives rise to a morphological transition which depends on the temperature gradient at the solidification front; from a *planar* front in a high temperature gradient to a *cellular* front in a low temperature gradient (Fig. 6.16). We assume a phase diagram with a *partition coefficient* less than one (*higher* solute solubility in the melt), in which case the *rejection* of solute from the solid into the melt gives rise to a region of high solute concentration in the neigh-

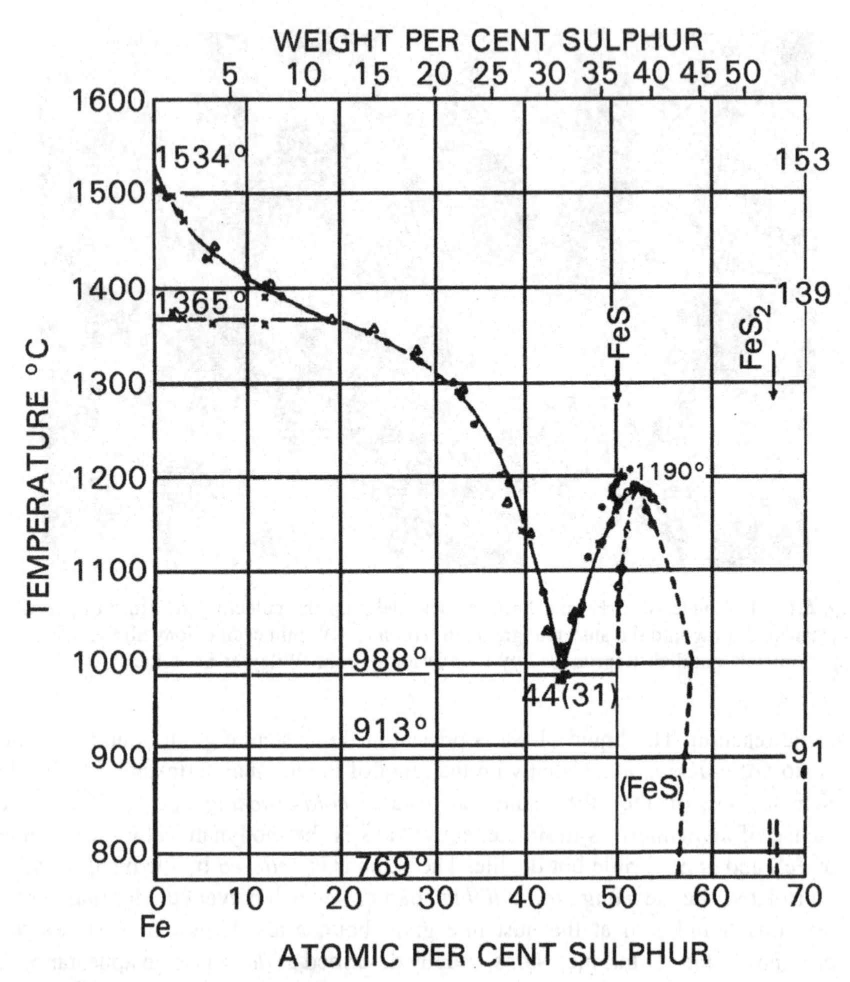

Fig. 6.13. The low eutectic melting point of the Fe–S binary system can lead to *hot-cracking*, since the molten eutectic *wets* the austenite grain boundaries. Reproduced from *Introduction to the Physical Metallurgy of Welding*, Easterling (1983), Buttterworths.

bouring melt (Fig. 6.17a). Solidification is initiated at the *liquidus* temperature appropriate to the melt composition T_L and is completed at the corresponding *solidus* temperature T_S. The rejection of solute from the solid *raises* the value of T_S, but solidification can *only* occur on a planar front if the temperature gradient is high enough to maintain T_L and T_S at the same position. That is, the *temperature* gradient in the melt must either equal or exceed the gradient in T_L imposed by the *concentration* gradient in the melt (Fig. 6.17b). At a *lower* temperature gradient, any random irregularity in the solidification front will tend to *grow*, with lateral rejection

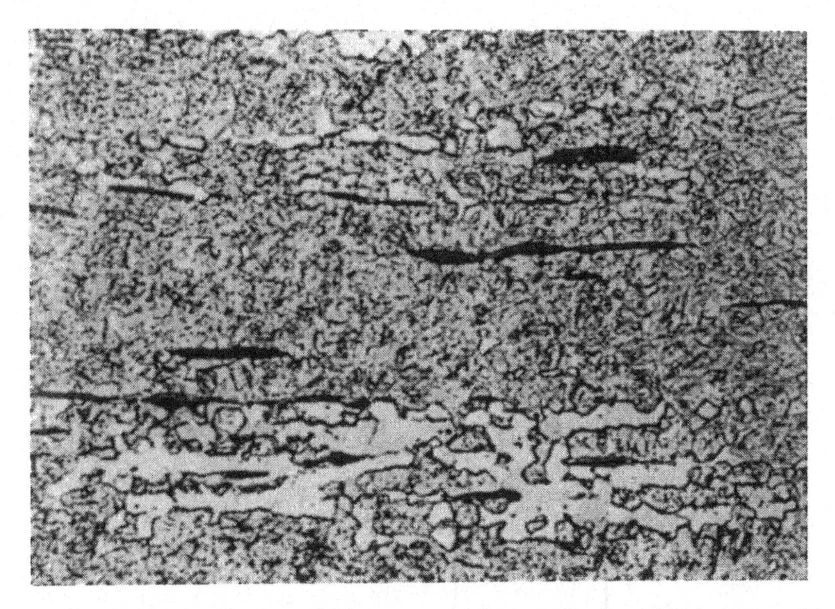

Fig. 6.14. Manganese alloying additions trap sulphur as ductile, *MnS inclusions* (dark, elongated regions). These reduce ductility but have no other detrimental effect on the properties of weld metal. Reproduced with permission from *Metals Handbook, 8th Edition, Vol 7 (1972)* ASM International (formerly American Society of Metals) Materials Park, OH, Fig. 1158, p. 143.

of solute into the surrounding melt. This will leave regions of high solute-content liquid between growing columnar crystals, resulting in a strongly segregated *cellular structure*.

It is the *combination* of the rate of mass transfer (diffusion) in the melt together with the rate of growth (the rate of solute rejection) of the solid which determines the *critical temperature gradient* necessary to prevent cellular growth. Each region of the weld bead experiences a different thermal history, leading to a microstructure which varies with distance from the weld centre line (Fig. 6.18). The *first* metal to solidify does so in a high temperature gradient, leading to a *planar growth* zone which is nucleated epitaxially at the edge of the weld. This is followed by a region of *cellular growth* which results in columnar crystals growing towards the centre line. Finally, the last region to solidify does so under near *isothermal* conditions, and nucleation of solid ahead of the growth front forms large, randomly-oriented *equiaxed crystals* showing evidence of *dendritic* branching.

6.2.4 Phase Transformations

The phases and their morphology in the weld metal depend on *alloy composition* and *thermal history*. In steels, both the amount of pro-eutectoid ferrite and the

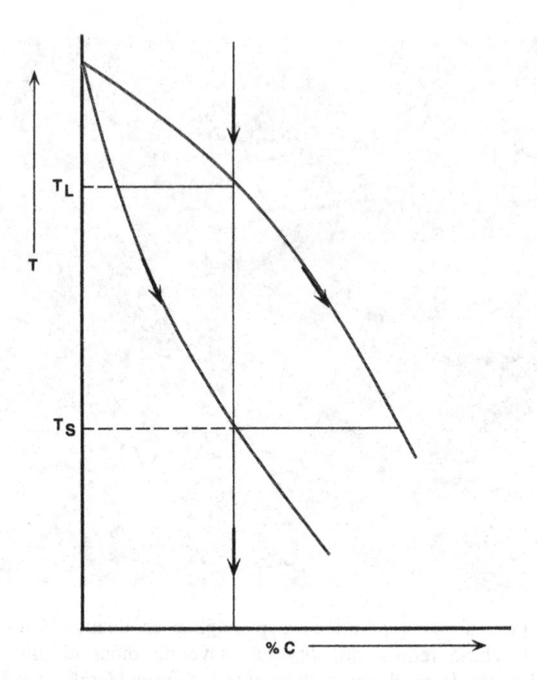

Fig. 6.15. *Solidification* of a binary alloy takes place over the solidification range $T_L - T_S$, the equilibrium compositions of the solid and liquid phases changing as the partially solidified metal cools.

interlamellar spacing in the pearlite depend strongly on the cooling rate from the austenitic region. The kinetics of *isothermal* phase transformation are determined by *quenching* fully austenitic samples to various holding temperatures and plotting the *degree of transformation* as a function of *time* and transformation *temperature*, to yield a *TTT diagram (time–temperature–transformation)* (Fig. 6.19). The diagram shows that as the transformation temperature is lowered below the phase boundary, the time required for the transformation at first decreases. This is a direct consequence of the very strong dependence of *nucleation rate* on *supercooling*. At transformation temperatures close to the stability limit of the, austenitic zone the critical nucleus size is large and few nuclei are available. At higher supercooling, the much smaller critical nucleus size results in a *faster* transformation and *finer* microstructure.

Two factors complicate this simple picture. *Firstly*, transformation below the eutectoid temperature is affected by *metastable extensions* of the phase boundaries which alter the relative proportions of ferrite and austenite which are in equilibrium, both with each other and with cementite (Fig. 6.20). The *lower* the transformation temperature, the *smaller* the quantity of pro-eutectoid ferrite which is formed, and the *larger* the volume fraction of pearlite. *Secondly*, the lamellar spacing in the pearlite *decreases* steadily at lower transformation temperatures, approaching zero in

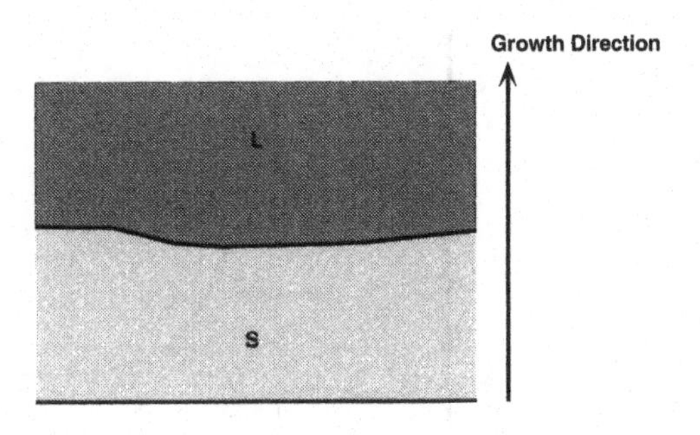

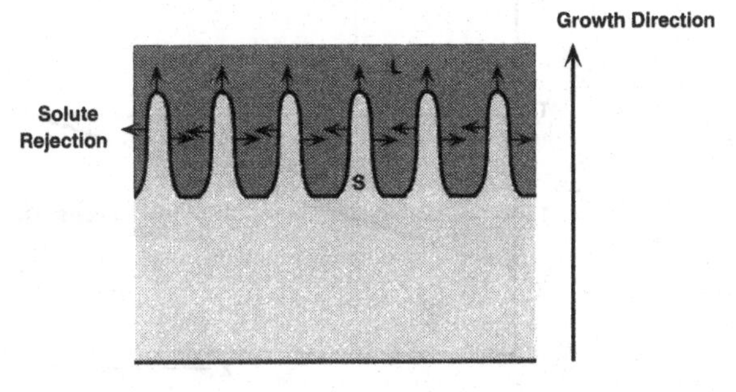

Fig. 6.16. Two morphologies can form during *solidification*, depending on whether a *planar* solidification front is formed normal to the temperature gradient, or a wavy, *cellular* growth front is formed associated with solute rejection into the melt.

the region of 550°C for a plain carbon steel. Below 550°C the nature of the transformation changes. The ferritic phase is no longer formed by nucleation and growth, but rather by a *shear transformation* which is accompanied by *carbide precipitation.* The resultant microstructure is termed *bainitic,* after the American metallurgist Edgar C. Bain. The bainitic microstructure is still two-phase, but with a transformation rate which is determined by the rate of carbide precipitation. Below this *nose* of the TTT diagram the rate of transformation is limited by the rate of *diffusion*, and *not* by the rate of nucleation. The rate of transformation then decreases rapidly with decreasing temperature, reflecting the exponential dependence of the diffusion rate on temperature.

Finally, at a sufficiently low temperature the bainite reaction is completely supressed, but a diffusionless transformation can *still* proceed by lattice shear, to form *martensite* (named this time after the German metallurgist Martens). Martensite has a *distorted* body-centred cubic structure, which arises from the excess carbon and results in a *tetragonal* symmetry. Since the martensite transformation is *diffusionless*, it does not require thermal activation. *Martensitic reactions* can

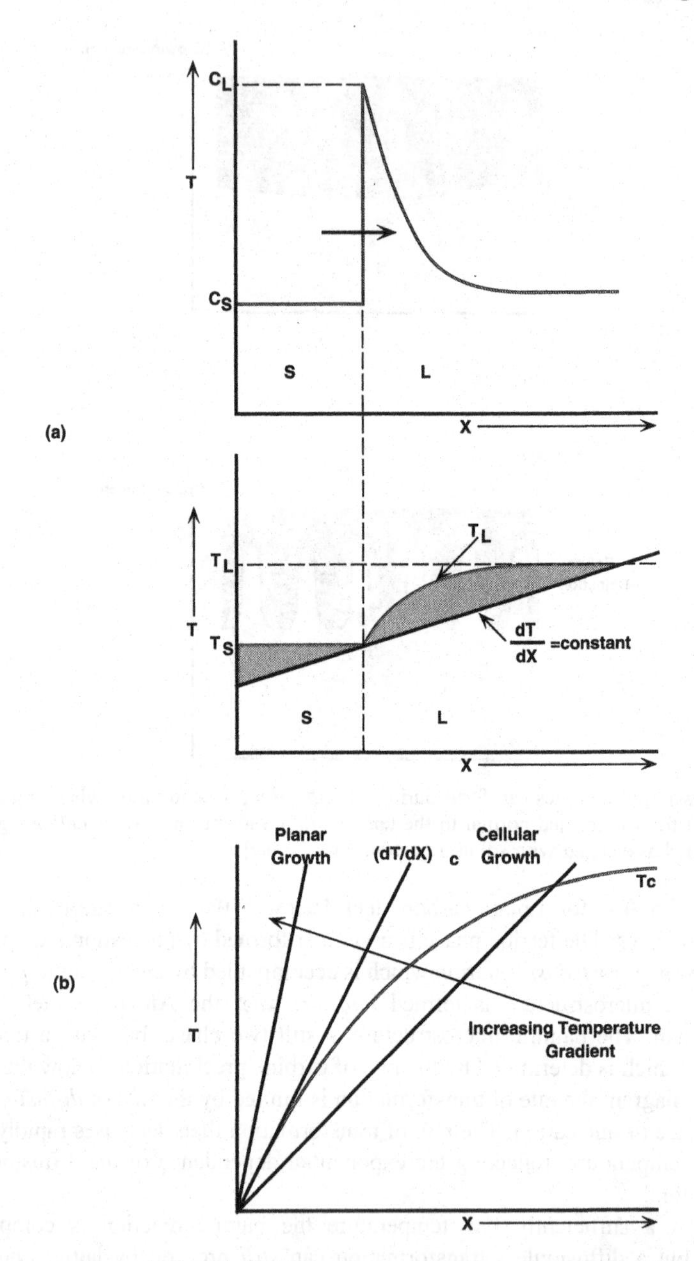

Fig. 6.17. (a) *Solute rejection* leads to a concentration gradient in the liquid phase and hence to a variation in the *liquidus temperature* as a function of position in the melt. Solidification *cannot* proceed in the region where solute enrichment results in a liquidus temperature *below* the temperature of the melt at that point. (b) For low temperature gradients solute rejection leads to a *cellular* growth front, but beyond a *critical* gradient *planar* growth is expected.

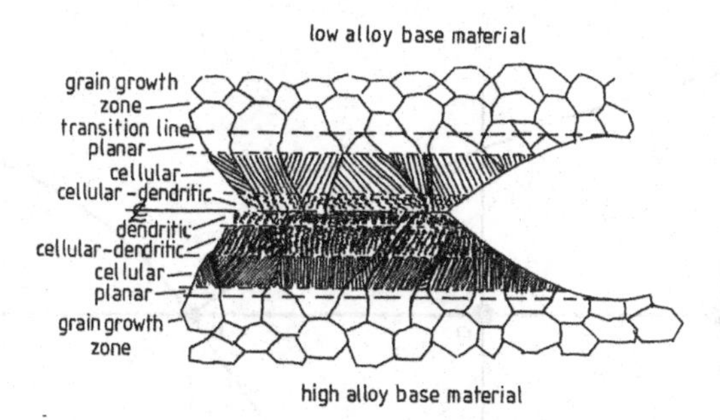

Fig. 6.18. The factors controlling the microstructure of the weld metal should now be apparent and the three morphologies observed (Fig. 6.6) can be understood in terms of the *monotonic reduction in cooling rate* of the weld metal as the weld bead solidifies. Reproduced from *Introduction to the Physical Metallurgy of Welding*, Easterling (1983), Butterworths.

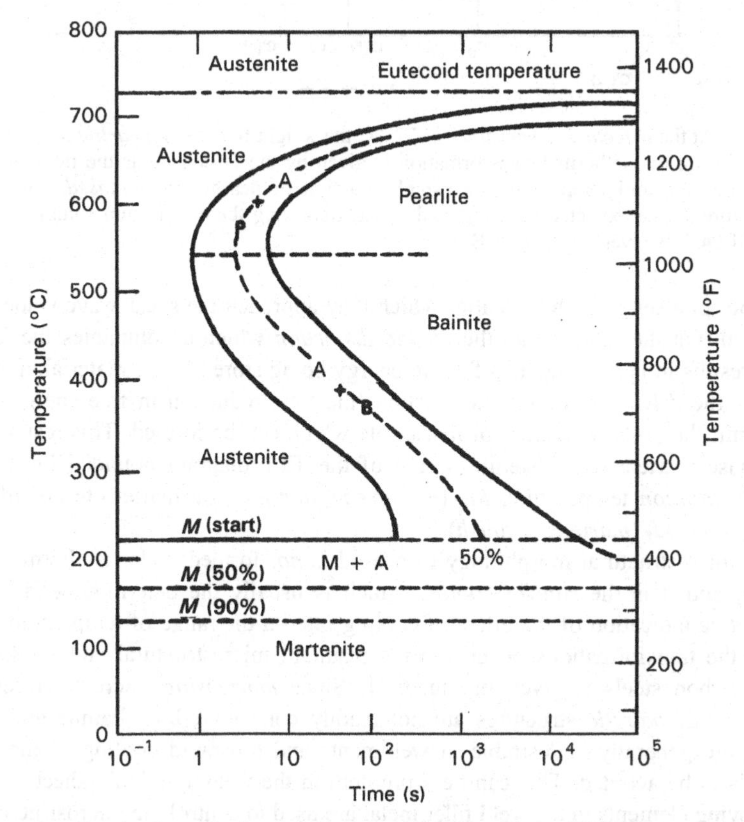

Fig. 6.19. The isothermal time–temperature–transformation *(TTT)* curve shows the transformation sequences which are to be expected for a given steel composition transformed from austenite at any *constant* temperature. Reproduced by permission of John Wiley & Sons Inc.

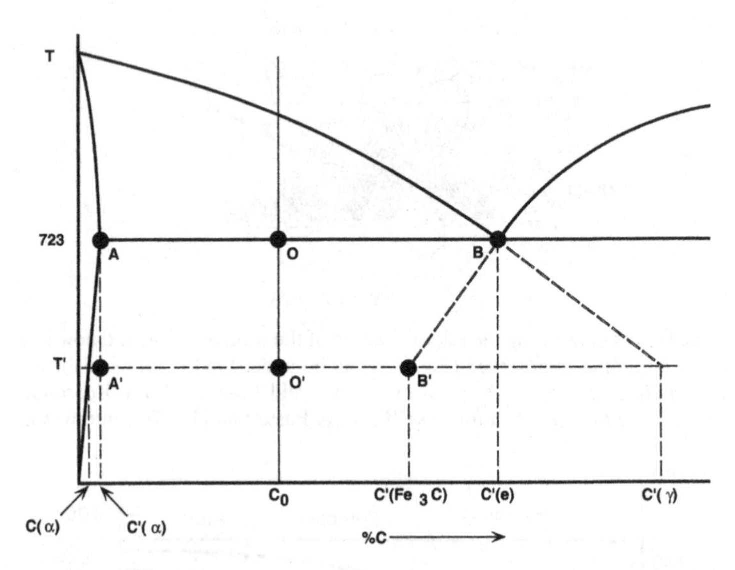

Fig. 6.20. At the *eutectoid temperature* of 723°C the weight fraction of *pearlite* formed in the microstructure by isothermal transformation is given by the *lever rule* as the ratio AO/AB. *Below* the eutectoid temperature the weight fraction of pearlite is *increased*, and at the temperature T' the expected ratio, derived by *extrapolating* the equilibrium solubility limits AA′ and OO′, is given by A′O′/A′B′.

therefore proceed at high velocities which may approach the shear wave velocity in the solid. On the other hand, the *elastic distortion* which accompanies the lattice shear results in large amounts of strain energy being stored in the material, and this reduces the *driving force* for the reaction (the total reduction in free energy), and may limit the volume fraction of martensite which can be formed. This results in a two-phase martensite + austenite region of the TTT diagram bounded by a martensite *initiation* temperature, M_S (*martensite start*), and a martensite *completion* temperature, M_f (*martensite finish*).

The microstructural morphology of a weld is *not* formed under isothermal conditions, and it is the *rate of cooling* which is usually the determining factor. A *qualitative* indication of the effect of cooling rate on the range of temperature over which the transformations occur and the resultant microstructural end product in plane carbon steels is given in Fig. 6.21. Since *martensitic* microstructures are brittle, while *bainitic* structures are not readily controlled, both bainite and martensite are generally undesirable in weldments and too rapid cooling of the weld metal is to be avoided. This can be a problem in the welding of thin sheet.

Alloying elements in the weld filler metal are used to control the microstructure of the weld bead, and of these alloy additions *chromium* and *nickel* are the commonest. Chromium is both a *carbide former* (precipitating either Cr_7C_3 or $(Cr,Fe)_{23}C_6$ under nearly all cases of practical importances) and a *ferrite stabilizer*. Nickel is a *grain*

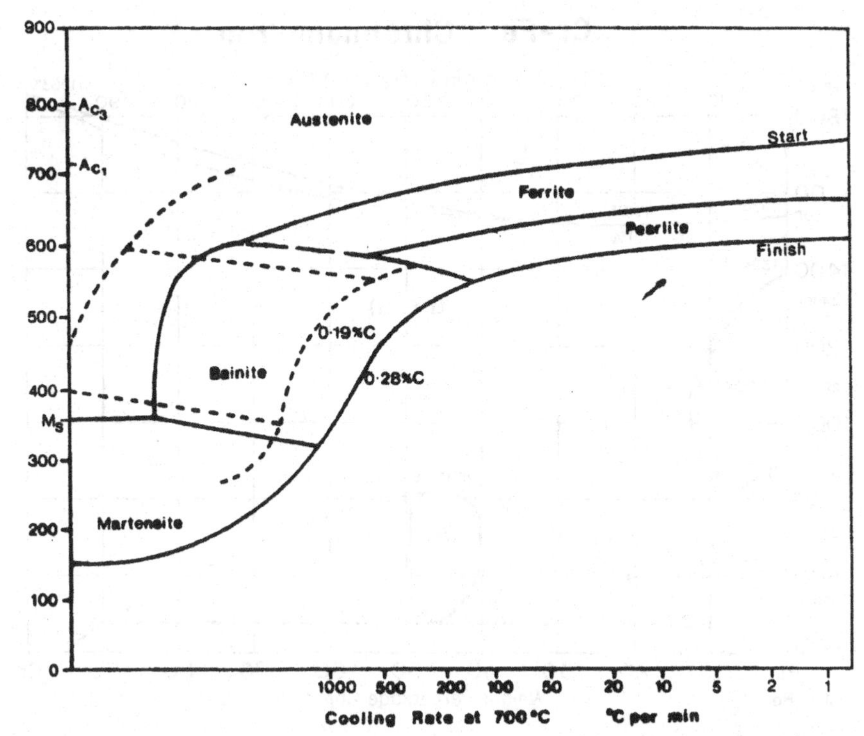

Fig. 6.21. For a *plain carbon steel* the temperature range of the phase transformations and their expected dependence on cooling rate can be summarized on a single diagram. Reproduced from *Introduction to the Physical Metallurgy of Welding*, Easterling (1983), Butterworths.

refiner and an *austenite stabilizer*. The binary phase diagrams for Fe–Cr and Fe–Ni are given in Fig. 6.22.

6.2.5 Nickel & Chromium Equivalents

A simple arithmetical device can be used to summarize, reasonably accurately, the effect of alloy composition on the final weld microstructure. The total alloy content is expressed as *nickel* (austenitic stabilizer) and *chromium* (ferritic stabilizer) *equivalents*. A number of equations have been proposed, depending on which alloying elements were considered and what experimental conditions were used to obtain the final microstructure, but the following definitions of nickel and chromium equivalents have proved widely acceptable and the predicted microstructures are plotted in Fig. 6.23:

$$Ni_{equiv} = \%Ni + 30x\%C + 0.5x\%Mn \qquad (6.4)$$

$$Cr_{equiv} = \%Cr + \%Mo + 1.5x\%Si + 0.5x\%Nb \qquad (6.5)$$

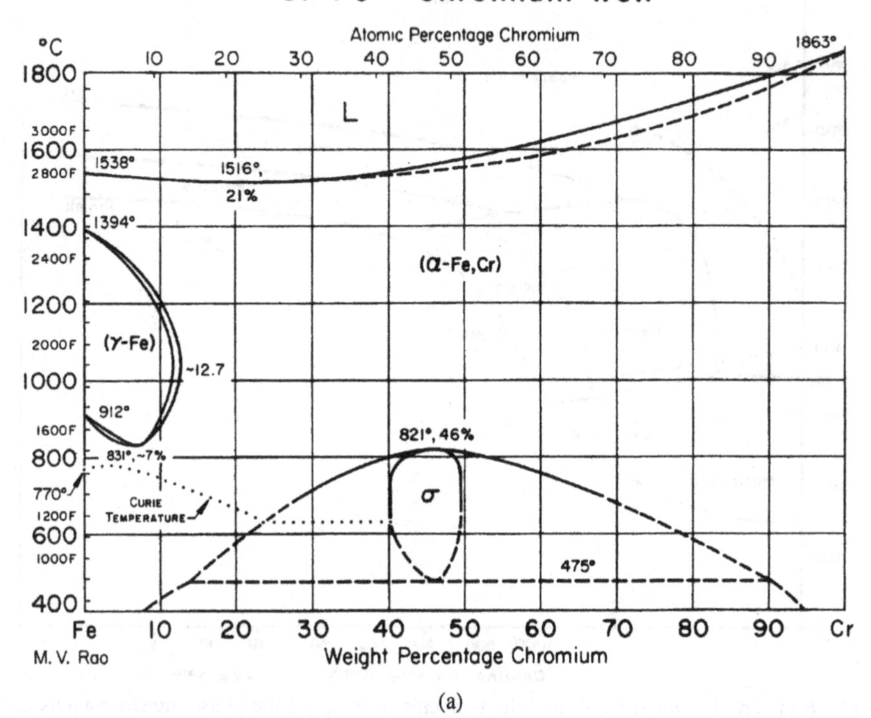

(a)

Fig. 6.22. The commonest alloying additions used to modify the properties of a steel filler alloy are *chromium* and *nickel*: the former stabilizes the BCC α-phase and the latter stabilizes the FCC γ-phase. Reproduced with permission from *Metals Handbook, 8th Edition, Vol 8, Metallography, Structures and Phase Diagrams*, Taylor Lyman Ed. (1973). ASM International (formerly American Society for Metals) Materials Park, OH, Fig. Ag-Cu, P. 253.

The range of resultant microstructures includes *austenitic* (A) *martensitic* (M) and *ferritic* (F) morphologies (the latter including *pearlite*), as well as more complicated morphologies, most notably those containing *retained austenite* in the weld metal.

6.3 THE HEAT-AFFECTED ZONE (HAZ)

The *heat-affected zone* (commonly shortened to *HAZ*) is a unique feature of fusion-welded structures and a direct consequence of the need to heat the junction between the two components above their melting point. The HAZ region *adjacent* to the fusion zone is heated to temperatures below the fusion point, and is the region where microstructural changes occur, either side of the central solidification zone.

In low carbon, structural steels the maximum rate of conversion from austenite to pearlite occurs at a temperature of about 550°C (Fig. 6.19), and this is also approximately the case in low alloy steels. For quenched and tempered steel com-

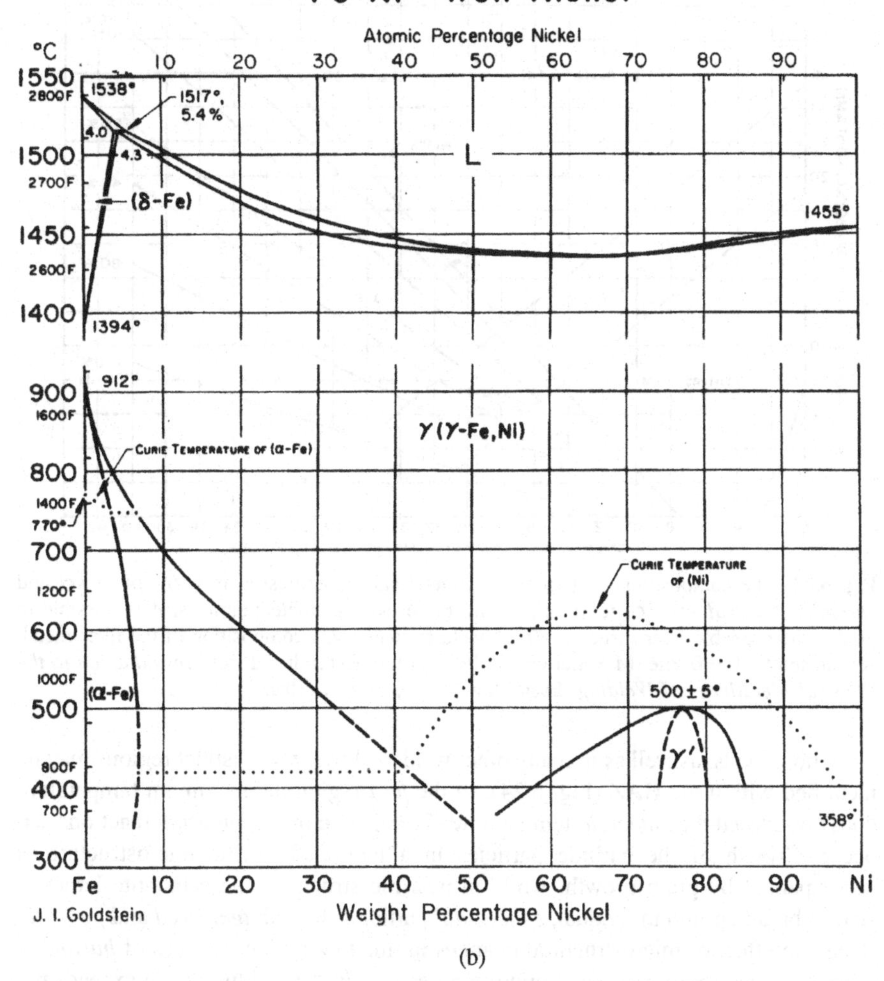

(b)

Fig. 6.22. (*continued*)

ponents, on the other hand, the tempering temperatures required to confer adequate toughness on the product are seldom below 350°C. It follows that the micro-structural changes which are responsible for the HAZ in a steel weldment are only likely to be found in regions of the material where the maximum temperature *exceeded* 350 to 550°C. The structure of the HAZ will depend on the initial microstructure of the unwelded components, so that *hot-rolled* steels, *normalized* components, and *quenched and tempered* structures will to some extent behave differently.

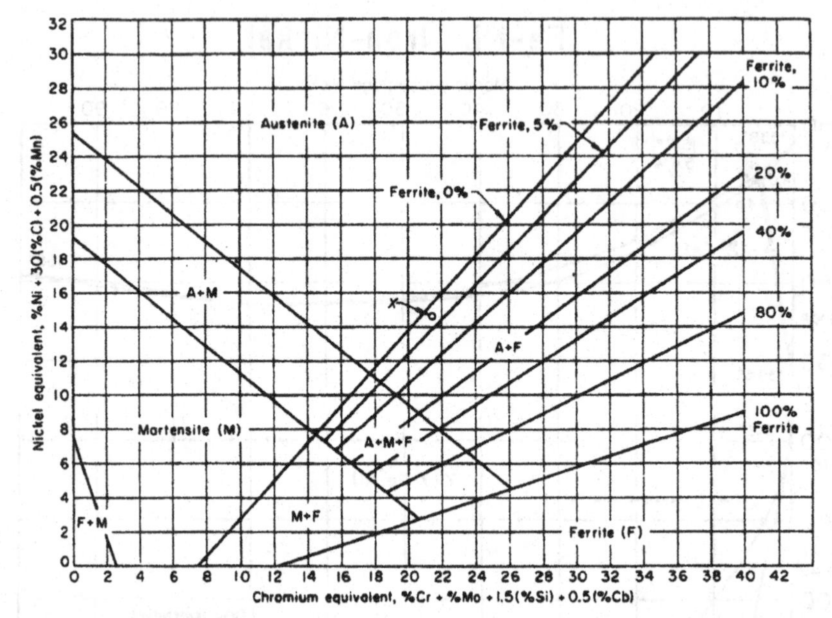

Fig. 6.23. The composition of a steel weld metal can be expressed in terms of *nickel* and *chromium equivalents* (see text), and, using these two parameters as axes, it is possible to predict the probable microstructure of the weld for any given composition of the filler metal. A = austenite; F = ferrite; M = martensite; P = pearlite. Reproduced from *Introduction to the Physical Metallurgy of Welding*, Easterling (1983), Butterworths.

In most steels, as well as in many other welded alloys, *four* distinct regions may be identified within the HAZ (Fig. 6.24). In the *first* region the maximum temperature does not exceed the *eutectoid* temperature (723°C) and only *tempering* reactions can occur. Growth of the carbide particles in a tempered ferritic microstructure is accompanied by grain growth, while in pearlitic structures the cementite lamellae start to break up into individual, *spheroidal* particles. In both *tempered* and *pearlitic* steels the effect of microstructural changes in this first region is a loss of *hardness*.

In the *second* region the maximum temperature in the welding cycle exceeds the eutectoid temperature. *Partial transformation* to austenite occurs and the carbides start to *dissolve* in the austenitic phase. On cooling back into the ferritic region a complex microstructure is formed which contains both undissolved carbide particles and, possibly, some ferrite grains which did not have time to transform. The carbon content of the transformed austenite in this region may be very variable.

In the *third* region the steel is fully transformed to austenite during the welding cycle, but the extent of carbide dissolution depends on both the concentration of carbide formers as well as on the actual thermal cycle. In many non-ferrous alloys *recrystallization* takes place in this region, relieving residual elastic strains and drastically reducing dislocation densities associated with prior plastic work. This

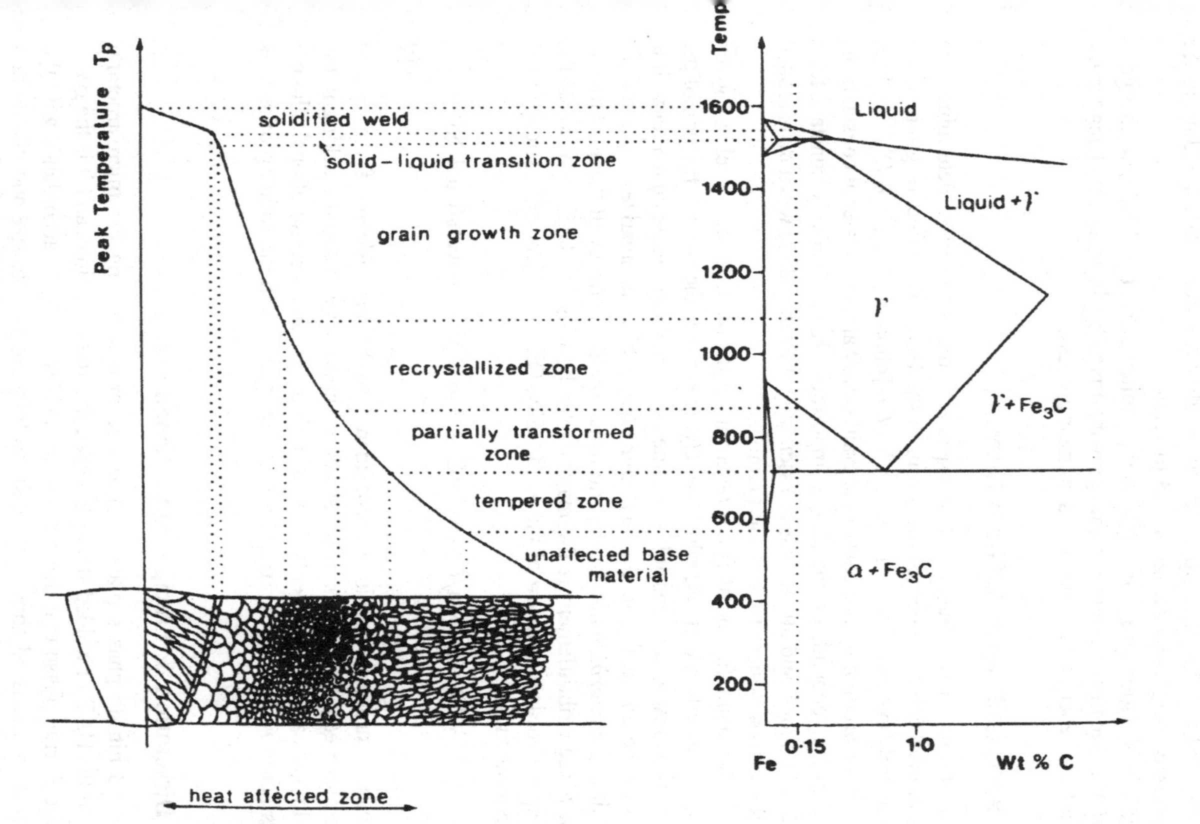

Fig. 6.24. The microstructure in the *heat-affected zone* of welded low-alloy steel can be arbitrarily divided into four *subzones* depending on the peak temperature reached in each region during welding. Reproduced from *Introduction to the Physical Metallurgy of Welding*, Easterling (1983), Butterworths.

region corresponds to a temperature regime in which diffusion is sufficiently rapid for the material to be able to approach its equilibrium state, corresponding to the minimum free energy. *Annealing* is said to occur.

Finally, in a region near the edge of the molten zone the dominant change is *coarsening* of the microstructure. Grain growth, driven by the resultant change in the total interfacial energy, becomes the dominant process.

6.3.1 Recrystallization & Grain Growth

The scale of the microstructure, and especially the *grain size*, determines the changes in mechanical properties that occur in the HAZ. The effect of grain size is most commonly expressed in terms of the *Petch relation*: $\sigma_y = \sigma_0 + k_y D^{-1/2}$, which gives the *yield strength* σ_y in terms of the grain size D and two material constants, σ_0 and k_y. While σ_0 depends on the alloy composition, k_y is effectively constant for a given class of alloys, and all steels give essentially the same slope when σ_y is plotted against the inverse square root of the grain size.

On the other hand, the major factor determining the *ductility* of a steel at a given grain size is the carbon content, which largely determines the volume fraction of the carbide phases. Just as it is possible to define *nickel* and *chromium equivalents*, it is useful to define an analogous *carbon equivalent* which summarizes the effect of alloy additions on *carbide formation*, and hence on the tendency of the steel to lose ductility. Since embrittlement is the major factor limiting weldability, in steels the carbon equivalent is a *measure* of the weldability. One empirical expression for the carbon equivalent is:

$$C_{equiv} = \text{wt\%C} + \text{wt\%Mn}/6 + \text{wt\%(Cr + Mo + V)}/5 + \text{wt\%(Cu + Ni)}/15 \quad (6.6)$$

In general, structural steels are considered to be *unweldable* if the carbon equivalent exceeds 0.4. That is, it is considered impossible to avoid cracking in the HAZ at higher carbon equivalents. Note that all the common substitutional alloying elements act to *reduce* the limiting carbon content at which the welding of steel is feasible.

6.3.2 Dissolution & Precipitation Phenomena

Carbide and nitride phases play a major role in determining the microstructural stability of the HAZ in welded steel. However, this role is indirect and it is the *grain size* of the ferritic phase that determines the *yield strength*, in accordance with the Petch equation discussed above. The volume fraction, particle size and distribution of the *carbides* (and, in some cases, the *nitrides*) act by limiting the rate of ferrite grain growth. The situation may appear less complex when it is realized that the *complete* processing history must be considered, provided that it is accepted that our understanding of the HAZ and its properties is at most semi-quantitative.

The initial microstructural condition of the components is important, especially the *particle size* of the carbides, which determines their rate of dissolution in austenite, and the *degree of cold work*, which determines the microstructural stability of ferrite at low temperatures. These two factors indirectly affect the rate of austenite formation and the subsequent austenite grain size. During the weld cycle it is the thermal history of each volume element which is important, that is, the *temperature* of the element as a function of *time* as the weld pool is formed, and the weld bead solidifies and cools. During this process, the particle size of the carbides first *increases* (by Ostwald ripening, Section 6.3.4), as the austenite transformation temperature is approached, then *decreases*, as the particles are dissolved in the austenite. Then, as the steel cools back into the ferritic region, *reprecipitation* occurs.

6.3.3 Precipitate Stability

The temperatures at which *carbides* and *nitrides* dissolve in the austenite and subsequently reprecipitate are strongly dependent on the *thermodynamic stability* of these interstitial compounds. Of the various alloying elements found in commercial steels, *Si*, *Mn*, *Ni* and *Co* do not form carbides. In the case of silicon, this is perhaps a little surprising, since SiC is a thermodynamically stable and refractory carbide, but the precipitation of SiC in steel is prevented by the high heat of solution of silicon in iron. Fe_3C is marginally more stable than Mn_3C, so that it is always Fe_3C which is precipitated in low manganese alloy steels. Of the carbide forming elements, chromium is the commonest, usually precipitating as either Cr_7C_3 or the mixed carbide $(Cr, Fe)_{27}C_6$.

Rather more stable carbides are formed by *molybdenum* and *tungsten*. Tungsten carbides are important in tool steels, which have high carbon contents (and are unweldable). Weldable alloy steels frequently contain molybdenum which usually precipitates as Mo_2C. Still more stable are the carbides of *tantalum* and *niobium*. *NbC* is only dissolved at the highest temperatures in the austenitic range, and inhibits austenite grain growth at lower temperatures. Niobium alloy contents need only be 0.1–0.2%, and the niobium-containing steels are referred to as *microalloyed*. Finally, the most stable carbides are those of *titanium* and *vanadium*, while *TiN* is also an important grain growth inhibitor in some alloy steels. The carbides V_4C_3 and *VC* are only dissolved if the austenite is annealed at high temperatures for long times, and these carbides are always present in vanadium-containing alloys throughout the thermal history of the HAZ. It is these, the most stable carbides, which are important in *heat-resistant* steels, which are designed to retain a stable precipitate distribution at temperatures approaching the eutectoid temperature. The thermodynamic stabilities of a variety of borides, carbides and nitrides are compared in Fig. 6.25 and is a function of the position of the carbide forming element in the *periodic table*. Only the relative free energy of formation of the carbides and nitrides is shown, and the heat of solution of the element in iron is ignored.

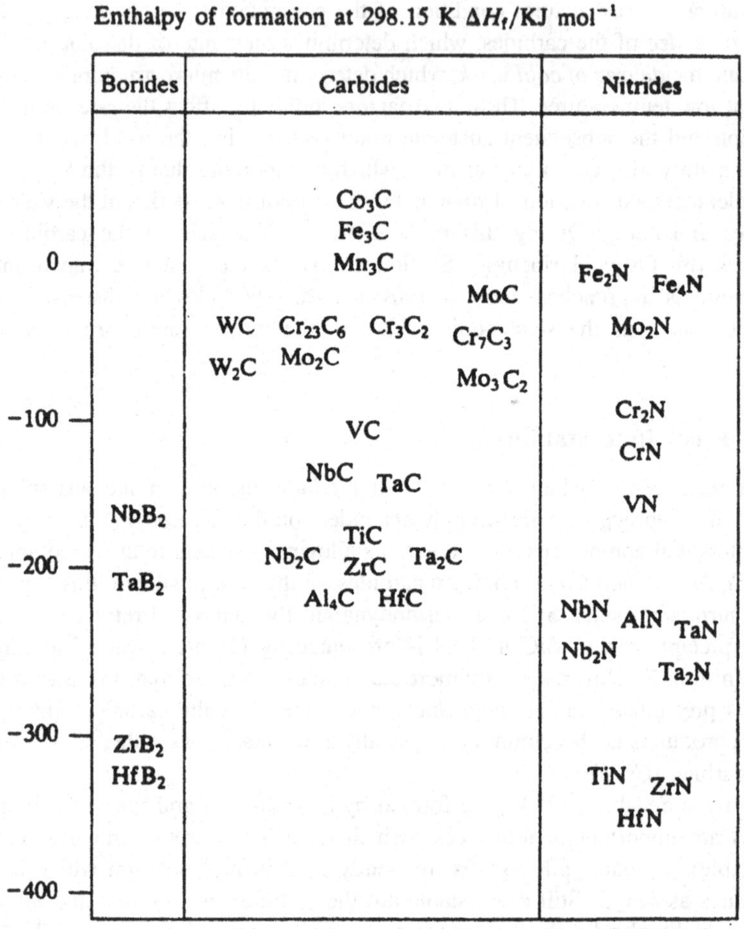

Fig. 6.25. The *enthalpy of formation* of the borides, carbides and nitrides which are commonly found in steel depend on the position of the metal component in the *periodic table*. However, in steels these enthalpy values need to be corrected for the *heat of solution* of the interstitial compound in the steel. Reproduced from *Introduction to the Physical Metallurgy of Welding*, Easterling (1983), Butterworths.

6.3.4 Particle Coarsening

The growth of the carbide particles in the HAZ is controlled by a process of particle coarsening or *Ostwald ripening* in which the smaller particles have a higher solubility limit than the larger particles, so that a concentration gradient is established between small and large particles in the matrix. Diffusion down the concentration gradient leads to loss of material from the *smaller* particles and growth of the *larger*, which thus further increases the driving force for particle coarsening. It is important to realize that Ostwald ripening occurs at a *constant volume fraction* of the carbide phase, and is therefore quite different from *precipitation* by nucleation and growth, in which the volume fraction of the carbide *increases monotonically* by precipitation from a supersaturated solid solution. The driving force for Ostwald ripening is the *dissolution pressure*, P_s: $P_s = \gamma(1/r_1 + 1/r_2)$, where γ is the surface energy and r_1 and r_2 are the principal radii of curvature of the particle. The factor $(1/r_1 + 1/r_2)$ is the *curvature* of the particle. *Thermodynamic equilibrium* in alloy phase diagrams is defined only for *zero* curvature of the phase boundaries. Fine particles are inherently *unstable* and will coarsen if diffusion (mass transport) can occur.

6.3.5 Particle Dissolution & Grain Growth

Since diffusion rates are exponentially dependent on temperature, the time required for *carbide dissolution* in austenite also shows an exponential dependence on temperature. The less stable carbides (starting with Fe_3C) dissolve at the lowest austenite temperatures (about 850°C), followed by carbides such as Cr_7C_3, then Mo_2C and finally NbC (at temperatures of 1550°C, close to the melting point). The carbide phases which inhibit austenite grain growth will depend on the type of steel: $Cr_{23}C_6$ inhibits grain growth during welding of *austenitic stainless steels*, while Mo_2C plays the same role in the welding of *pressure vessel steels*. Figure 6.26 shows some examples of the time and temperature dependence of dissolution and demonstrates the strong dependence of the dissolution time on the solubility of the interstitial compound, which is inversely related to the thermodynamic stability.

Grain growth is also driven by the rate of change of boundary curvature, now best defined as *boundary surface area per unit volume*. The two-dimensional case of a material for which the boundary energy is *isotropic* (independent of orientation) is straight-forward (Fig. 6.27). The equilibrium angle of contact at a boundary *triple junction* is 120°, so that 2-D six-sided grains should neither grow nor shrink. Grains with *fewer* sides *must* have boundaries which are *convex* (positive curvature), and such grains therefore shrink, while those with *more* than six sides have *concave* boundaries (negative curvature) and hence grow. In three dimensions the *larger* grains will have more neighbouring grains than the average, and hence tend to possess *negative* curvature. *Smaller* grains have fewer neighbours and are associated with nett *positive* curvature of the boundaries.

A graphic representation of the changes to be expected in the HAZ of an alloy steel has been developed by *Easterling and Ashby* who plotted the HAZ microstructure as a function of the *time* of the heat pulse in a spot-weld and the *maximum temperature* reached in a volume element of the steel. Two examples are given in Fig. 6.28. The time of the heat pulse is proportional to the *total energy input*. Precipitation and dissolution take place over a narrow range of temperatures, and this range moves to *lower* temperatures at *longer* pulse times (longer precipitate dissolution times). *Grain growth* initiates as soon as precipitate dissolution is complete, the grain size increasing with both maximum temperature and pulse time. The

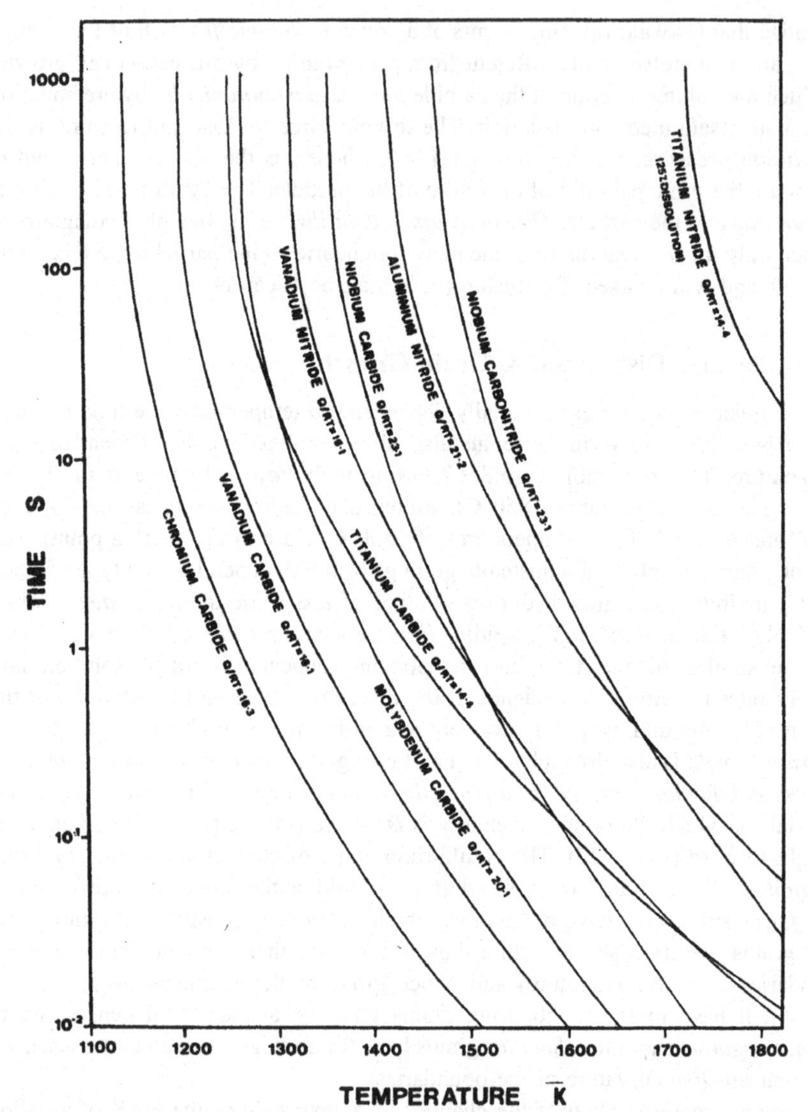

Fig. 6.26. *Dissolution* of an interstitial precipitate takes time, but the time–temperature dependence of dissolution accurately reflects the *thermodynamic stability* of the interstitial compound. Reproduced from *Introduction to the Physical Metallurgy of Welding*, Easterling (1983), Butterworths.

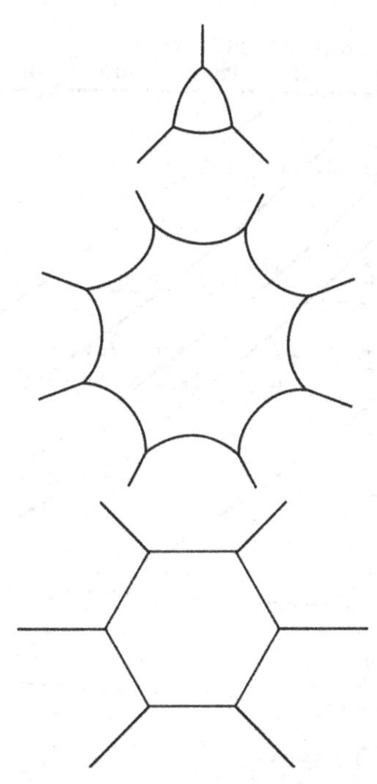

Fig. 6.27. Grain boundary *curvature* controls grain growth. Small grains *shrink* because they have few neighbours and the curvature is *convex*. Large grains *grow* because they have many neighbours and the curvature is *concave*. In the isotropic, *two-dimensional* case *six-sided* grains are expected to be stable, since the boundaries then have zero curvature.

position corresponding to a given maximum temperature moves further from the centre of the weld bead as the pulse time increases, so that the *width* of the HAZ *increases* with the total energy input. As can be seen by comparing the two diagrams in Fig. 6.28, the *alloy content* of the steel has a large effect on the precipitate dissolution history, and hence on the final grain size in the HAZ.

For the HAZ in a non-ferrous weld similar considerations lead to a qualitative plot of the *generalized* HAZ structure as a function of the *energy input* (Fig. 6.29). The fusion line borders a zone of grain growth, which is separated from the bulk microstructure by a region in which recrystallization (or phase transformation) is expected.

6.4 WELD PROPERTIES & PROBLEMS

Both the liquid slag and the molten metal in the weld pool may react with the bulk components and lead to some additional problems. These have been discussed under the heading of *weld defects*, Section 5.1.3, and will only be summarized here.

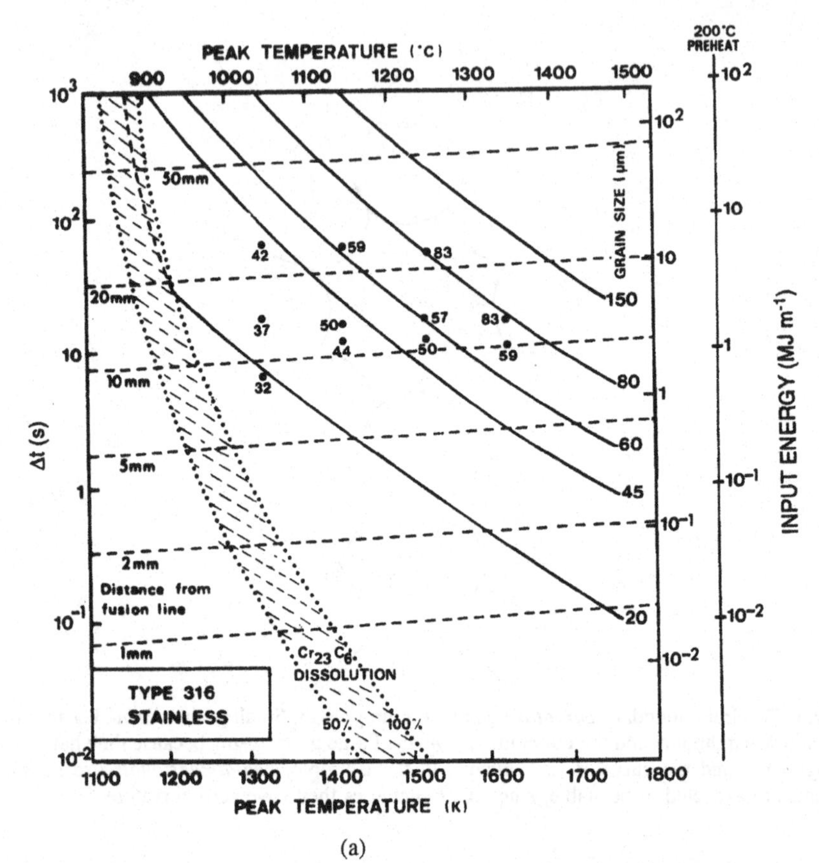

(a)

Fig. 6.28. (a) *Austenitic grain growth* calculated for a welded *stainless steel* half-space. The near-horizontal dotted lines give the expected HAZ width. Chromium carbide ($Cr_{23}C_6$) dissolution controls the onset of grain growth. (b) *Austenitic grain growth* calculated for a welded *pressure vessel steel.* Molybdenum carbide (Mo_2C) dissolution controls the onset of grain growth. Reproduced from *Introduction to the Physical Metallurgy of Welding*, Easterling (1983), Butterworths.

In low carbon steels ($<0.3\%$C) changes in composition of both the filler and the base metal are expected as a result of slag–metal reactions. If the slag is *basic* manganese will be transferred to the weld pool, while with *acid* slags it is silicon which is transferred:

$$MnO + Fe \rightarrow Mn + FeO \quad \text{or} \quad SiO_2 + 2Fe \rightarrow Si + 2FeO$$

Sulphur may embrittle the weld, but when 'gettered' by manganese will precipitate as harmless *MnS. Any* liquid phase which wets the austenite grain boundaries may lead to *hot-cracking*. This includes not only *FeS*, but also some oxide phases, such as B_2O_3. *Hydrogen*, arising from the presence of *moisture*, can lead to *hydrogen*

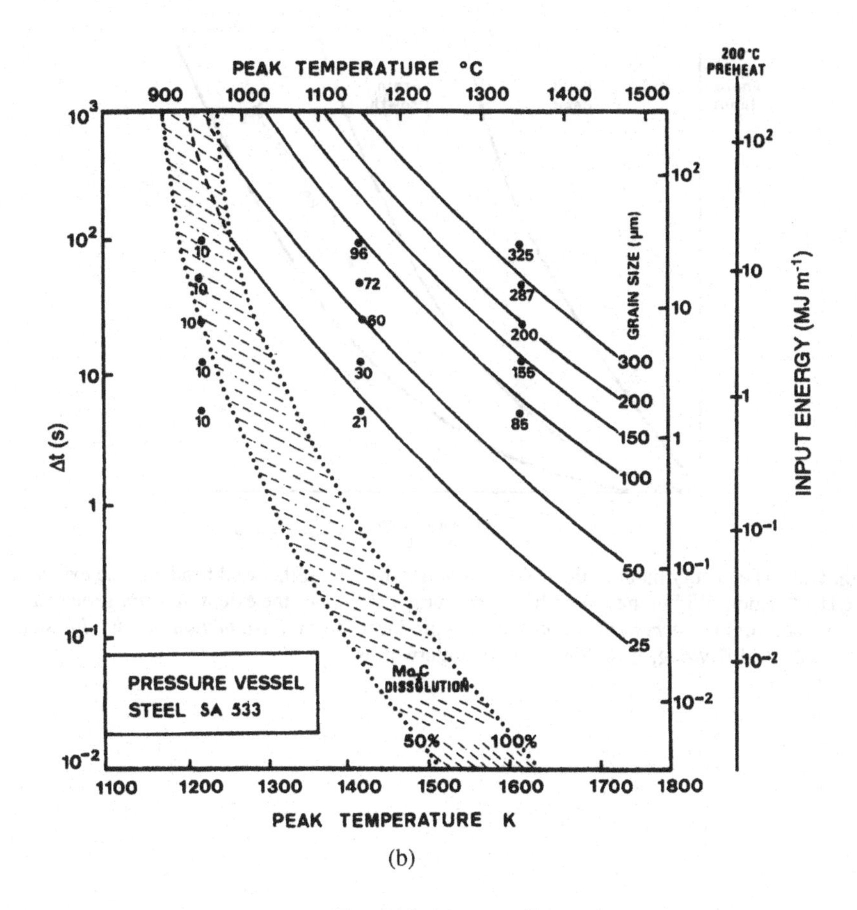

(b)

Fig. 6.28. (*continued*)

embrittlement. All these effects are sensitive to *residual thermal stresses*, which, in turn, depend both on *weld geometry* and the overall *thermal history*.

6.4.1 Mechanical Properties of Welds

The *hardness* of a welded component usually varies appreciably across the HAZ. In steels the changes are usually minimal below a maximum HAZ temperature of 900°C unless the components were cold-worked, but between 900 and 1500°C the hardness changes are significant. A hypothetical example is given in Fig. 6.30. The final *ferrite grain size* may be *refined* (up to 900°C, in the region where austenite

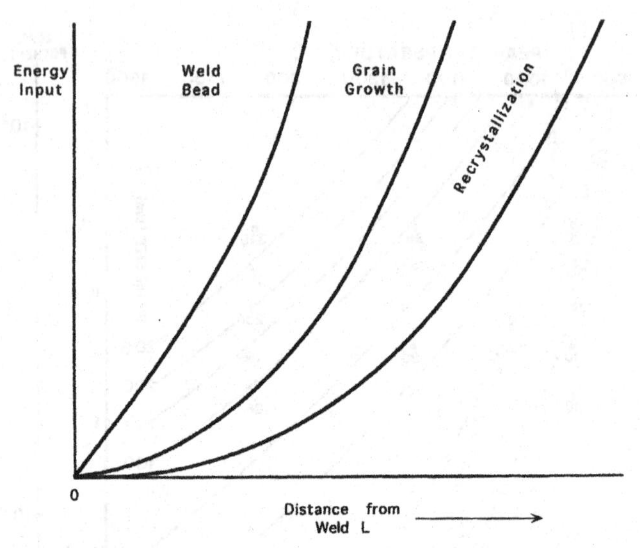

Fig. 6.29. The energy input to the weld determines the *size* of the weld bead and the *extent* of the HAZ region. Within the HAZ, the input energy determines the extent of *grain growth* and the extent of the *recrystallization* zone. Reproduced from *Introduction to the Physical Metallurgy of Welding*, Easterling (1983) Butterworths.

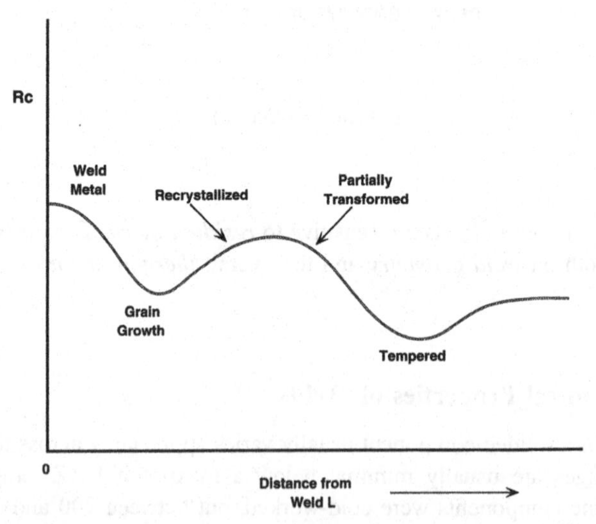

Fig. 6.30. The *hardness* is expected to vary across the weld and HAZ, and the microstructural changes may lead to more than one hardness minimum (or maximum).

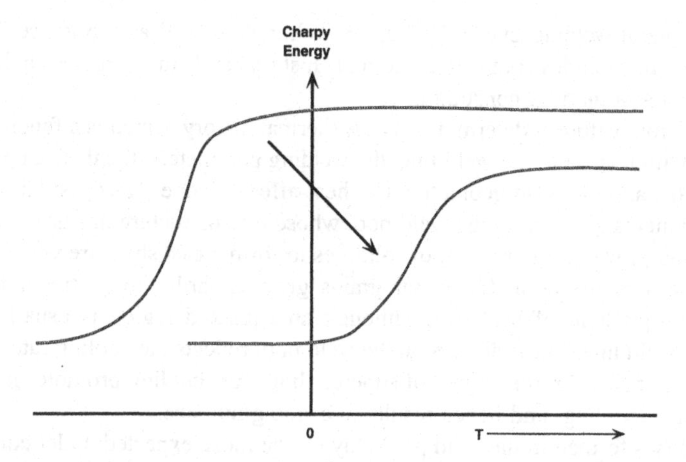

Fig. 6.31. *Charpy tests* performed on specimens taken from the HAZ generally show degradation of the *notch impact energy* and an increase in the *ductile–brittle transition temperature* with respect to the bulk steel.

grain growth is *inhibited* by the presence of precipitated carbide) or *coarsened* (above 900°C, as a result of austenite *grain growth*).

The Charpy impact strength of steel is adversely affected by the microstructural changes in the HAZ, and this is seen both in a general reduction in the *Charpy energy* at room temperature, as well as in an increase in *Charpy transition temperature* (Fig. 6.31). The Charpy test is often used as a major indicator of the degradation of mechanical properties associated with welding.

Many welded structures are *heat-treatable*, and the mechanical properties can be ameliorated by a post-welding heat treatment. A *residual stress relaxation treatment*, at a temperature of the order of 350°C for carbon steels, does not result in any microstructural changes. In *high-alloy* maraging and PH (precipitation hardening) steels (which contain no carbon), heat treatment of welded components may include a full *solution treatment and aging* schedule.

Finally, high alloy steels, especially the stainless steels, may be either *ferritic* or *austenitic*. *Segregation* of the alloying additions, either austenite (Ni, Co) or ferrite (Cr, Mo) formers, can drastically modify both the phase content and the phase distribution in the HAZ of these materials.

Summary

The art of the metallurgist is used to develop a wide range of properties from a single alloy composition, primarily through the sensitivity of the microstructure to plastic deformation and heat treatment. Similarly, the properties of a welded joint are also a sensitive function of the microstructure resulting from the weld cycle. The processing history of each volume element of the welded components is complex, and may include several welding cycles (multipass welding) as well as a post-welding heat treatment. It includes both the thermal history and the stresses induced by mechanical and thermal constraint.

The microstructure is determined by the thermal history, which is a function of the position with respect to the weld line, the welding parameters, the thermal properties of the alloy, and the weld geometry. The heat-affected zone (HAZ) is that region in the components adjacent to the weld pool whose microstructure has been altered by the welding process. The weld pool solidifies to form a cast structure which initiates at the edges of the pool. Columnar grains grow towards the centre of the pool, against the gradient of heat flux, although an equiaxed region is usually present along the weld line. The grain size in the weld pool reflects the cooling rate, which is in turn responsible for the degree of supercooling. Fast cooling promotes nucleation at high supercooling, and hence results in a fine grain size.

The phases formed in the weld pool may not be those expected under equilibrium conditions, since alloy segregation may occur. Impurities may also lead to undesirable effects, as in the hot-cracking of steel welds associated with the presence of sulphur forming a low temperature intergranular eutectic. The transition from a planar growth front to a cellular growth front depends on the temperature gradient, which thus determines the morphology of the cast weld metal, but phase transformations in the solid state may determine the final microstructure. In ferrous weld metals alloy additions may act as carbide formers, or either austenite or ferrite stabilizers, as well as affect the rate of the pearlite transformation (the time–temperature–transformation, TTT, diagram).

The HAZ is a unique feature of welded structures. The width of the HAZ is determined by the minimum temperature required to alter the microstructure in the time available during the welding cycle. In steels, this minimum temperature is between 350 and 550°C, and the HAZ itself is subdivided into four distinct regions. Below the eutectoid temperature (723°C) only tempering reactions can occur. Above the eutectoid there is some transformation to austenite. In the fully austenitic region there is no appreciable growth until carbide dissolution is complete. At the edge of the weld pool complete carbide dissolution permits coarsening of the austenite grain size.

Embrittlement in the HAZ is a major factor limiting the weldability of carbon steels, and the weldability is related to the carbon equivalent, an empirical parameter based on the carbon concentration and alloy content of the steel. The presence of stable carbides and nitrides in the region transformed to austenite limits microstructural coarsening and improves weld properties, but some particle dissolution and particle coarsening are to be expected. Qualitative analysis of the expected features can be summarized by plotting the structure as a function of energy input.

Problems associated with weld metallurgy are usually connected with either the presence of impurities (for example, sulphur or hydrogen in steels) or the control of grain size. Post-welding heat treatment is commonly limited to promoting stress relaxation, but may also be used to refine the microstructure.

Further Reading

1. K. Easterling, *Introduction to The Physical Metallurgy of Welding*, Butterworth, London, 1985.
2. J. E. Lancaster, *Metallurgy of Welding*, Fourth Edition, Allen & Unwin, London, 1987.
3. O. F. Devereux, *Topics in Metallurgical Thermodynamics*, Robert E. Krieger Publishing Company, John Wiley & Sons, New York, 1983.
4. Metals Handbook, Eighth Edition, Volume 7: *Atlas of Microstructures of Industrial Alloys*, American Society For Metals, Metals Park, Ohio, 1972.

Problems

6.1 What material properties are involved in heat transfer calculations? When applying a heat transfer calculation to a welding process, there are a number of simplifying assumptions which can lead to error. List these sources of error and explain how they affect the heat transfer calculation.

6.2 Estimate the weld-pool radius in 1040 steel (0.4 wt%C) and the distance from the heat source at which martensite would be expected to form. Assume a static electrode heat source supplying 400 A at a voltage of 18 V. For the 1040 steel assume a heat transfer coefficient of $K = 41$ W m^{-1} K^{-1}. List any assumptions you have made in arriving at your estimate.

6.3 Sketch the microstructure expected in a welded 0.15 wt% carbon steel as a function of distance from the centre of the weld-pool, including the HAZ.

6.4 How would welding be expected to affect the presence and size of precipitates in the HAZ (including carbides, in the case of welded steels)?

6.5 Explain how the extent of plastic deformation (cold work) affects the process of recrystallization, and how recrystallization may affect the mechanical properties of a welded joint.

6.6 Steel containing Mn and Si is welded. Based on the following reactions, explain why the concentration of Mn and Si is likely to be lower at the surface of the weld region. (Assume a temperature of 1600°C in the liquid metal and use the Ellingham diagram for data.)

$$SiO_2 + 2Fe \leftrightarrow Si + 2FeO$$

$$2Fe + O_2 \rightarrow 2FeO$$

$$Si + O_2 \rightarrow SiO_2$$

$$2MnO + 2Fe \leftrightarrow 2Mn + 2FeO$$

$$2Fe + O_2 \rightarrow 2FeO$$

$$2Mn + O_2 \rightarrow 2MnO$$

6.7 List and explain the potential problems involved in the welding of stainless steel, with special attention to the microstructure. How can these problems be avoided?

6.8 Two 5.0 mm thick nickel plates are to be welded together in plane. Choose a specific weld process and geometry, and explain your reasoning. Compare the advantages of the geometry and process you have chosen to other possible geometries. Outline the types of weld defect and the weld microstructure which you expect.

6.9 You are required to specify a welding process for stainless steel plates to be used in a pressure vessel which will operate at cryogenic temperatures. The plates are 18 mm thick and no overlap of the plates is permitted. Choose an appropriate welding process and geometry. Using handbook information, specify which alloy should be chosen for the steel, as well as the approximate welding conditions.

6.10 Explain the term partition coefficient. How does the partition coefficient affect the formation of the microstructure in the weld region?

6.11 'There is an inherent problem in attempting to predict the phases and microstructure in the weld region based on phase diagrams alone.' Explain this statement and list other sources of information which can help to provide more accurate information.

6.12 A moving point source of heat causes an asymmetric temperature distribution in the workpiece. How do the following factors affect the temperature distribution profile?

a. The velocity of the heat source with respect to the workpiece.
b. The thickness of the workpiece.
c. The heat transfer coefficient of the workpiece material.

6.13 Welding usually leads to the formation of residual stresses in the weld region (both the weld bead and the HAZ). What are the primary sources of the residual stresses during welding? Sketch the magnitude and type (compressive or tensile) of residual stresses expected across a completed weld in thin plate.

6.14 Aluminium alloy plates are to be welded in the construction of an airplane frame. The stringent engineering requirements include: minimum contamination, maximum strength and fracture toughness, and a small HAZ. Of the following techniques, which in your opinion would be most applicable? Explain your reasoning!

a. TIG.
b. Laser welding.
c. Electron beam welding.
d. Resistance spot welding.

7

Soldering & Brazing

The terms *soldering* and *brazing* are frequently linked, and there is often some confusion involved in their usage. Both are bonding processes which involve the use of a filler alloy with a melting range well below that of the components to be joined. These components are usually *metallic*, although *ceramic* components can also be bonded within the framework of these technologies. *Solders* have lower melting points than brazes, and the alloys used for solder compositions owe their low melting points to the large atomic radii which result from the high atomic numbers of their constituents, which therefore have high densities.

Brazes have higher melting points than solders, and the melting point of the braze may approach that of one or other of the components to be bonded. The constituents of brazing alloys have low heats of formation for their oxides, and are generally less readily oxidized than other metals. They include the lower melting point noble metals (*silver* and *gold*) and some of their near-neighbours in the periodic table (*copper*, *zinc* and *nickel*). The exceptions are the Al–Si eutectic compositions which are used to braze aluminium alloys, and, to some extent, the low melting point eutectic alloys of nickel and copper with phosphorus. The most important constituents for solder and braze alloys are identified by their relative position in the *periodic table* of the elements (Fig. 7.1).

7.1 GENERAL PRINCIPLES

All soldering and brazing operations are performed in a three-phase system consisting of the solid components (considered as a single phase) and two liquid phases, the metallic *braze* and the non-metallic *flux*. There is no real distinction between a flux and a *slag*. Slags are low density, non-metallic melts which float to the surface of a molten metal and separate it from the atmosphere. They form by preferential oxidation of selected constituents and, in steels, have compositions based on the mixed oxides of silicon, manganese and iron. Liquid slags form in

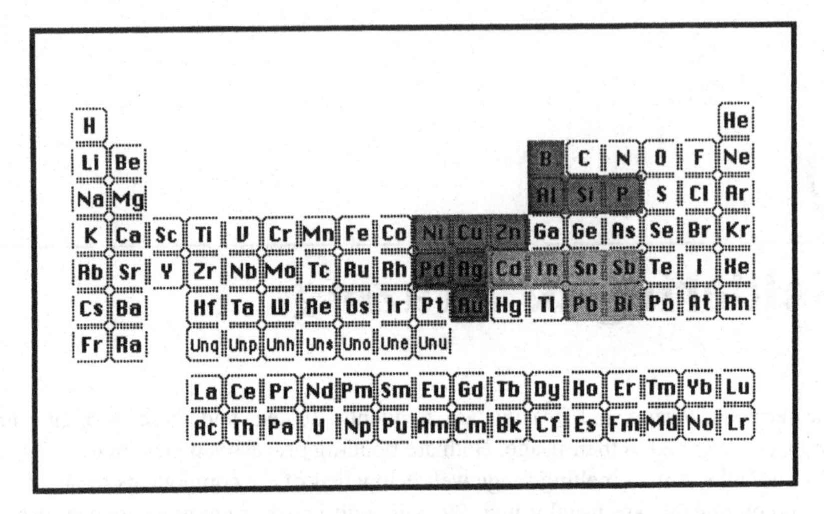

Fig. 7.1. The common constituents of *solder* and *braze* alloys fall into well-defined positions in the *periodic table* of the elements.

fusion welding operations, where they *flux* away surface oxides and help to protect the weld pool from atmospheric contamination. In the welding of ferrous alloys this fluxing action is provided by a mixed-oxide *silicate glass*. The low density slag subsequently solidifies on the surface of the weld *after* solidification of the weld bead. In the welding of aluminium alloys, no low melting point oxide slag can form and a protective non-metallic liquid phase has to be formed by *fluxing* additions. For the aluminium alloys the welding fluxes are halide-based and the oxyhalides formed by dissolution of surface oxide have low melting points, below the melting temperatures of the aluminium alloy components (550 to 600°C). The fluxes which are to be used for *soldering* and *brazing* operations must also have low melting temperatures, to ensure that solidification of the bonding alloy precedes that of the flux.

7.1.1 Stages of Joining

The *soldering* or *brazing* process must thus take place in several well-defined, sequential steps (Fig. 7.2):

1. As the temperature is raised, the *flux* melts and *wets* the components to be joined, *dissolving* any contaminant (oxide) film which is on the surface.
2. At higher temperatures, the *solder* or *braze* alloy melts, and *preferentially* wets the components, displacing the flux, while *itself* being wetted by the flux.
3. At the final brazing (soldering) temperature, the *triple junction* between the (solid) component, the (non-metallic liquid) flux and the (metallic liquid) solder

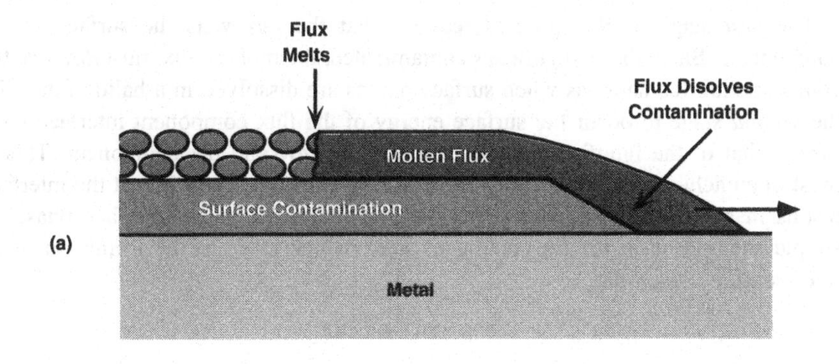

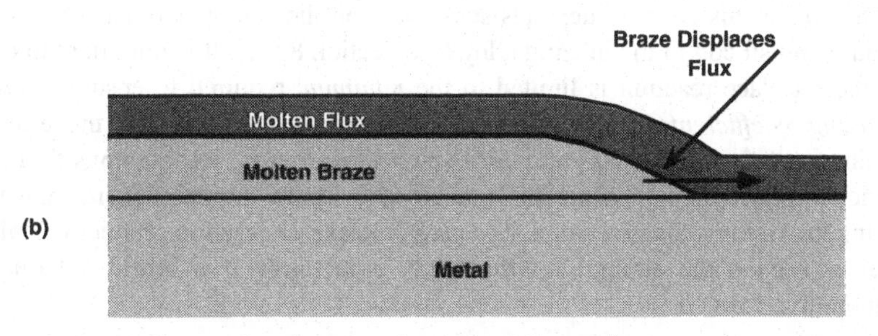

Fig. 7.2. All soldering and brazing procedures depend on a *flux* to wet the surfaces of the components and dissolve surface films and contamination, before being displaced by the molten braze or solder alloy (which must therefore also wet the components). (a) The flux *melts* and dissolves the film of surface contamination, completely *wetting* the cleaned surfaces of the components. (b) The molten braze or solder *displaces* the molten flux layer to wet the surfaces of the components, while itself being protected from the atmosphere by the molten flux.

or braze migrates, to *increase* the area of the contact interface between the component and the molten alloy (solder or braze).

4. Once the area to be joined is *fully* wetted by the molten alloy the source of heat can be removed, when the alloy cools and *solidifies* to form the bond.
5. Further cooling then results in solidification of the (oxide-contaminated) *flux* (now termed *dross*), which must subsequently be completely removed before the joint is put into service.

The *first* stage of this process requires that the *flux* wets the surface of the components. Since these are *always* contaminated, the molten flux must *dissolve* the contamination (as happens when surface oxides are dissolved in a halide flux). For the *second* stage to occur the surface energy of the flux component interface must *exceed* that of the interface between the molten alloy and the component. This is most often achieved by *reactive wetting*, that is, a chemical reaction at the interface results in a *positive spreading force* and the formation of an interface phase. A simple example is in the tin coating of steel, which leads to the formation of an *intermetallic compound*

$$Fe + 2Sn \rightarrow FeSn_2$$

Another example is in the brazing of a silicon nitride component to a ferrous alloy. In this case the addition of a small quantity of *titanium* to the Cu–Ag brazing alloy results in reactive wetting through an *exchange reaction*

$$4Ti + Si_3N_4 \rightarrow 4TiN + 3Si$$

where the silicon is dissolved in the molten brazing alloy. The *equilibrium coefficient* for this reaction depends sensitively on the activities (concentrations) of titanium and silicon in the brazing alloy (see Section 8.5.1). It is important to ensure that the interface reaction is limited to the *minimum* required to ensure a *positive spreading coefficient*. If the interface reaction is excessive, then the effect on bonding is likely to be *deleterious* for two reasons. *Firstly*, a continuous film of the reaction product may *separate* the brazing alloy from the component, drastically altering the wetting characteristics. *Secondly*, the excess reaction product will almost certainly *reduce* the strength of the bond, particularly if a brittle, intermetallic compound is formed.

7.1.2 Surface Finish & Roughness

Surface roughness may have a significant effect on the wetting process, as was discussed in Section 2.3. During spreading the roughness imposes a periodic change in the *contact angle* which may either *promote* or *inhibit* wetting. Fig. 7.3 illustrates three possible situations. In the *first* case grinding marks *parallel* to the advancing wetting front *inhibit* the wetting by imposing a periodic force component *parallel* to the spreading force which may, in extreme cases, trap regions of non-wetting. In the *second* case the grinding marks are *perpendicular* to the wetting front. The periodic component of the spreading force now acts to deflect the wetting front into a *periodic curve* and tends to channel the molten metal into the grooves. *Finally* two sets of grinding marks at right-angles (or random abrasion) will leave a *network* of channels, isolating individual *high*-spots. Wetting may occur *faster* (the molten metal is 'sucked' into the channels by the *capillary forces*), but wetting may be *incomplete*, leaving regions of free surface on the high-spot 'islands'.

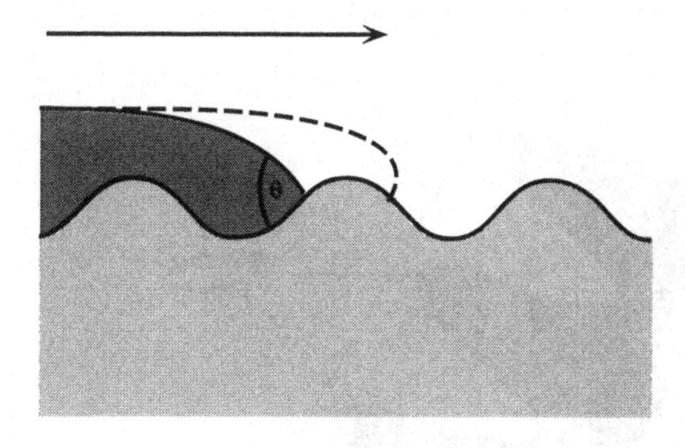

(a)

Fig. 7.3. The *surface* roughness has a large effect on the extent and rate of wetting. (a) Machining or grinding grooves *parallel* to the advancing front of the melt *inhibit* wetting by introducing a periodic variation in the *wetting angle* and hence in the spreading force. (b) Grooves *perpendicular* to the wetting front tend to *promote* wetting by introducing additional *capillary forces*. The steady-state wetting front is then *wavy*, with the melt 'sucked' into the grooves. (c) A *network* of grooves or scratches generally *promotes* wetting as a result of the additional *capillary forces*, but may leave behind *unwetted islands* ('high spots') if the roughness is extreme.

7.1.3 Failure Tolerance

Many commercial processes involve the mass-production of brazed or soldered joints. Soldered printed circuit boards (*PCBs*) for consumer electronics constitute an important example of the *simultaneous* production of a *large* number of electrical contacts on a single board. In this case *defective contacts* can be identified by automated inspection techniques (and sometimes repaired by manual soldering). In a complex circuit a single '*dry*' (incompletely wetted) joint will affect the performance of the PCB. If the *probability of incomplete wetting* is P_f, then the probability of making a successful join is $1 - P_f$, and the probability that a PCB containing n soldered joints will pass inspection is $(1 - P_f)^n$. It follows that the probability that the board will *fail* inspection is $[1 - (1 - P_f)^n]$, which is a very sensitive function of n. For *mass-produced* items for a consumer market, a 1% chance of failure to pass inspection would probably be considered high, while 100 or more contacts may have to be made on each PCB in a single pass over the molten solder bath. For this particular case it follows that: $1 - (1 - P_f)^{100} = 0.01$, which gives an allowed value

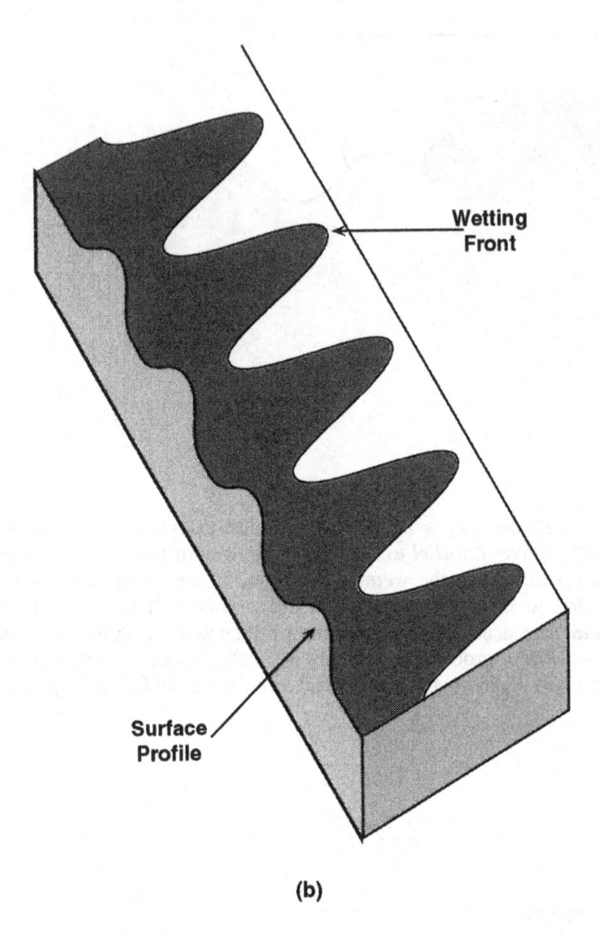

Fig. 7.3. (*continued*)

of P_f of only 10^{-4}. Attaining this level of reliability is no trivial challenge for the process engineer.

Precleaning of the components prior to bonding is essential to ensure reproducible wetting, and for many applications grinding or abrasion may be the acceptable, economically effective cleaning methods. However, the resultant *surface*

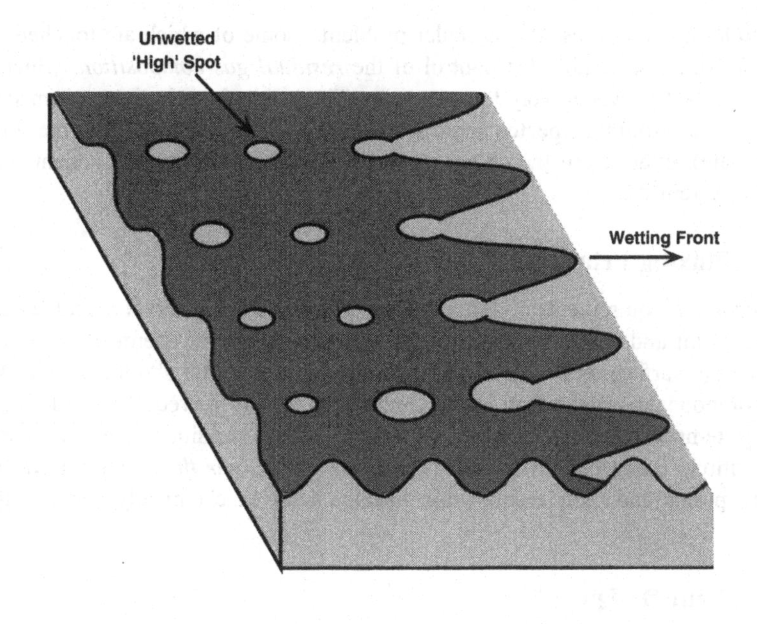

(c)

Fig. 7.3. (*continued*)

roughness must also be controlled within close limits if both the extent and rate of wetting are to be kept within the working tolerances for the bonding process.

7.1.4 Protective Atmospheres

Failure probabilities can often be *reduced* by bonding in a *protective atmosphere*. This is particularly true of *batch processes* like furnace brazing, in which the components are assembled and the complete assembly is then heated to the brazing temperature. Protective atmospheres are either *reducing*, so that oxide formation is thermodynamically unfavourable, or *neutral*, preventing oxygen access to the surfaces to be bonded. *Vacuum*, *argon*, *hydrogen*, *nitrogen* and *carbon monoxide* are all used as brazing atmospheres, as are gas mixtures, such as forming gas ($96\%N_2 + 4\%H_2$). In many cases the control of a *gas dissociation* reaction determines the degree of protection afforded by the atmosphere. Two common examples are:

$$N_2 + 3H_2 \leftrightarrow 2NH_3 \text{ and } O_2 + 2H_2 \leftrightarrow 2H_2O$$

Vacuum bonding presents particular problems, some of which are touched on in Chapter 10. Most notably the control of the *residual gas composition*, which is a significant factor, even at very low pressures. The residual gas in the vacuum system may originate from unexpected sources: *outgassing* (gas evolution from the solid or liquid state), either from the charge in the furnace or from the components of the brazing system itself).

7.1.5 Fluxing Principles

The *chemical* requirements of a flux are rather demanding. The flux must have a low melting point and a low viscosity, combined with sufficient chemical reactivity to dissolve the surface oxide films. On the other hand, the flux should not attack the base components, at least not at any appreciable rate at the required soldering and brazing temperatures. In *brazing* operations these conditions can be met by compositions based on *borax* and the *halides*, but *solder fluxes* must have lower melting points and their composition usually involves chemically active *organic resins*.

7.1.6 Joint Design

Brazed and soldered joints cannot normally match either the strength or the corrosion resistance of welded joints, since they employ bonding materials whose mechanical, physical and chemical properties differ drastically from those of the components being joined. In particular, *mismatch* of the *elastic properties*, the *thermal expansion coefficients* and the *electrochemical potentials* are all expected to limit the performance of the bonded assembly.

A common method of attaining adequate strength in a brazed or soldered construction is by employing a *lap-joint*, in which the load is transferred from one component to the other predominantly in *shear* (Fig. 7.4). The *strength* of the joint then depends on the *length* of the overlap over which the shear stresses operate. The strength will also be affected by the *thickness* of the braze and the *elastic modulus mismatch*, as well as by the *residual stresses* associated with differential shrinkage across the joint. In particular, it is common to find that there is an *optimum length* of overlap, beyond which the *integrity* of the join is difficult to ensure, because bonding tends to be incomplete and differential shrinkage leads to excessive *thermal stresses*. Good design of brazed or soldered joints typically minimizes the normal stresses by using a lap joint construction which employs *sleeves* wherever necessary. Figure 7.4 illustrates this principle with some examples.

7.2 SOLDERS

Solder alloy systems are all based on the low melting point, high atomic weight metals to be found towards the end of the periodic table (Fig. 7.1). The constituents

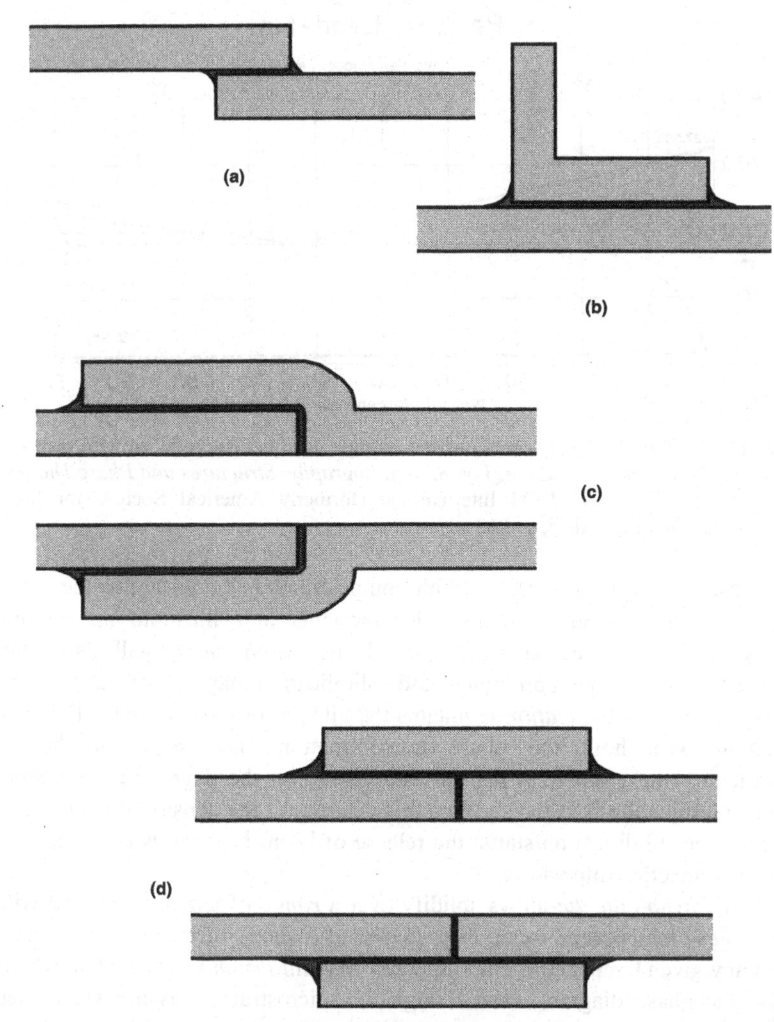

Fig. 7.4. Soldered and brazed joints tend to be weak in *tension*. Designing with *lap-joints* reduces tensile stresses to a minimum. (a) Single lap-joint. (b) Lap-joint for a bracket or T-junction. (c) Lap-joint for a mated tube connection or elbow. (d) Butt-joint reinforced by a lap-jointed sleeve.

of these alloys possess a *range* of melting points from the order of 40°C up to about 220°C, but the classic *lead–tin* solders all melt in the range 180 to 220°C. The Pb–Sn eutectic diagram is given in Fig. 7.5. Two classes of Pb–Sn solder alloy are important: those based on the eutectic temperature (183°C), which melt *isothermally*, and those precipitating proeutectic tin prior to solidification of the eutectic, which melt over a range of temperature. These two classes of alloy, *eutectic*

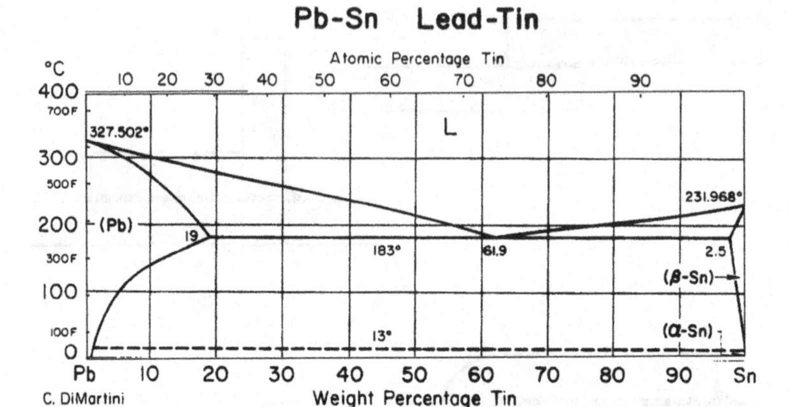

Fig. 7.5. The *lead–tin* binary system contains a single eutectic. Reproduced with permission from *Metals Handbook, 8th Edition, Vol. 8, Metallography, Structures and Phase Diagrams*, Taylor Lyman, Ed. (1973) ASM International (formerly Americal Society for Metals) Materials Park, OH, Fig. Al-Si, p. 263.

and either *hyper-* or *hypo-eutectic* (with compositions below or above the eutectic composition), are found in many other systems and illustrate an important metallurgical principle: eutectic alloys which melt *isothermally* will also solidify over a very narrow temperature range, and solidification may be difficult to control. The *latent heat of solidification* maintains the alloy at or close to the solidification temperature throughout the phase transformation. The *scale* of the final microstructure (the grain size) depends sensitively on the degree of *supercooling* at which solidification is initiated, since this determines the density of nuclei for the solid, but once solidification starts, the release of latent heat raises the temperature back to the eutectic temperature.

Hyper- or hypo-eutectic alloys solidify over a *range* of temperature, and within this *two-phase* temperature range they consist of a *slurry* of the solid metal in the melt. At any given temperature this slurry has an *equilibrium* solids content which is fixed by the phase diagram, even though the microstructure is not stable (large particles tending to grow at the expense of the small ones by *Ostwald ripening*, see Section 6.3.4). Nevertheless, there is a tendency to form a *metastable* plastic mass which can be worked like clay. This offers an opportunity to *mold* the solder into and around a joint: a familiar process when domestic 'plumbing' really did involve the use of lead (*plumbum*) piping and the joints were '*wiped*' using a 50/50 Pb–Sn alloy.

7.2.1 Compositions & Properties

Lead tin solders are commonly available in the compositions 19, 61.9 and 97.5% Sn, and their melting and solidification response can be predicted from the Sn–Pb phase diagram. Plumber's solder (the 50/50 composition) is now required relatively rarely, since domestic plumbing has long replaced lead piping with galvanized iron or copper, and these are, in their turn, being rapidly replaced by filled polymers, which can be readily hot-formed on site.

The 61.9% Sn *eutectic* alloy is the material of first choice for electrical applications where *automation* is commonly employed. The other two alloys have a wide melting range and their selection depends on whether a *lead-based* or a *tin-based* alloy is preferable.

The two major disadvantages of the Pb–Sn solders are the *high cost* of the constituents and the *low creep strength* of the alloys. Antimony (Sb) is commonly added to improve the creep strength of *lead* alloys, for example in lead–acid accumulators and car batteries, as well as in lead sheets for roofing and guttering. While Sb additions do inhibit creep at temperatures up to 100°C, they also *embrittle* copper alloys, especially the Cu–Zn alloys (the brasses), and are not suitable for soldering copper-based alloys.

The *toxicity* of lead and the need to find more *corrosion-resistant* solder alloys has led to the development of several alternative tin-based alloys. The Sn–Ag system (Fig. 7.6) has a eutectic temperature of 221°C and a eutectic composition of 3.5% silver. This eutectic alloy is commonly used in jewellery manufacture, where it is known as '*hard solder*'. The Sn–Zn system (Fig. 7.7) has a eutectic temperature of 198°C and a eutectic composition of 8.5% Zn. Sn–Zn solders can be used for soldering *aluminium alloys* using a suitable flux, but the *corrosion resistance* of the joint is poor. Compositions in the range 10 to 30% Zn are commonly available.

Lower melting point solders exist in multi-component alloy systems containing the constituents *Sn*, *Pb*, *Bi*, *Cd*, *In* and *Sb*. These range from *Wood's metal* (5%Bi, 25%Pb, 12.5%Sn, 12.5%Cd, 70°C), which is used for novelties (teaspoons that melt in hot tea!), to the Sn–In solders (Fig. 7.8) which can be used to bond to glass.

7.3 BRAZES

Brazing as a process was originally restricted to bonding noble metal alloy systems (copper, silver and gold), and has a history that goes back over 2000 years. Today, the range of available materials has expanded to include compositions suitable for joining anything from *aluminium alloys* (maximum use temperatures of the order of 300°C) and stainless steels, to high temperature *superalloys* for heat-engine and aerospace applications at temperatures approaching 1000°C.

7.3.1 Copper-based Brazing Alloys

The classic brazing alloy system is the *copper–silver* alloy whose binary phase diagram is shown in Fig. 7.9. This is a *eutectic* system with a *eutectic* temperature of 780°C and a eutectic composition of 72% silver. The commercial brazing

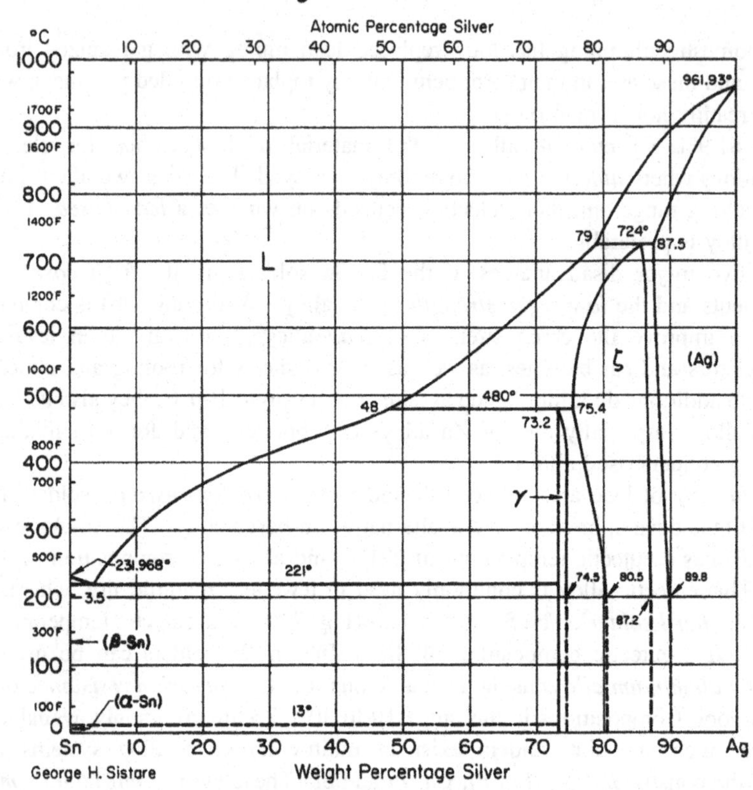

Fig. 7.6. The *silver–tin* binary system contains a eutectic transformation involving the γ-phase intermetallic Ag_3Sn. Reproduced with permission from, *Metals Handbook, 8th Edition, Vol. 8, Metallography, Structures and Phase Diagrams*, Taylor Lyman, Ed. (1973) ASM International (formerly American Society of Metals) Materials Park, OH, Fig. Au-Nl, p. 267.

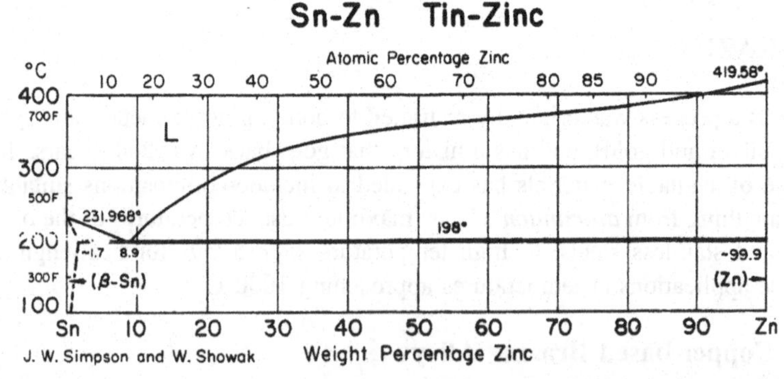

Fig. 7.7. The *tin–zinc* binary alloys are another example of a simple, single eutectic system. Reproduced with permission from, *Metals Handbook, 8th Edition, Vol. 8, Metallography, Structures and Phase Diagrams*, Taylor Lyman, Ed. (1973) ASM International (formerly American Society of Metals) Materials Park, OH, Fig. Cr-Fe, p. 291.

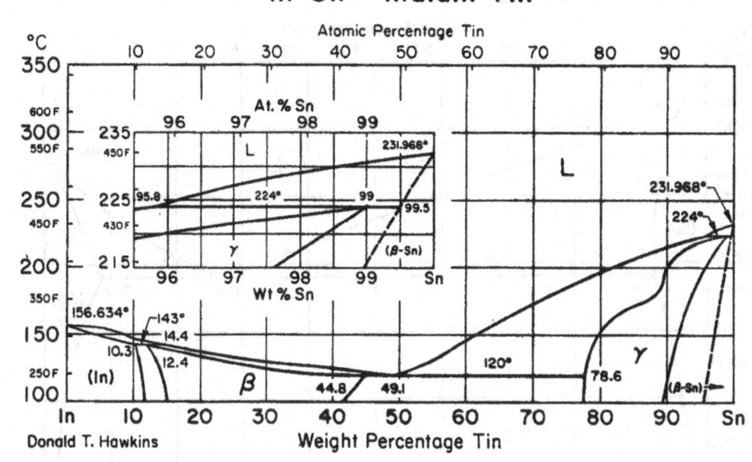

Fig. 7.8. Low melting point eutectic *tin–indium* solders have specialty applications, although this binary system is quite complex. Reproduced with permission from, *Metals Handbook, 8th Edition, Vol. 8, Metallography, Structures and Phase Diagrams*, Taylor Lyman, Ed. (1973) ASM International (formerly American Society of Metals) Materials Park, OH, Fig. Fe-Ni, p. 304.

compositions have excellent corrosion resistance. This is exploited in the assembly of *stainless steel* systems for the petrochemical and food industries, and in the construction of high-vacuum and ultra-high vacuum systems used in the processing of microelectronic components. Ag–Cu brazing alloys are often the compositions of first choice in the jewellery industry. Major considerations in this application are not only the *corrosion resistance* of the join (perspiration at body temperature!), but also the matching of the joint *appearance* to that of the assembled components. Alternative binary copper-based brazing alloys are based on the Cu–Zn (a white appearance) and the Cu–Au systems (a deep yellow finish). Both systems again have applications in the jewellery industry, where *appearance* is a primary consideration.

Cadmium additions to a Cu–Ag alloy reduce still further the temperature for the onset of *solidification* (the completion of melting) to 640°C. In this case the improved microstructural stability of the components at the lower brazing temperature must be balanced against some sacrifice of the *corrosion resistance* and *strength* of the joint. Alternatively, the *Cu–Ag–Zn* system (Fig. 7.10) has a *ternary eutectic* at 677°C with a composition of 25%Zn, 55%Ag and 20%Cu. This alloy therefore has some advantage in the joining of components whose microstructures would be too unstable when heated to higher temperatures.

7.3.2 Brazes for Aluminium Alloys

The *aluminium silicon* system is the basis of brazes for joining aluminium alloys (Fig. 7.11). The *eutectic* is at 12.6% Si and the eutectic temperature is 577°C. In the

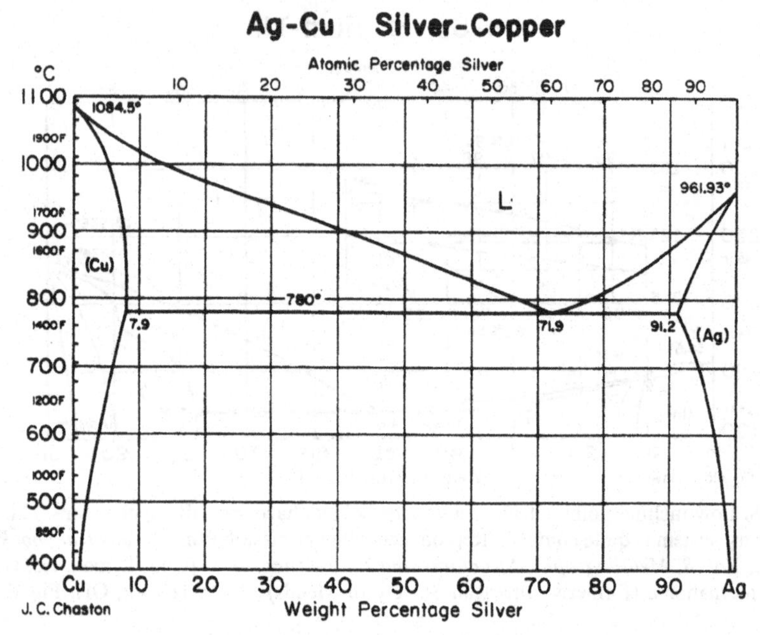

Fig. 7.9. The simplicity of the *silver–copper* binary eutectic system makes these alloys the brazes of first choice for many applications. Reproduced with permission from, *Metals Handbook, 8th Edition, Vol. 8, Metallography, Structures and Phase Diagrams*, Taylor Lyman, Ed. (1973) ASM International (formerly American Society of Metals) Materials Park, OH, Fig. In-Sn, p. 312.

binary alloy the eutectic microstructure tends to be rather coarse, leading to poor mechanical properties, but small additions of a third constituent (*iron*, for example) '*modify*' (refine) the microstructure.

The adherent, cohesive alumina film, which forms on aluminium alloys at all temperatures and oxygen partial pressures, can be dissolved by suitable *halide* (especially fluoride) fluxes, but adequate fluxing of this stoichiometric surface oxide is still a major problem in producing acceptable brazed joints between aluminium alloy components.

7.3.3 Brazing Stainless Steels & Heat-resistant Alloys

The brazing of *stainless* and *heat-resistant* alloys places severe demands on the *corrosion resistance* (stainless steels), the *oxidation resistance* (both stainless steels and heat-resistant alloys) and the *elevated temperature mechanical properties* (heat-resistant alloys) of the braze compositions to be used. The materials to be joined are all highly alloyed, and are mainly iron, cobalt or nickel-based alloys. These alloys also include the *superalloys*, which have temperature capabilities that approach and in some cases exceed 1000°C (which no braze can match).

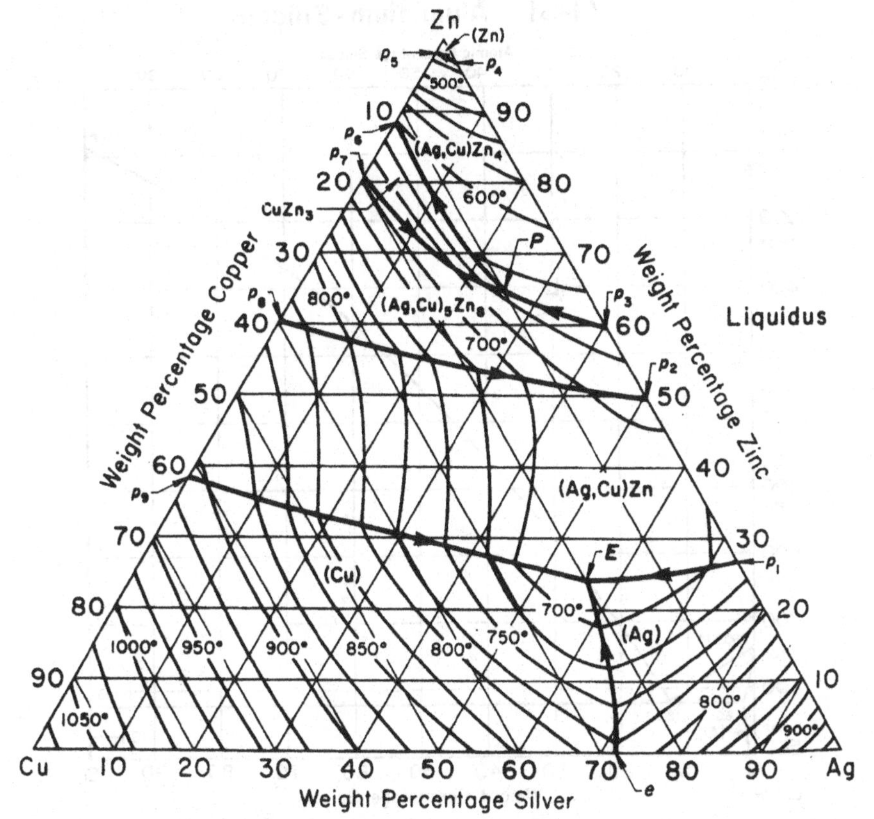

Fig. 7.10. The low melting point of the *ternary eutectic* in the *silver–copper–zinc* system is a major advantage for some brazing operations. Reproduced with permission from, *Metals Handbook, 8th Edition, Vol. 8, Metallography, Structures and Phase Diagrams*, Taylor Lyman, Ed. (1973) ASM International (formerly American Society of Metals) Materials Park, OH, Fig. Ni-Pd, p. 324.

The *gold–nickel* and *gold–copper* systems (Fig. 7.12) are used to bond nickel-based *superalloy* turbine blades to the turbine disc. The actual turbine blades may reach temperatures of the order of the melting point of the braze, but the temperature gradient across the join normally limits the *operating temperature* at the braze in service to the order of 700°C. The blades can be *debonded* and *replaced* at the end of their service life (and the gold recovered!). The *nickel palladium* eutectic system (Fig. 7.13), with a eutectic temperature of 1235°C at 40% Pd, has a similar working temperature and oxidation resistance.

For applications where *corrosion resistance* is a dominant requirement *stainless steel* and *superalloy* components can also be brazed using a *nickel–phosphorus* eutectic (Fig. 7.14; the eutectic temperature is 875°C at a composition of 11%P).

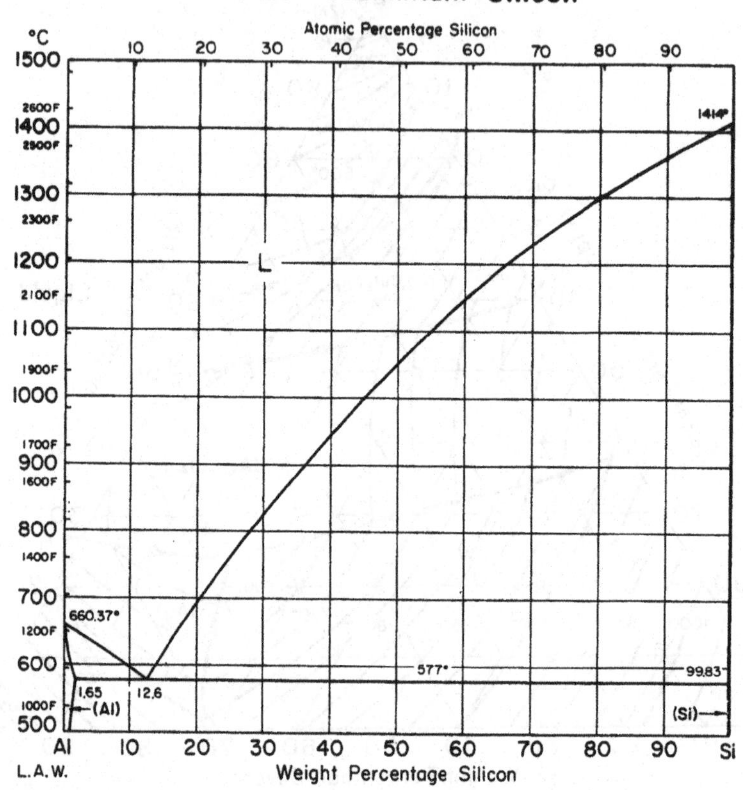

Fig. 7.11. The *aluminium–silicon* binary eutectic system is the basis for most aluminium casting and brazing alloys. Reproduced with permission from, *Metals Handbook, 8th Edition, Vol. 8, Metallography, Structures and Phase Diagrams*, Taylor Lyman, Ed. (1973) ASM International (formerly American Society of Metals) Materials Park, OH, Fig. Pd-Te, p. 330.

Finally, the brazing of *carbide* tool inserts to a *carbon steel* tool holder involves the joining of two *dissimilar* materials. The brazing alloy composition commonly used in this case is the *copper–phosphorus eutectic* at 7.5%P.

7.3.4 Selection Criteria

Which particular brazing alloy (or class of brazing alloys) should be selected for a particular application depends in each case on the *engineering requirements* and their *relative importance*. Typical criteria are listed in Table 7.1. To these factors should be added the *costs* of both the *brazing materials* (alloys and fluxes) and the *bonding operation* (initial preparation of the components, their assembly, the heating process, and post-bonding cleaning and inspection).

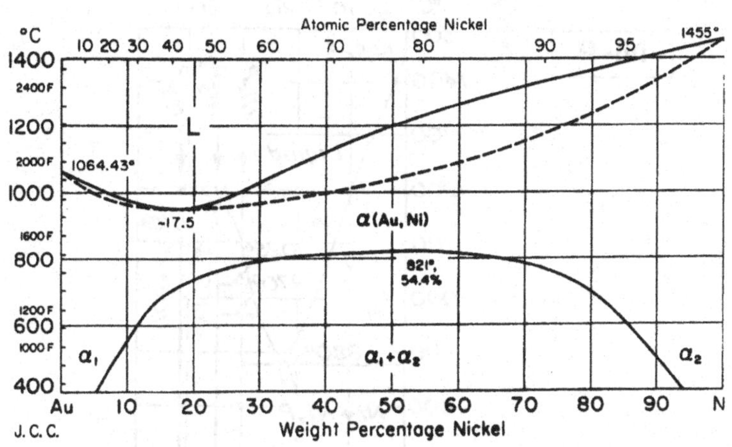

Fig. 7.12. The *gold–nickel* binary system is 'almost' eutectic, with a melting point minimum and solid state phase separation at lower temperatures. Reproduced with permission from, *Metals Handbook, 8th Edition, Vol. 8, Metallography, Structures and Phase Diagrams*, Taylor Lyman, Ed. (1973) ASM International (formerly American Society of Metals) Materials Park, OH, Fig. Sn-Zn, p. 236.

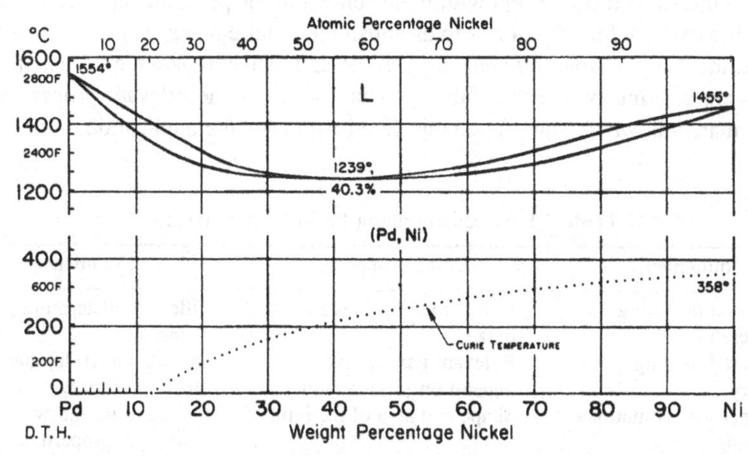

Fig. 7.13. *Nickel–palladium* binary alloys show complete solid state solubility, but they do have a melting point minimum at 1239°C. Reproduced with permission from, *Metals Handbook, 8th Edition, Vol. 8, Metallography, Structures and Phase Diagrams*, Taylor Lyman, Ed. (1973) ASM International (formerly American Society of Metals) Materials Park, OH, Fig. Au-Cu-Zn, p. 380.

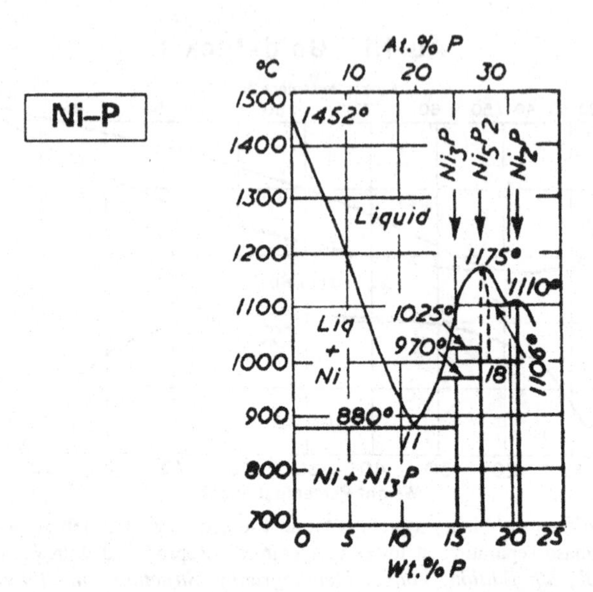

Fig. 7.14. The *nickel–phosphorous* binary system shows an exceptionally steep melting point dependence on composition which is associated with the non-metallic properties of the alloying element, *phosphorus*. Reproduced with permission from *Smithells Metals Reference Book*, Brandes and Brook (1992) Buttterworth-Heinemann Ltd.

Of the factors listed in Table 7.1 *none* can be classed as more important or less important than the others, except within the context of the particular application. The *only* viable strategy for selection is to establish the requirements as *quantitatively* as possible, identify the *candidate bazing systems*, and then compare the values of their relevant engineering properties. Table 7.2 lists some of the relevant properties of brazing materials which should be considered *whatever* the application.

Table 7.1. Selection criteria for brazing systems

System requirements	Braze properties	Comments
1. Required operating temperature	Limits melting range of braze	Often the determining criterion
2. Expected loading history	Relevant parameters depend on joint design	Include creep, fatigue and stress corrosion
3. Geometrical limitations on design	Shape and area of the joint	Optimize design with respect to properties
4. Corrosion/oxidation resistance over life	Electrochemical potential and oxygen access	Joint resistance to pitting and crevice attack
5. Reactivity and bonding to components	Chemical reactivity with respect to bulk	Includes wetting and spreading characteristics

Table 7.2. Relevant properties of brazing materials

Properties	Parameters
1. Melting range and wetting characteristics	Solidus, liquidus and eutectic temperatures, contact angles
2. Strength characteristics	Yield strength, toughness, creep strength, stress corrosion
3. Corrosion/oxidation performance	Electrochemical potential, rates of oxidation
4. Chemical reactivity	Thermodynamic parameters for potential reactions

The *final* stage in selection is the most difficult, since it involves *subjective* decisions about the relative weight to be attached to each of the factors which have been considered. Selection is an *iterative* process, in which modifications of the joint *design* can change the order of preference for the candidate brazing alloys. This is particularly the case if, for example, the cost of corrosion protection can be saved by employing a more expensive, but less corrosion-susceptible braze. The economic factors noted previously do not make the decision any easier, especially when potential improvements in the *performance* and *life* of the system must be balanced against increased materials and processing *costs*.

7.4 FLUXES

Fluxes are often critical for the success of both soldering and brazing operations, and hence merit a separate section. The flux must:

1. *Wet* the materials to be bonded.
2. *Dissolve* surface (oxide) films present on the components.
3. Be readily *displaced* by the molten braze.
4. Be readily *removed* after joining is complete.

Fluxes for *solders* are often *organic-based resins*, which attain the required *fluidity* below the melting point of the solder. Brazes may be *self-fluxing* (see below), or the appropriate fluxes may be based on low melting point *ionic compositions*.

7.4.1 Requirements

In addition to the basic requirements for a flux, listed above, successful fluxes must also fulfil the following additional criteria:

1. Once applied to the surface, the flux should remain *adherent* during the heating of the component up to the melting point of the flux. It does not help for the

molten flux to wet the component if most of the flux has flaked off the surface well below this melting point.

2. The flux should be *chemically stable* at the braze temperature, at least for the length of time required to accomplish the braze. In practice, some products of the surface reactions with impurities are likely to be *solid phases*, known as *dross*, which are *insoluble* in the flux. Such dross must be carried free of the brazed components by the flux, that is, the flux should fully wet any dross which is formed.

3. The flux should be easy to apply (as a *paste* or *solution*, for example), and the residual flux, dross or slag must be removable after brazing (by mild *abrasion* or by *dissolution* in hot water).

The *boron-* and *phosphorus*-containing brazing alloys are to a large extent *self-fluxing*. Low melting point borates and phosphates are formed from the molten braze by reactions of boron or phosphorus with the surface oxides and the atmosphere. These self-fluxing reactions also help to ensure excellent spreading characteristics for the molten braze.

7.4.2 Compositions

Adequate fluxing for solders is possible using aqueous solutions of $ZnCl_2$ or NH_4Cl. These can be brushed, dipped or sprayed onto the surfaces to be bonded. Breakdown to HCl on heating and subsequent reaction with the surface oxide leads to the formation of molten flux at temperatures of the order of 200°C. Replacement of water by *glycerol* or *propanol* helps to prevent splashing at the boiling point of the solvent and loss of flux from the surfaces to be bonded. Admixtures of other halides (*bromides* and *fluorides*) are also common.

Organic acid fluxes can replace the halides in the soldering of electrical components, thus avoiding problems of corrosion associated with incomplete removal of halide residues after soldering. *Rosin*, the clear, amber residue left after distilling turpentine from tree sap, can be dissolved in an *organic solvent* and used as a flux for PCB (printed circuit board) manufacture. The active agents reacting with surface oxides are the organic acid groups in the rosin.

Higher melting point fluxes are used for the higher temperature brazing alloys and *stainless steels* are usually fluxed with an *orthophosphate-based flux*. Alternative high melting point fluxes include *alkaline* systems of borates and fluorides, which are also used for fluxing *aluminium* alloys.

In all cases the *selection* of a flux is a compromise between the need to react with the surface contamination and remove it from the components, and the need to avoid flux residues, which could lead to subsequent corrosion of the system in service.

Summary

Soldering and brazing employ filler alloys with melting points well below those of the components to be joined. Solders are alloys of reactive, high atomic number metallic elements with very low melting points, while brazes are much less reactive alloys, typically of the noble metals, but with higher melting points. Soldering and brazing operations take place in a three-phase system comprising the molten filler metal, the solid substrate and a molten, non-metallic flux which fulfils the twin functions of dissolving surface contamination from the substrate and protecting the molten filler metal from oxidation.

It follows that the flux must both wet the substrate (to dissolve surface contamination) and be displaced by the molten filler metal (to form the join). In many cases this requires reactive wetting of the substrate by the filler, which may lead to the formation of intermetallic compounds at the interface. Surface roughness generally promotes wetting of the substrate, but may also lead to incomplete coverage of the interface by the molten filler. For complex assemblies, such as a printed circuit board, the tolerance for such joining defects may be exceptionally low. Under some circumstances it may be justified to employ a protective atmosphere during joining in order to improve product reliability.

The strength of brazed and soldered joints is generally far inferior to that of a welded joint, both as a result of the residual stresses associated with thermal and elastic mismatch between the filler metal and the components and due to the limited strength of the filler metal. Replacing tensile loading of the joint by a design involving shear (a lap-joint) is an effective method for assuring mechanical reliability through good design.

The commonest solder alloy system is that based on the lead–tin alloys, melting in the range 180–220°C. Large quantities are employed in the automated soldering of printed circuit boards. Other solder compositions have specific application niches in jewellery assembly (the Sn–Ag eutectic), vacuum applications (Sn–In solders) and the soldering of aluminium alloys (the Sn–Zn system).

Most brazing alloys are copper-based, the commonest being in the Cu–Ag–Zn system with a typical melting range of 650 to 800°C. The aluminium–silicon eutectic is the basis for aluminium alloy brazes, while heat-resistant steels and nickel alloys (superalloys) are commonly brazed with gold or nickel alloys. These latter have much higher melting temperatures (of the order of 1250°C) and service temperatures that may exceed 700°C. Phosphorus-containing alloys have low melting point eutectics and are used in the brazing of stainless steels and superalloys (the Ni–P system), as well as in the assembly of carbide tool inserts (the Cu–P system).

When selecting a soldering or brazing system for a specific application it helps to consider the engineering requirements of the system and then compare these to the relevant properties of the available filler metal alloys. The possible alternatives then have to be weighed against the materials and processing costs.

Successful soldering or brazing depends very much on the prior cleaning and preparation of the surfaces to be joined and on the flux used to ensure wetting. The flux must be easy to apply and remove, surface oxides on the components must be soluble in the molten flux, and the flux must be displaced by the molten filler metal.

Typical flux compositions are based on halide salts or organic acids, the latter commonly being employed for electrical solders. Higher melting point fluxes are used for the higher temperature brazing alloys. The selection of a flux composition is a compromise between the requirements for adequate reactive fluxing and the need to be able to remove the flux residues.

Further Reading

1. J. E. Lancaster, *Metallurgy of Welding*, Fourth Edition, Allen & Unwin, London, 1987.
2. C. J. Thwaites, *Capillary Joining—Brazing and Soft-Soldering*, John Wiley & Sons, Chichester, 1982.
3. D.R. Frear, W. B. Jones and K. R. Kinsman (eds.) *Solder Mechanics—A State of The Art Assessment*, TMS, Warrendale, Pennsylvania, 1991.
4. P. Shewmon, *Diffusion in Solids*, Second Edition, TMS, Warrendale, Pennslyvania, 1989.
5. Metals Handbook, Eighth Edition, Volume 6: *Welding and Brazing*, American Society For Metals, Metals Park, Ohio, 1971.
6. Metals Handbook, Eighth Edition, Volume 8: *Metallography, Structures and Phase Diagrams*, American Society For Metals, Metals Park, Ohio, 1973.

Problems

7.1 Outline the differences between soldering, brazing and welding. To what extent are these processes limited to the joining of metallic components?

7.2 Describe in detail the role of fluxes in soldering and brazing. What are the required properties of a successful flux?

7.3 How does the contact angle between a drop of liquid and a solid surface depend on the surface tensions and interface energy? In soldering or brazing a liquid flux is involved in the process. What are, qualitatively, the desired relative values of the surface tension for the flux, the liquid metal, the solid metal substrate, and their respective interfaces.

7.4 Two components of a pure metal (A) are brazed together in air with a binary braze alloy containing only the elements (A) and (B). During the brazing process oxides are formed at the joint interface. Why has the binary system been transformed into a ternary? Suggest conditions under which a quaternary system could evolve. How would these transformations be expected to affect the spreading of the braze alloy?

7.5 Soldered and brazed joints are usually designed for loading under shear rather than tension. Explain the primary advantage of this design concept.

7.6 To what extent is there an optimal value of the thickness for a soldered or brazed joint? Consider in your answer:
   a. The strength of the joint.
   b. Production of the joint.

7.7 Ni–B braze alloys are often used for brazing special Ni-based alloys. Based on the fact that the ionic radius of B is small (~ 0.2 Å;), so that B enters the Ni lattice as an interstitial atom, explain which *properties* of the Ni–B braze would assist the brazing *process*.

7.8 A wide range of brazing alloys are based on eutectic compositions (for example in the Cu–Ag–Zn system).
   a. What is the primary reason for selecting a eutectic composition?
   b. What is the composition and approximate melting point of the ternary eutectic in the phase diagram (Fig. 7.10)?

c. A common brazing alloy is 45%Ag, 30%Cu, 25%Zn. Indicate this composition on the phase diagram.
d. What is the approximate liquidus temperature of this alloy?
e. Zinc has a high vapour pressure at the brazing temperatures used for this alloy. How would you expect this to affect the process?

7.9 A suitable flux for brazing copper tubing with the brazing alloy given in the previous question might be a eutectic mixture of chloride salts. Outline the engineering requirements for this flux in terms of the following factors:
a. The melting point of the flux.
b. Its solubility in water.
c. The solubility of copper oxide in the flux.
d. The relative interfacial energies in the flux/braze/metal system.
e. The reactivity of both the braze and the flux with respect to the copper tubing.

7.10 Three important criteria for the design of soldered joints are:
1. Minimize the tensile stresses.
2. Spread the distribution of forces over the joint area.
3. Use mechanical locking to supplement the soldered joint.

Explain how each of these criteria helps to improve the reliability of the joint.

8

Metal–Ceramic Joints & Diffusion Bonding

The integrity of *metal–ceramic* joints is particularly difficult to guarantee, not just because of the problem of securing adequate wetting and chemical attachment to the ceramic, but also because of the brittle nature of the ceramic component and the mismatch in thermal expansion coefficient and elastic stiffness across the interface. Of course *mechanical bonding* may be feasible. In this case a *flexible interlayer* (an elastomeric sealing gasket) will help to prevent failure associated with unwanted assembly stresses or localized stress concentrations. Alternatively, a *mechanically ground seal* may improve the dimensional tolerances sufficiently to reduce local stress concentrations to an acceptable level, as in the assembly of glass cup-and-cone joints. There is no reason why one component of a ground joint should not be a metal which has been given a suitably ground mating surface with a glass or ceramic. Of course, neither a flexible gasket nor a ground seal is a possible solution at elevated operating temperatures.

Another common strategy for *mechanical bonding* is the *shrink-fit* assembly, used to place the *brittle* component in *compression*, with a surrounding ductile ring or sleeve, which is placed in *tension* (Fig. 8.1). In a shrink-fit assembly the heated, ductile component is placed over the cold, brittle component, and *thermally contracts* to seal the join. Improved sealing can be achieved by using an even more ductile *gasket* as an intermediate layer, and ensuring that plastic flow only occurs in the *interlayer*, while the ductility of the tensile sleeve serves only to ensure adequate *toughness* of the assembly and guard against brittle failure.

For glass components, three alternatives to a simple mechanical assembly are possible:

1. *The use of a sealing wax or adhesive*. Adhesives can be regarded as the *non-demountable* alternative to the use of a *grease* in the assembly of ground joints, with sealing waxes an intermediate possibility. In the case of a *wax* the joint has to be

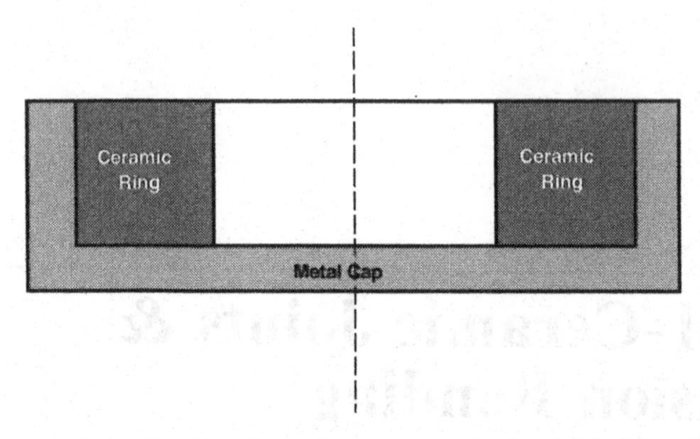

Fig. 8.1. A *ceramic* bearing-ring clamped in *compression* by shrink-fitting the ring into a heated metal cup.

heated above the *softening point* of the wax, while the solid joint can be disassembled by *reheating.*

2. *Soldering using an appropriate alloy composition.* Indium alloys are particularly effective in the soldering of *glass*, both to other glasses and to some metals (particularly copper and its alloys). The *soldering alloys* have low melting points (typically in the range 70 to 150°C), so that thermal shock of the glass is not a problem, and bond strongly to the glass, reflecting the strength of the *oxide bond* with indium.

3. *Using a high melting point metal of matched properties.* In this case bonding is achieved *above* the softening point of the glass, but *below* the melting point of the metal, while the *thermal expansion coefficient* of the metal is matched, within reasonable limits, to that of the *glass* or *ceramic.*

The *engineering requirements* for a bond of adequate strength between *elastically stiff* and *brittle* components may vary widely, but must include those requirements dictated by *geometry* (shape and dimensions), as well as those determined by the *operating temperature* and the *environment* (which might also include vacuum requirements). There is no such thing as an 'ideal' bond, but in a later section *transient liquid-phase* (TLP) bonding is described. In TLP joints the microstructure of the bonded region, after completion of the liquid to solid transition, is practically indistinguishable from the microstructure of the bulk components. Unfortunately, there are relatively few metallurgical systems whose physical properties lend themselves to the TLP process, and in the present chapter we seek a more general solution to the problem of forming a *stress-free*, high-integrity bond. In what follows, we first consider the bonding of *similar* components, having comparable composition and microstructure, and then introduce the bonding of *dissimilar*, metal and ceramic components, with a pronounced mismatch in properties across the interface.

For *similar* materials it is *sometimes* possible for the discontinuity at the bonded interface to be all but undetectable, both in its *properties* (chemical, physical and mechanical), and also in the *microstructure* (composition, phase content and morphology). For *dissimilar* materials there is a discontinuity in the properties of the two components to be joined, so that a *transition* at the interface is inevitable. In this second case the basic engineering objective is for a *smooth transition* between the properties of one component and those of the other, and the avoidance, as far as practicable, of any abrupt interfacial discontinuity. The bonded interface then becomes a *transition zone*, in which the relevant properties change gradually from those of one component to those of the second component.

The first task is to specify those properties which determine the integrity of the joint. The properties commonly considered when discussing the feasibility of joining a pair of dissimilar materials are the *elastic moduli* and the *thermal expansion coefficients*, and further simplification is then achieved by choosing just the *tensile modulus* (ignoring any mismatch in Poisson's ratio), and an *average thermal expansion coefficient* (ignoring variations in thermal expansion coefficient with either temperature or crystallographic orientation).

The options for approximating a *smooth transition* in properties across the interface region can be analysed in a number of ways, but a clear appreciation of the *geometrical* possibilities is a prerequisite. Figure 8.2 illustrates three of these options. The simplest is to introduce one or more *interlayers* of intermediate properties, so that the transition is accomplished in a sequence of steps (Fig. 8.2a). An alternative approach is to use a *surface profile* (Fig. 8.2b), to introduce a measure of *mechanical interlock*, increase the *area of contact* between the components, and establish bonding over a region of controlled thickness. Both the *periodicity* (wavelength) and the *amplitude* of the surface profile are important, since they control the balance between the local *stress concentration* associated with the profile and the improvement in the strength of the bond associated with the increase in *contact area* (compare the geometry of a *lap-joint*, Section 9.3). Finally, in some cases it has proved possible to *blend* the components of the two materials being bonded to provide a *continuous* transition region (Fig. 8.2c). Such a region is the distinguishing feature of a *functionally graded material*, in which the *planar* interfacial discontinuity is replaced by a *transition zone* over which the properties (and microstructure) change continuously. The *dimensions* of the transition region have to be chosen carefully in order to reduce the gradient in the critical engineering properties to an acceptable, safe level for the *in-service* performance requirements.

In what follows we will explore some of the ways in which a reasonable approximation to the above joining principles can be implemented, both for similar and for dissimilar ceramic-ceramic and metal-ceramic bonding pairs, despite the extreme mismatch in properties which is commonly observed in the engineering properties of ceramic and metal components.

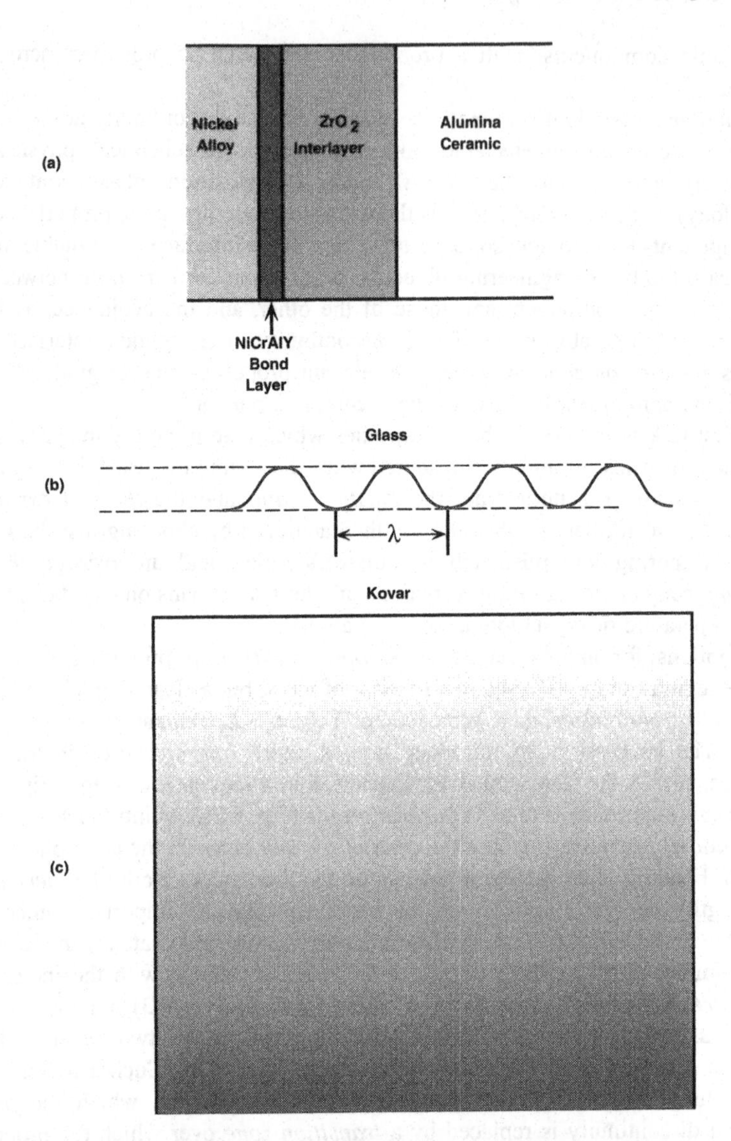

Fig. 8.2. Some strategies for bonding *unlike* materials. (a) *Plasma-spray* deposited thermal insulation for a nickel alloy component is built up in stages. A thin bonding layer of an *Ni-Cr-Al-Y* alloy is first laid down. This is followed by a *zirconia coating* (which has a thermal expansion coefficient similar to the bond coat). Finally, *alumina* is deposited to the thickness required to maintain the temperature gradient. (b) For some systems controlled *surface roughness* can serve to improve the strength of the bond, despite the local stress concentration associated with the amplitude of the roughness. (c) *Graded coatings* are the subject of current development. The phase content at the joint varies *continuously* over a controlled distance. The scale of the *microstructure* and the thickness of the *graded interlayer* are selected to reduce the in-service failure probability to an acceptable value.

8.1 THERMOELASTIC MISMATCH & MATERIAL ANISOTROPY

Thermoelastic mismatch between components has already been identified as a major source of undesirable *residual stress* in the bonding of brittle ceramics and other components which have a *low compliance*. Table 8.1 lists the *coefficients of thermal expansion* of some common engineering materials. These coefficients of thermal expansion are *average* values for a specific range of temperature. Figure 8.3 shows the thermal expansion coefficient of *alumina* as a function of temperature. Clearly, the value of this coefficient which is appropriate for a particular application will depend on the range of temperature of interest, and may easily vary by a factor of two. From Table 8.1, values for the thermal expansion coefficient of *metal* alloys vary from about 20 to $10 \times 10^{-6}\ °C^{-1}$, with *aluminium* alloys at the high end of this range, and *stainless steels* and the refractory metals at the low end of the range. Some transition metal alloys may have exceptionally low expansion coefficients outside the normal range: *Kovar* (54%Fe, 29%Ni, 17%Co) has an expansion coefficient matched to specific glass compositions over the appropriate range of temperature required for bonding the metal to glass. *Invar* (64%Fe, 36%Ni) has an expansion coefficient of approximately zero over a range of temperature from −100 to +200°C, and has been used for the manufacture of measuring standards.

Glasses, also, have coefficients of thermal expansion which can be controlled over a wide range through control of the composition, and certain *glass ceramics* can have a negligible thermal expansion. Most of the latter are based on *lithium aluminium silicate* compositions, and precipitate a crystalline, ceramic phase on heat treatment (hence the term *glass ceramic*). Glass ceramics may be either opaque or

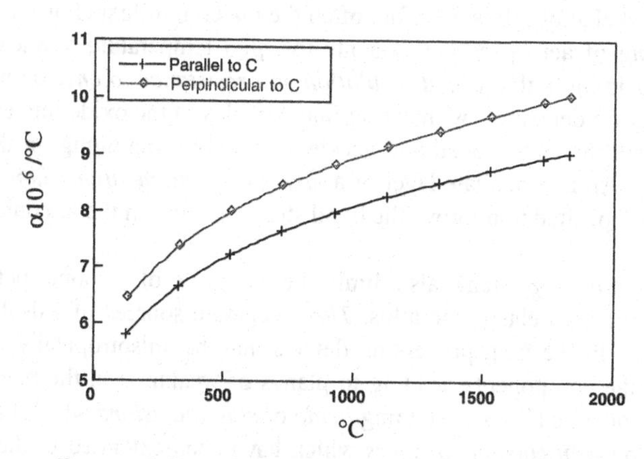

Fig. 8.3. The *coefficient of thermal expansion* is frequently a strong function of both orientation and temperature. The curves show the temperature dependence of the thermal expansion coefficient for an *alumina single crystal* (sapphire) *parallel* and *perpendicular* to the axis of the rhombotedral unit cell.

Table 8.1. Coefficients of Linear Thermal Expansion for Selected Pure Metals at 25°C.

Metal	α ($\times 10^6$) [°C]$^{-1}$
Al	25
Cr	6
Co	12
Cu	16.6
Au	14.2
Fe	12
Pb	29
Mg	25
Mo	5
Ni	13
Pt	9
Si	3
Ti	8.5

transparent. Some compositions have found important applications in ovenware, where the absence of thermal expansion makes them immune to thermal shock, associated with *transient temperature gradients* due to rapid heating and cooling. Other oxide ceramics may also have exceptionally low coefficients of thermal expansion. Both *mullite* (aluminium silicate) and *cordierite* (magnesium aluminium silicate) come into this category.

By contrast, *zirconium oxide* has both a higher thermal expansion coefficient and a lower elastic modulus than most ceramics, closely approximating those of some steels and nickel alloys. It is therefore often the material of first choice in the *plasma-spray coating* of aerospace components for high temperature applications. These coatings providing both *thermal insulation* and *resistance to environmental attack.* The coatings are deposited by injecting fine particles of the oxide into an electric arc plasma, where they are melted and accelerated before impinging on the surface of the substrate. An intermediate layer of a *nickel–chromium–aluminium* alloy, a *bond coat,* is first deposited to improve the bond strength between the zirconia coating and the substrate.

Material anisotropy may also limit the integrity of a bond between brittle components of high elastic modulus. *Three* separate sources of anisotropy may be distinguished. In the *first*, processing defects may be anisotropically distributed in the bulk of the components, leading to planes of weakness in the material. This is common in metallic alloys containing *oxide* or *slag inclusions* which have been hot-rolled, leading to *inclusion stringers* which have been extended in the direction of rolling, with a resultant loss of ductility in the *transverse* direction.

The *second* form of anisotropy is related to the lattice orientation of the *grains* in crystalline materials. Many engineering alloys develop strong *preferred orientation,* also known as *crystalline texture*, during processing. For example, during *plastic forming* there is a strong tendency for the active slip vectors in the crystals to align *parallel* to the direction of maximum tensile stress. This leads to a strong *cold-worked texture*, which may be observed by studying the diffracted X-ray intensity for the different reflecting planes as a function of the sample orientation. *Hot-working* the material leads to alternative crystalline textures, as does *recrystallization* after cold-work. In all cases the preferred orientation has an appreciable effect on *all* the properties of the material: physical, chemical and mechanical. In particular, both the *elastic modulus* and the *tensile yield strength* are affected by preferred orientation, resulting in an appreciable variability of the elastic compliance and plastic response of the components to be bonded.

The *chemical activity* of the surface also depends on the crystal orientation, with the close-packed surfaces of a metal generally less active than other orientations. It follows that *preferred orientation* can affect *wettability*. But perhaps the most important effects arise from *thermal expansion anisotropy* in materials having non-cubic crystal symmetry. In this case the *residual microstresses* arising from thermal expansion anisotropy are in competition with the *residual macrostresses* at the bonded interface. Simple analysis of the residual stress state is then invalid.

Thirdly and finally, microstructural anisotropy may arise from the *grain shape* and its distribution in the material. If the grains can be approximated as ellipsoids of revolution, then the *aspect ratio* may be defined as the ratio of the dimension normal to the plane of mirror symmetry to that in the symmetry plane. An *oblate spheroid* would then have an aspect ratio *less* than unity, while an *ellipsoid*, formed by rotation about the major axis, would have an aspect ratio *exceeding* unity. By extension of this definition, *plate-like* grains have aspect ratios *less* than one, while *needle-like* whiskers have an aspect ratio *greater* than one. On a micrographic section through the material, only the *apparent aspect ratio* can be measured, based on the maximum and minimum caliper dimensions of each grain section. It is not possible to determine from the sample section whether the 'elongated' grain is a section through a platey or a whisker-like grain, so that in *two dimensions* the distinction between plates and whiskers is lost. In other words, $R \equiv 1/R$, where R is the *apparent aspect ratio*.

When whiskers and platelets are *aligned* in the microstructure of a brittle material, they will interact strongly with a propagating crack, tending to *deflect* the crack into the plane containing the whiskers or platelets. As a result, *microstructural anisotropy* is of concern in the bonding of *brittle ceramics*.

In *ductile* metals and alloys elongated grains are generally the result of *plastic work* and their presence is often accompanied by appreciable *residual stress*. Attempting to bond the components in the cold-worked state is very likely to lead to shape changes which are related to the residual stress pattern, as much as to any microstructural asymmetry. The best way to improve *dimensional stability* is to *anneal* the components prior to bonding, and an *equiaxed* grain structure is to be preferred, at least in the plane of the bonded surface.

8.2 TRANSIENT LIQUID PHASE BONDING

One specific *isothermal* bonding technology, which has been developed for the high-performance superalloys, manages to approach the concept of an *'ideal'* bond: the bonded zone has engineering properties, composition and microstructure which closely approximate those of the components. The process is based on the use of *chromium* or *nickel interlayers* alloyed with several percent of *boron*, which drastically reduces the melting temperatures of the pure metals (Fig. 8.4). The rate of boron diffusion in the solid state is extremely rapid at temperatures near the melting point, since the boron is present in *interstitial* solid solution. The *Cr-B* alloy interlayer is chosen for bonding superalloys based on the *BCC* phase (chromium- and iron-based alloys), while the *Ni–B* interlayer is selected for the predominantly *FCC* superalloys (nickel- and cobalt-based alloys). The interlayer is compressed between the components to be bonded and the assembly is then heated in a protective atmosphere to a temperature above the melting point of the interlayer. The molten interlayer wets the components and the boron diffuses rapidly into the bulk components. As the boron diffuses away, so the melting point of the molten interlayer rises, until it exceeds the bonding temperature. The whole joining process occurs *isothermally* and is known as *transient liquid-phase (TLP) bonding*. The residual boron dissolved in the surrounding metal is present in very small concentrations, and may help to restrict unwanted grain growth at the receding liquid–solid interface. The final *microstructure* is generally indistinguishable from that of the bulk components, and the mechanical properties closely approximate those of the bulk.

8.3 GLASS–METAL SEALS

Glass–metal bonding has a long history, rooted in *craftsman skills* associated with the decoration of glassware and the production of silvered mirrors in the 15th and 16th centuries. Development of these technologies has been continuous, leading to the preparation of electrical lead-throughs for experiments on electricity and the generation of 'cathode' rays at the end of the 19th century. The *geometry* of the bond between the metal and the glass is a major factor in determining the mechanics of bonding, and a distinction should be made between the essentially *planar* symmetry associated with the *coating* of glass and the *cylindrical* symmetry associated with the *bonding* of a thin wire, through the wall of a glass-vessel. *Silvered mirrors* can be prepared by a purely *chemical* process: the deposition of finely divided metallic silver onto clean glass from a solution of silver nitrate under controlled conditions of pH and temperature. Silver bonds well to glass, forming *oxide bonds*, but must be protected from further oxidation by a paint layer of an inhibitor. Silver has the highest reflection coefficient of any metal in the visible range of electromagnetic radiation, and the result is a high quality mirror. *Gold* is also readily bonded to glass,

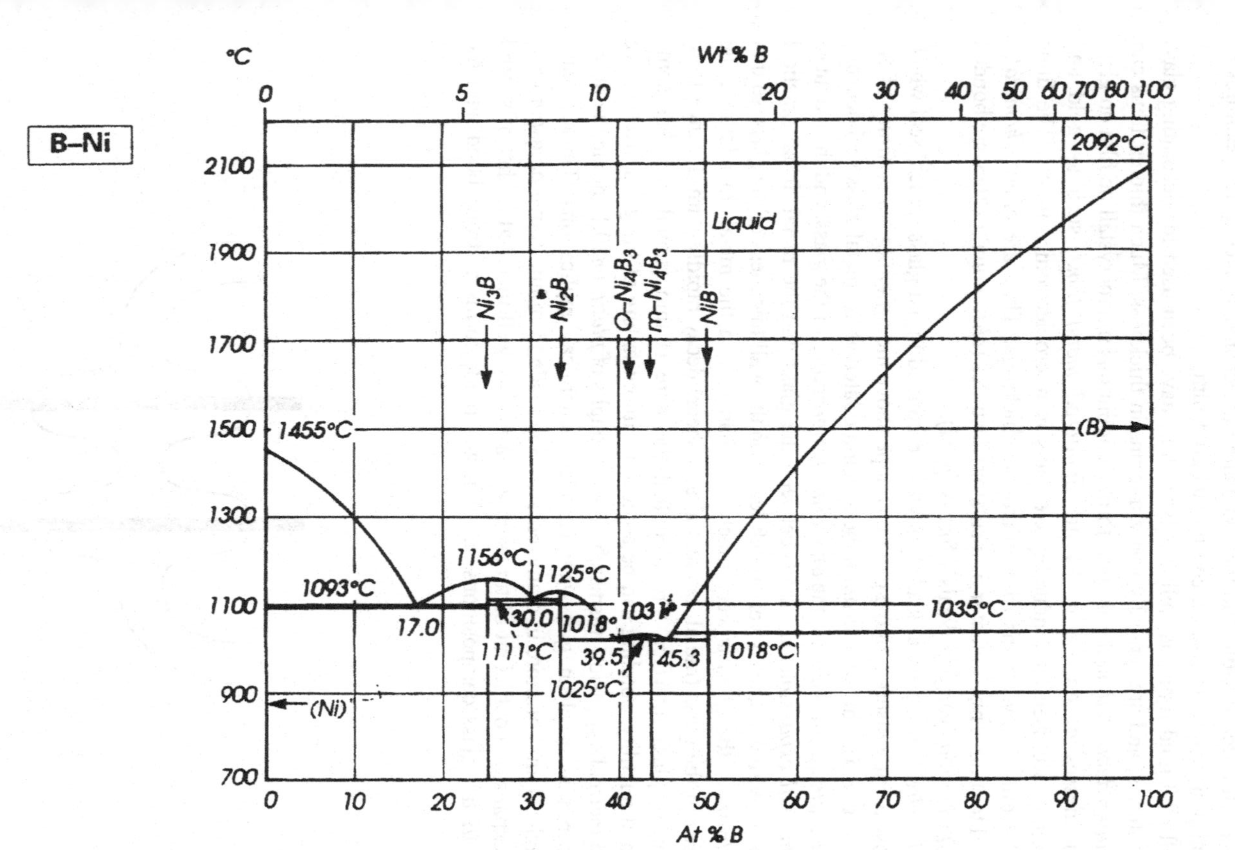

Fig. 8.4. The *nickel–boron* binary phase diagram. The addition of the non-metallic, interstitially soluble *boron* drastically reduces the melting point of nickel. Reproduced with permission from *Smithells Metals Reference Book*, Brandes and Brook (1992). Butterworth-Heinemann Ltd.

but in this case the film is generally formed by painting the clean glass surface with a *colloidal solution* of gold and then annealing the coated surface at temperatures of several hundred °C to form a continuous gold film.

Colloidal gold films, as well as silver films, have been used for decorating glass for centuries, and are typically many microns in thickness. Much thinner films can be *sputter-deposited*, and gold will form a continuous microcrystalline film on glass at thicknesses of only 5 to 10 nm. *Sputtered metal films*, such as gold and chromium, are used as conducting elements in microelectronic circuits. The films may be only a few tenths of a micron in thickness. They are either *chemically* bonded to their semiconductor or insulator substrates (*chromium*) or they are bonded by *polarization* (van der Waals') forces (*gold*).

Thin wires for electrical lead-throughs are bonded into a glass 'pinch' seal which is subsequently inserted into a glass envelope above the *softening temperature* of the glass (Fig. 8.5). The *total strain* to be accommodated at the metal/glass seal depends on the *diameter* of the wire, the *annealing temperature* of the glass, and the extent of the *thermal expansion mismatch* between the glass and the metal. In all practical cases the wire diameter is very much less than the wall thickness of the glass, so that the *strain* in the glass is very much less than that in the metal. In addition, the *stresses* in the glass (radial, tangential and shear) decay rapidly with distance from the metal–glass interface. Provided that the wire is thin enough and has some minimal ductility, bonding can be achieved without failure of the glass, irrespective of the thermal expansion mismatch. Lead-throughs of *platinum*, 0.1 mm in diameter, are common in clean glassware for electrochemical applications. Thicker lead-throughs, 1 mm or more in diameter, are often made from hot-drawn *tungsten* wire. The expansion coefficient of tungsten is reasonably well matched to that of several *borosilicate* glass compositions and tungsten oxide forms a strong bond with the glass.

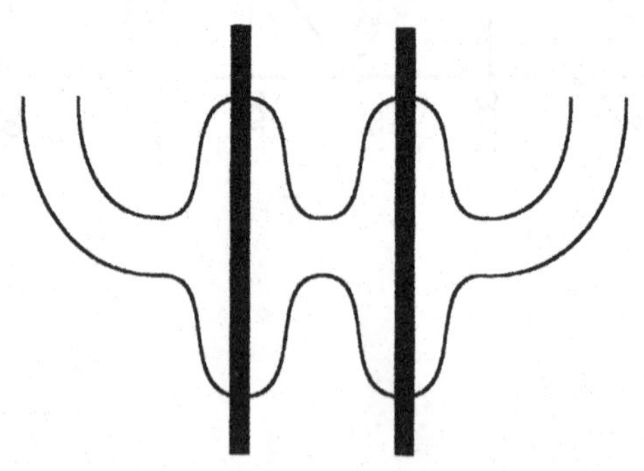

Fig. 8.5. *Electrical feed-throughs*, consisting of a metal wire embedded in a glass envelope, are familiar from the *glass pinch* in an electric light bulb.

In many cases, however, the seal geometry and dimensions are such that a much closer match of the thermal expansion characteristics is required, and this can only be achieved through the use of *graded seals*.

8.3.1 Graded Seals

Graded seals reduce the level of the *thermal stress* by spreading the mismatch over a sequence of bonds, using materials whose *thermoelastic* properties differ incrementally between each couple in the sequence (Fig. 1.5). It follows that the bond is now far from being a planar interface, and that the region of the bond may take up considerable space. Indeed, it is common for the graded seal between tubes of *stainless steel* and *borosilicate glass* to have linear dimensions of the order of the tube diameter, compare Fig. 8.6.

If the two end members of the bonded materials in a graded sequence differ in *thermal expansion coefficient* by $\Delta\alpha$, then the *total strain mismatch* to be accommodated is $\Delta\alpha\Delta T$, where ΔT is the difference between the annealing temperature of the glass and room temperature. For *borosilicate glass* and *stainless steel* the appropriate values are about $5 \times 10^{-6}{}^{\circ}C^{-1}$ for $\Delta\alpha$ and 850°C for ΔT, leading to a total strain mismatch of just over 0.4%. With a tube diameter of 50 mm this amounts to a displacement of over 0.2 mm which has to be accommodated over the length of the graded seal.

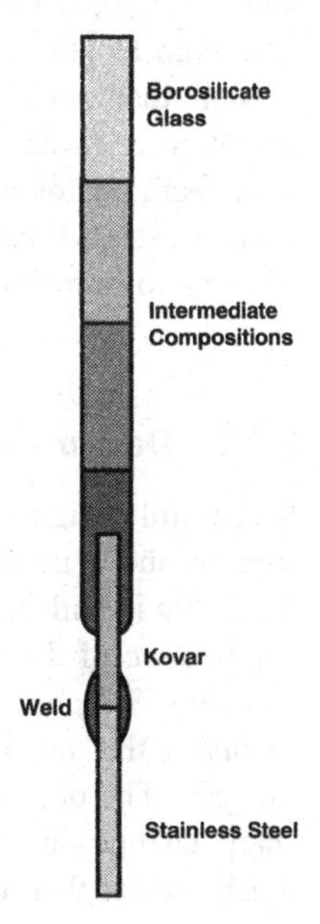

Fig. 8.6. Each component of a *borosilicate glass–stainless steel seal* has properties that differ slightly from those of the neighbouring components on either side. The dimensions are chosen to ensure that this mismatch is insufficient to cause failure.

The *metal* components of the seal have a much higher *toughness* than the *glass* components, so that it is possible to accommodate a large part of the mismatch in the *weld* between a Kovar sleeve and the stainless steel tube. Let us assume that the *shear stresses* developed by the mismatch arise from a *uniform* shear strain in the graded seal. If we further assume an *average* value for the elastic shear modulus, say 100 GPa, then the *average* shear stress in the graded seal corresponding to a 0.2 mm shear displacement is $\tau = 100 \times 0.2/L \times 10^3$ MPa, where the length of the seal L is in mm. Glass has a poor shear strength, and inserting an optimistic value of 50 MPa for τ_f, the shear strength, we obtain a *minimum* value for L of 40 mm, that is, of the order of the *tube diameter*, as noted above.

More realistically, the Kovar sleeve has a *thermal expansion coefficient* of about 5×10^{-6}°C^{-1}, compared to about 10×10^{-6}°C^{-1} for stainless steel. It follows that the largest residual shear stress is developed in the *weld zone* between the Kovar sleeve and the stainless steel. Kovar has a ductility of the order of 30%, and *plastic deformation* of the Kovar should reduce the residual stress at the weld to the level of the yield strength of Kovar, typically 400 MPa or less. The *residual strain* which remains to be accommodated arises from a $\Delta\alpha$ of the order of only 1×10^{-6}°C^{-1}, and corresponds to the expansion mismatch between Kovar and glass. That is, less than 0.1% strain for a temperature difference of 850°C. Assuming the *same* elastic shear modulus (100 GPa) and shear strength (50 MPa) as before, we now obtain a minimum L of 10 mm for the required linear dimensions of the *glass* components in the graded seal. To this length must be added the *length* of the Kovar sleeve and the *width* of the welded zone in the stainless steel, probably another 10 to 15 mm. The total, 25 mm, is still appreciably less than the previous estimate of 40 mm for the minimum length of a graded seal on a 50 mm diameter tube.

More than one intermediate glass composition is commonly required to make the transition from Kovar to borosilicate glass. In practice, the graded seals which are manufactured for assembly consist of a *Kovar sleeve* (ready for *arc welding* to stainless steel) bonded to a *graded glass* component (ready for assembly by *glass blowing* to borosilicate glassware).

8.3.2 Design & Geometry

Successful design of *glass–metal* seals is based on the avoidance of residual *tensile* stresses and is accomplished at two levels. On the scale of the components, *elastic flexibility* is built into the system to ensure that any misfit is accommodated by the *compliance* of the system. *Metal bellows* and *glass spirals* or *helices* accomplish this (Fig. 8.7), although always at the cost of increases in the space requirements (*volume*), the wall area of the system (*surface*) and the path length within the system (*length*). The objective is always to place the *glass* in mild *compression* and avoid sharp changes in cross-section. *Glass* components always have wall thicknesses much greater than those of the *metal* components, which are readily encastrated in

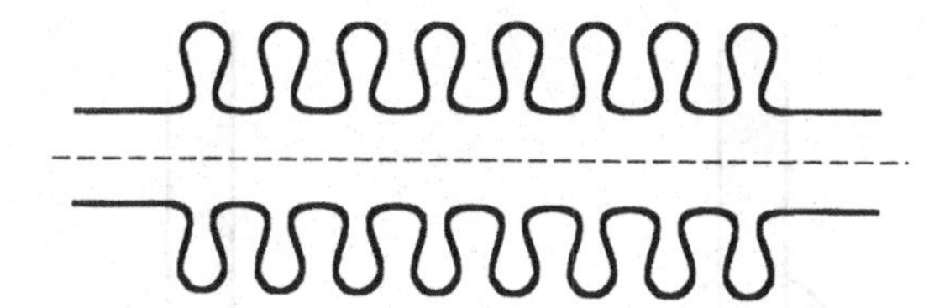

Fig 8.7. *Strains* associated with either *mechanical mismatch* (tolerances) or mismatch of the *physical properties* (thermal expansion coefficient) can frequently be accommodated by incorporating a component of *high compliance* in the system, such as a metal bellows or helically wound glass spiral.

the glass. A good metal/glass seal is *symmetrical*, has no *reentrant angles*, and has a length to diameter ratio of the order of *2 to 4* (Fig. 8.8a). Poorly constructed seals are asymmetrical, have reentrant angles or have an insufficient bond length for the thickness of the glass (Fig. 8.8b–d).

The same criteria apply to *electrical lead-throughs*. The diameter of the lead-through should have a cross-section much less than that of the glass through which it passes, while the length of the glass pinch should be about 4 times the thickness of the glass envelope (Fig. 8.9). Lead throughs, including *multiple lead-throughs*, are supplied bonded to a suitable intermediate *glass pinch* for direct insertion by glass-blowing into a glass vessel (Fig. 8.5).

8.4 DIFFUSION BONDING

Solid state bonding processes are based on the combined application of *pressure* and *temperature*, and involve *thermally activated creep deformation* of one or other of the components, usually over extended *times*. All such processes are termed *diffusion bonding*. In addition, both *friction welding* and *explosive welding* are, strictly speaking, solid state bonding processes. Friction welding relies on *frictional heating* to raise the temperature at the interface to the bonding temperature. Very extensive *plastic flow* makes this process akin to hot-forging rather than creep.

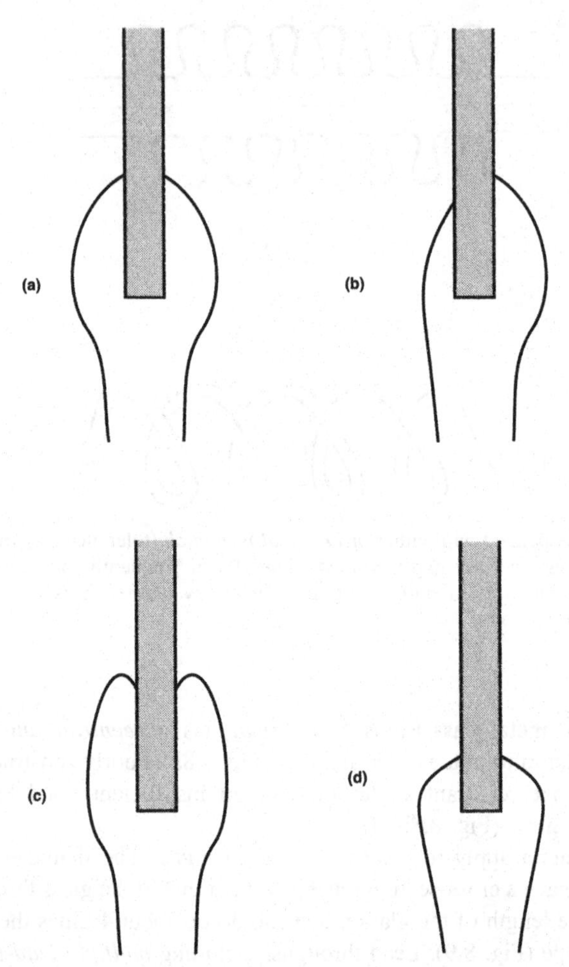

Fig. 8.8. (a) A *metal–glass seal* should be *symmetric* with no reentrant angles and of sufficient length to ensure adequate load transfer from the metal to the glass component. (b) An *asymmetric* seal. (c) Incomplete wetting by the glass results in a *reentrant angle*. (d) Too short a *sealing length* also reduces the strength of the joint.

Explosive welding, by comparison, takes place in a fraction of a second. The periodic pattern of intense *hydrostatic shear* which occurs in the bond zone is largely determined by *shockwave physics*, although *adiabatic heating* plays an important part in controlling the final microstructure of the bond.

Diffusion bonding processes depend for their success on a combination of three factors:

1. Absence of *contamination* at the mating surfaces and an adequate *surface finish*. The surfaces must be cleaned and, in many cases, polished before bonding.

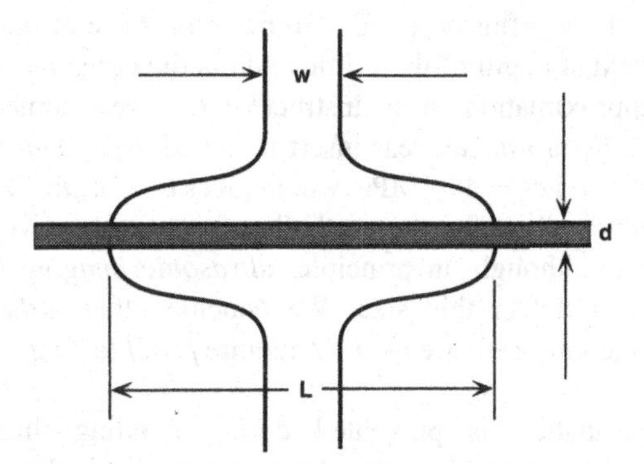

Fig. 8.9. The mechanical performance of an electrical feed-through in a *glass pinch* depends on the relative dimensions of the components, principally the *wire diameter d*, the *thickness* of the glass *w* and the *length* of the seal *L*.

2. The ability of at least one component to undergo sufficient *plastic flow* (creep) in order to develop complete contact across the interface between the two components.
3. Sufficient time for *diffusion* to occur in the interface region in order to eliminate microstructural instability and establish an adequate bond strength.

8.4.1 Thermal Mismatch

To these three factors should be added the essential condition that thermal stresses in the assembled joint should be *minimized* through adequate matching of the thermal expansion characteristics of the components over the temperature range from room temperature up to the temperature required for diffusion bonding. When one component is *ductile*, the level of residual shear stress at the interface will be of the order of the yield strength of the ductile component. If the second component is *brittle*, then the fracture strength of the bond may be reduced to a value given by: $\sigma_b = K_c/\sqrt{\pi c} - \sigma_y$, where K_c is the *fracture toughness* of the brittle component, c is the dimension of the largest inherent *interfacial defect* and σ_y is the *yield strength* of the ductile component (the residual stress in the joint). *Elastic mismatch* between the two components will affect this relation (by determining the extent of *mode mixity*), while the actual *bond strength* will depend on the *loading* (tensile or shear, uniaxial or biaxial), but the equation does give a clear indication of the likely influence of the three parameters: *defect size, fracture toughness* of the brittle component and *yield strength* of the ductile component.

Since there is no point in making the bond strength greater than the yield strength of the ductile component, we can set $\sigma_b = \sigma_y$, to derive a criterion for ensuring that the bond strength is *equal* to this yield strength: $K_c/\sqrt{\pi c} = 2\sigma_y$. In practical terms this reduces to $\sqrt{\pi c} < K_c/2\sigma_y$, which is less than the *maximum* allowed size of interfacial defect if the diffusion bond is to make *best* use of the respective fracture toughness and yield strength of the brittle and ductile components. Although this is a very rough approximation, it is instructive to insert some values, say those appropriate to a *silicon nitride* wear insert mounted on a *heat-resistant steel* base: $K_c = 8$ MPa $m^{-1/2}$ and $\sigma_y = 400$ MPa, which gives $c < 32$ μm. This is on the border line of the *detection limit* for the methods of non-destructive evaluation (NDE) currently available, although, in principle, *ultrasonic imaging* (C-scan) should be able to visualize a defect this size. We conclude that *diffusion-bonded* joints involving a *brittle* component are likely to require *proof testing*—NDE alone may be insufficient.

Surface contamination is prevented during bonding through a *controlled atmosphere*, usually a *reducing* atmosphere (wet or dry hydrogen, or a mixture of 96% nitrogen with 4% hydrogen—'forming gas'), although bonding is also performed in *vacuum* or dry *argon.*

8.4.2 Surface Preparation

All surfaces must be adequately prepared before bonding, and this includes both *chemical cleaning* (as well as degreasing with an organic solvent) and *mechanical abrasion.* A variety of abrasives are in use, ranging from sand (*silica*), used to *sandblast* metal components, to *diamond powder* pastes, use in submicron mechanical polishing operations. The commonest abrasive is probably *silicon carbide.* The particle *grit size* used for mechanical abrasion determines the *surface roughness* of the abraded components, and there is considerable variability in the *micromechanics* of abrasion, depending on the ductility of the surface, the hardness of the abrasive, and the type of carrier used for the abrasive during the abrasion process. The least ambiguous definition of the particle *grit size* is in terms of the *sieve dimensions* which bracket the range of particle sizes involved: that is, a *360/600 grit* will pass through the coarse sieve (360), but be retained by the finer sieve (600). The numbers themselves refer to the number of *mesh holes per linear inch* in the sieve.

The four terms *erosion, abrasion, lapping* and *polishing* refer to four distinct mechanisms for achieving a surface finish. Sandblasting is an *erosion* process in which the kinetic energy of the particles carried in an air jet does mechanical work when the particles strike the surface, gouging out and breaking away particles of the workpiece. It is commonly used to clean away rust and dirt from steel surfaces before welding. *Abrasion* makes use of grit particles which have been bonded to either a flexible or rigid *substrate.* The flexible substrate may be either *cloth* (for the coarser particles) or *paper* (for the finer grit sizes), as in the flexible silicon carbide grit papers commonly employed for metallographic preparation. Alternatively, a diamond grit may be bonded in a metal grinding wheel. The grinding wheel retains its *cutting efficiency* by the periodic fracture and pull-out of the grit particles, exposing fresh cutting surfaces. Most *abrasion* processes employ a fluid *coolant*, which may be water, paraffin or a commercial machining fluid (an emulsion).

Lapping processes differ from abrasion in that the grit particles are free to *rotate* on the carrier substrate, which is commonly a cast iron disc rotated with respect to the workpiece. Once again, new cutting surfaces are continuously exposed, this time by the rotating particles, but, for a given grit size, the *depth* of cut is generally reduced compared to abrasion, resulting in a slower *rate of material removal*, less *subsurface damage* and a better *surface finish*. The surface density of grit particles trapped between the workpiece and the lapping substrate has a large affect on the *lapping efficiency*. In general, the grit particles should be sufficiently widely-spaced so as not to interfer with their mutual freedom of rotation, although too few grit particles will drastically reduce the rate of lapping. The *optimum loading* of the substrate surface corresponds to perhaps 60% coverage. A common fault in lapping is the overloading of the lapping surface with excessive quantities of grit, resulting in both a poorer surface finish and wastage of the lapping compound. *Diamond* is the preferred material for the finer grits used in lapping.

Polishing refers to the preparation of a mirror finish, corresponding to a surface roughness below the wavelength of visible light (0.4 to 0.7 μm). The surface appears *shiny* and completely featureless to the unaided eye. *Mechanical polishing* requires *submicron* grit sizes, and *diamond* is usually the material of first choice. Polishing wheels may be based on *lapping* principles (allowing free rotation of the polishing grit), or they may employ a *cloth* (sometimes felted) to trap the particles and provide a flexible substrate. *Chemical polishing* employs chemical agents which form a *viscous layer* of the reaction products at the surface, resulting in the preferential attack of any protrusions. *Electropolishing* achieves the same result by forming a viscous layer of high electrical resistance at the anode, but is obviously only suitable for *metallic conductors*. *Mechanochemical* polishing is sometimes possible, in which a chemically active cooling fluid is used with the the polishing grit.

8.4.3 Plastic Flow

Diffusion bonding requires yielding and *plastic deformation* of at least one component to reduce *residual porosity* at the interface and increase the *true* contact area at the join until it is equal to the *nominal* area of contact: that is, porosity has been eliminated. As the original contact points develop and link up during the diffusion bonding process, the local stresses *decrease* in inverse proportion to the *true area of contact*. The rate of plastic flow in the region of the contact points is determined by the local stress and the bonding temperature, and will also decrease as the true contact area increases.

Ashby has identified the principal *mechanisms of plastic flow* in metals, alloys and ceramics, together with the governing equations for time-dependent plastic flow. Quite generally, little *creep relaxation* is expected to take place at temperatures less than $0.6\ T_m$, where T_m is the *melting point* in K, while *diffusion bonding* generally requires temperatures of the order of *0.8* T_m to ensure that sufficient plastic flow can occur (T_m is now the melting point of the *less refractory* component).

The *strain rate* generally increases exponentially with temperature, but follows a power law dependence on stress: $\ln(\varepsilon/\varepsilon_0) = -Q/RT$ (σ constant), $\varepsilon/\varepsilon_0 = (\sigma/\sigma_0)^n$ (T constant), where n is typically in the range 2 to 10.

Since the local stress *decreases* as the contact area *increases*, most of the *time* required for diffusion bonding is associated with the *final stages* of the bonding process, corresponding to the elimination of isolated, individual pores. In this final stage, the *radii of curvature* of the isolated pore provides a *hydrostatic driving force* for pore closure: $\Delta P = \gamma(1/r_1 + 1/r_2)$. *Two* populations of residual pores may develop: those *larger* pores which result from geometrical mismatch of the mating surfaces on a *mesoscopic* scale, and the *finer* pores which are present on a *microscopic* scale corresponding to the *original* surface topology. The former can be extremely difficult to eliminate. Somewhat paradoxically, the mesoscopic pores can often be reduced in scale and importance by employing a *coarser* grit for the final surface abrasion, in order to reduce the spacing of the original contact points.

It follows that optimum results are often obtained for a specific *surface roughness* (abrasive grit size) which *inhibits* formation of the *larger*, isolated pores, but still leaves behind *smaller* pores which are able to shrink under the influence of the hydrostatic *capillary pressure.* A corollary of this discussion of the influence of surface topology is that the *initial surface finish* may be a more important variable than the *applied bonding pressure*, since in the final stages of bonding the *capillary pressure* may exceed the applied pressure. *General yielding* at high temperatures (0.8 T_m) may occur at pressures *less* than 50 MPa, so that applied bonding pressures must be limited to of the order of 20 MPa, since the bonding pressure must be kept below the stress for general yielding, in order to avoid distortion of the components. A 600 grit surface finish is likely to give rise to residual pores of the order of *1* μ*m* in diameter, which for a surface energy of *2* $J\ m^{-2}$ will lead to a hydrostatic capillary pressure of *5 MPa*, an appreciable fraction of the bonding pressure.

8.4.4 Diffusion

Pore shrinkage depends on the rate at which material can be transfered in the solid state, and the *mechanism* of mass transfer determines the *activation energy* for the process. *Surface diffusion, boundary diffusion* and *bulk diffusion* are all possible contenders, as well as *pipe diffusion* along the line of a *dislocation core*. However, *boundary diffusion* is the most common limiting case, and involves the migration of *vacant lattice sites* from the pore surface into the boundary, where they are absorbed (the boundary acting as a *vacancy sink*). Shrinkage of the pores is thus accompanied by migration of the *centre of gravity* of the grain towards the plane of the *boundary* (Fig. 8.10).

The rate at which diffusion can occur is usually analysed in terms of Fick's two laws. The first controls the nett *flux of atoms* in a *concentration gradient*:

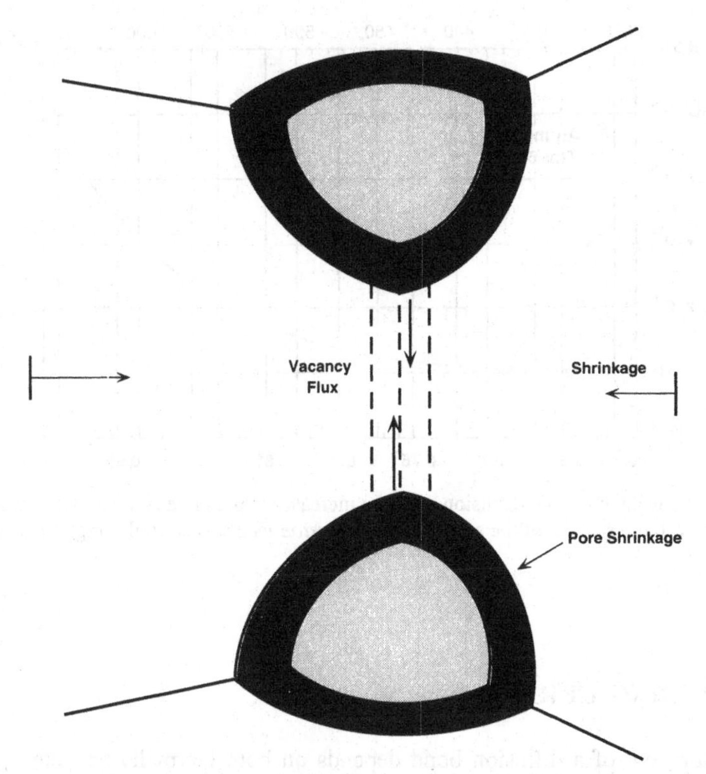

Fig. 8.10. The second stage of compaction by sintering is for the most part a function of the rate of *pore shrinkage*. Vacant sites at the pore surface are absorbed as *vacancies* into the grain boundaries. These migrate along the boundaries until they are annihilated, leading to a *decrease* in the distance between the centres of gravity of the grains.

$J_x = -D \cdot \mathrm{d}c/\mathrm{d}x$, where D is the diffusion coefficient. Fick's second law determines the *rate of change of flux*: $\mathrm{d}J/\mathrm{d}x = -D\mathrm{d}^2c/\mathrm{d}t^2$. A common result for planar diffusion is the *error function* solution in which diffusion distances are *normalized* by the parameter $2\sqrt{(Dt)}$. Bearing in mind that the diffusion coefficient is an *exponential function* of the temperature: $D = D_0 \exp(-Q/RT)$, the *normalized diffusion distance* must also depend sensitively on temperature. The *minimum* diffusion distances required to ensure pore closure will be some multiple of the pore diameter, while the *activation energy* will most probably approximate that for *boundary diffusion*. As in the previous discussion, it follows that *small* pores are readily absorbed at temperatures of the order of 0.8 T_m. Table 8.2 lists some values for the *diffusion coefficients* for both *boundary* and *bulk* diffusion, while Fig. 8.11 shows the value of the parameter $2\sqrt{Dt}$ as a function of the *homologous temperature* (T/T_m) for one typical case. As can be seen, it is not difficult to obtain diffusion distances of the order required for pore closure.

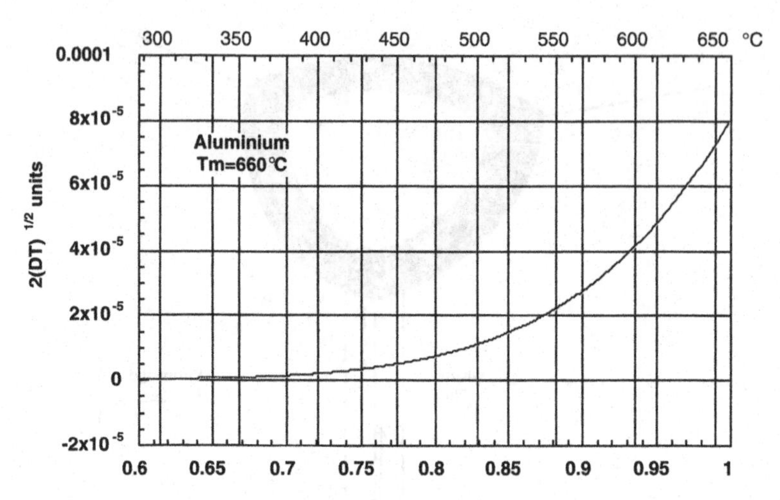

Fig. 8.11. The average atomic diffusion distance increases exponentially with temperature and depends on both the *species* of the atom and the *diffusion mechanism* (bulk or grain boundary, see text).

8.5 BONDING CERAMICS

The development of a diffusion bond depends on both thermally activated plastic flow and diffusion-controlled mass transfer. The growth of the contact areas in the *early* stages of bonding occurs by plastic flow of the more ductile component, while pore closure, corresponding to the *final* stages of bonding, is limited by the rate of diffusion, usually boundary diffusion. In the present section we will review some of the technological strategies which have been developed for *bonding ceramics*, as distinct from the scientific principles, which have been covered in the previous sections.

An important practical limitation of diffusion bonding technology is the absence of accepted *international standards*, both for the processes and for the *testing* of the finished joint. This may be compared to the present state of the art in *soldering, brazing* and *welding technologies*, for which national and international institutions have developed standards, not only for the *processes* and for the testing of the *joints*, but also for the *qualification* of the *welders*, who are responsible for maintaining the *reliability* and *reproducibility* of the final assembly.

This absence of standards is exacerbated by the *low fracture toughness* of ceramic components, which often reduces the size of the critical flaws responsible for brittle failure to below the resolution limit of *non-destructive evaluation*. The only alternative is then *proof testing* of the assembly, which must be performed on a *representative sample* of the product, very often every single assembly.

Table 8.2. Diffusion data for selected pure metals. $D = D_0 \exp\left(\frac{-Q}{RT}\right)$

Metal	Solute	Single or Polycrystal	Temp. Range [°C]	Q [kcal/mole]	D_0 [cm²/sec]
Al	Al	S	450–650	34.0	1.71
	Cu	S	433–652	32.27	0.647
	Ga	S	406–652	29.24	0.49
Cr	C	P	1200–1500	26.5	9.0×10^{-3}
	Cr	P	1030–1545	73.7	0.2
	Fe	P	980–1420	79.3	0.47
Cu	Ag	S,P	580–980	46.5	0.61
	Cu	S	698–1061	50.5	0.78
	Fe	S,P	460–1070	52.0	1.36
Au	Ag	S	699–1007	40.2	0.072
	Au	S	850–1050	42.26	0.107
	Fe	P	701–948	41.6	0.082
Fe(α)	Cu	P	800–1050	57.0	0.57
	Fe	P	809–889	60.3	5.4
	Ni	P	680–800	56.0	1.3
Pb	Ag	P	200–310	14.4	0.064
	Cu	S	150–320	14.44	0.046
	Pb	S	150–320	25.52	0.887
Mg	Fe	P	400–600	21.2	4×10^{-6}
	Mg	S	467–635	32.5	1.5
	Ni	P	400–600	22.9	1.2×10^{-5}
Mo	C	P	1200–1600	41.0	2.04×10^{-2}
	Mo	P	1850–2350	96.9	0.5
	W	P	1700–2260	100	1.7
Ni	Cr	P	1100–1270	65.1	1.1
	Fe	P	1020–1263	58.6	0.074
	Ni	P	1042–1404	68.0	1.9
Pt	Co	P	900–1050	74.2	19.6
	Cu	P	1090–1375	59.5	0.074
	Pt	P	1325–1600	68.2	0.33
Si	Cu	P	800–1100	23.0	4×10^{-2}
	Ni	P	450–800	97.5	1000
	Si	S	1225–1400	110.0	1800
Ti(β)	Cr	P	950–1600	5.1	5×10^{-3}
	Fe	P	900–1600	31.6	7.8×10^{-3}
	Ti	P	900–1540	31.2	3.58×10^{-4}

Proof testing assumes that failure in service can be avoided if the assembly is subjected to an *overload test* which results in failure of the *weakest* products. This is not completely true, since the very act of overloading during proof testing may result in *sub-critical crack growth* to the point at which a *second*, repeat proof test on the *same* component *could* cause failure. This is illustrated in Fig. 8.12. Figure 8.12a shows the expected distribution of *fracture strength* for an engineering assembly

Table 8.3. Diffusion data for self grain boundary diffusion.
$D = D_0 \exp\left(\frac{-Q}{RT}\right)$

Metal	Temp. Range [°C]	Q [kcal/mole]	D_0 [cm^2/sec]
Ag	350–480	20.5	0.12
Zn	75–160	14.3	0.22
Cd	50–110	13.0	1.0
Sn	40–115	9.55	0.06
Pb	214–260	4.7	0.81
Sb	370–548	23.1	5.87
Fe	530–650	40.0	2.5
Ni	850–1100	28.2	0.0175

whose failure is controlled by *Weibull statistics.* In general, there is no lower limit to the load below which the *probability of failure* falls to zero (in principle, a large enough flaw would cause failure close to zero load). The applied load *in service* is indicated by F_a, and would give rise to some *finite* probability of failure. *Proof-testing* at a higher load F_p will result in *rejection* of a proportion of the product and, in the first instance, eliminate these *'defective'* components from the distribution (Fig. 8.12b). However, the *weakest* remaining components will have experienced some additional damage (*sub-critical crack growth*) during proof-testing, *redistributing* the Weibull distribution for the remaining products, as indicated by the dotted line in Fig. 8.12b. The *Weibull modulus,* an inverse measure of the width of the failure probability distribution, is *increased* by proof testing, but a *finite* probability of failure below the applied load P_a still exists. A *second* proof test will further *increase* the Weibull modulus of the product, and hence further *reduce* the probability of failure in service (Fig. 8.12c) but only at the cost of further *rejections* of the weaker components.

Proof testing can be optimized, for example by first testing at a *lower* proof-stress and only then testing at the *higher* load level, but *acceptance criteria* will always remain a compromise, with the *expense of testing* to be set against the *cost of failure*. In many instances the *engineering requirements* specify exceptionally low failure probabilities because of the high *social* cost of failure. Two well-publicized examples are the sealing and disposal of nuclear waste and the reliability of space craft. Both these applications employ brittle *glasses* and *glass ceramics* for their assembly, and in both cases the acceptable probabilities of failure approach *zero*, leading to high costs of evaluation (approaching *infinity*!). This is a *political* issue in which the cost to the *tax-payer* should be weighed against the *social benefits* of the technology.

While adequate *national* and *international standards* are not yet available for *ceramic joining*, manufacturers certainly have *in-house* standards which offer some measure of *guarantee* to the customer, providing he or she is aware of the standards

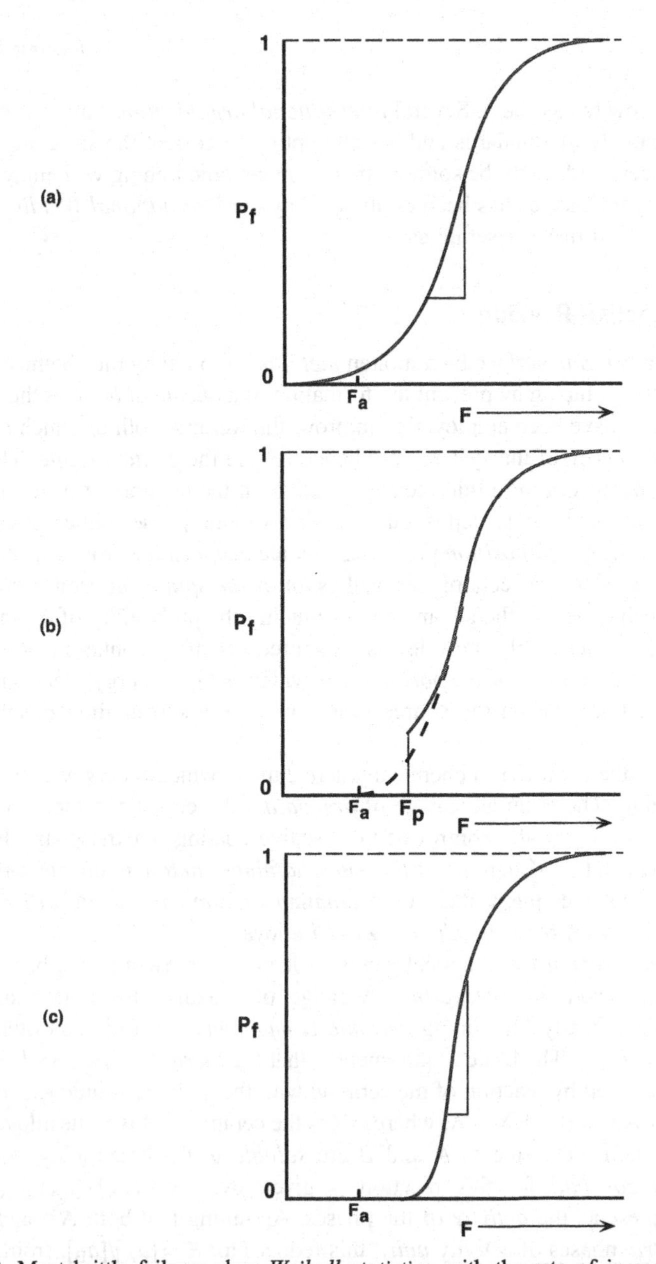

Fig. 8.12. Most brittle failures obey *Weibull statistics*, with the rate of increase in failure probability with load depending on the *Weibull modulus*, which is a measure of the *inverse* width of the distribution. (a) The shape of the curve suggests that a *proof test* at a load in the region of the tail of the distribution should eliminate the *weak* members of the sample population. (b) *Proof testing* at a load F_p eliminates the part of the population having strengths below the proof-stress cut-off, but damages some samples in the *remaining* population (dotted curve). (c) *Repeated* proof testing continues to *increase* the Weibull modulus of the remaining population, *reducing* the width of the probability function.

and knows how to use them. Several *international organizations* are now concerned about the paucity of standards and are attempting to correct the situation. This is a lengthy process and it will be some years before ceramic joining will enjoy the same reliability for the user as has been established by the *International Welding Institute* (IWI) for welded metal assemblies.

8.5.1 Reactive Brazing

Wetting of a *ceramic* surface by a molten *metal* is inhibited by the chemical stability of the ceramic, which may prevent the formation of a *chemical bond* at the interface. *Two* strategies have been employed to improve the wetting, both of which reduce the *surface free energy* of the system, and hence reduce the *contact angle*. The *first* of these is the deposition of a thin metallic coating on the ceramic prior to brazing. In most cases the coating is deposited from the vapour phase, either physically or chemically. *Physical deposition* processes include *evaporation* from a heated source and *condensation* on the ceramic, as well as *ion-beam sputtering* from a cold target. The latter technique has the advantage of reducing the probability of contamination from the source assembly. Low levels of source-assembly contamination are also achieved in *electron-beam evaporation*, in which a high energy electron beam is focused to a fine spot on the source target, while the surrounding metal remains unmelted.

In general, the coating is a chemically active metal which *reacts* with the ceramic during brazing. The main advantage of *precoating* the ceramic, rather than using a *reactive braze*, lies in the control of the reactive coating thickness, and hence the extent of the reaction. Coatings of *titanium, hafnium, tantalum* and *zirconium* have been used to enhance the wettability of *alumina, silicon carbide* and *silicon nitride* during brazing with both *Ag–Cu* and *Au–Ni* alloys.

In *reactive brazing* the chemically active element is present in the brazing alloy, obviating the need for *precoating*. A range of brazing alloys have now been developed specifically for joining *ceramic components*, both to each other and to *metallic substrates*. The basic requirement is that the *thermodynamic stability* of the compound formed by reaction of the ceramic with the additive should *exceed* that of the ceramic: $AX + B = BX + A$, where AX is the ceramic and B is the alloy additive. In general, both the elements A and B are *soluble* in the braze alloy, so that the *equilibrium constant* for this reaction is given by: $K = [a_{BX}] \cdot [a_A]/[a_{AX}] \cdot [a_B]$, where a represents the *activity* of the phases. Assuming that both AX and BX are *stoichiometric* phases of activity *unity*, this reduces to: $K = [a_A]/[a_B]$, from which it is clear that it is the *concentrations* of the two elements dissolved in the *braze* that control the reaction, together with the *equilibrium constant* for the reaction: $K = -RT\ln(\Delta G_0/RT)$, where ΔG_0 is the standard free energy for the reaction and R is the gas constant. *Titanium* has proved to be a most successful additive in ceramic brazes, both for brazing *alumina* and for brazing *silicon nitride* or *carbide*. The amount of titanium in commercial brazing alloys is of the order of 1 or 2%.

However, little attempt seems to have been made to control wetting by adding *aluminium* or *silicon* to the braze alloy, and commercial compositions appear to have a larger than necessary spreading force for wetting.

The *strength* of the joints obtained by *reactive brazing of ceramics* is frequently quoted in the literature without reference to the *method* or *condtions* of testing. Most of these tests were probably made in *four-point bending* on specimens of *standard size* prepared according to one or other of the *national specifications* for determining the *bend strength* of ceramics. *Bend strengths* of *less* than 100 MPa are considered poor, while those *exceeding* 200 MPa are considered good. However, the reported bend strength should not necessarily be taken at face value, for the following reasons:

1. The mismatch in *elastic modulus* across the interface commonly results in a *mixed-mode* fracture.
2. *Residual shear stresses* at the interface, associated with *thermal expansion mismatch*, reduce the bend strength.
3. Fracture is usually initiated from *surface defects* associated with the *surface finish* (grinding defects), unbonded *inclusions* (dross particles), or *residual porosity* (incomplete bonding).

It follows that a major problem in reactive brazing is the *reproducibility* and *reliability* of the joint formed, and scrupulous attention to surface preparation and the control of the brazing parameters are primary conditions for success.

8.5.2 Alumina-Metal Seals

Of all the high-performance ceramic–metal seals which have been developed, those associated with the sealing of *alumina* and its derivatives have received the most attention. One common example is the sealing of a *spark-plug insulator*, both to the high-current electrical conductor and into its metal socket. This technology, which has been developed over three quarters of a century, has also been adapted to high-voltage *vacuum lead-throughs* based on high purity, high density *alumina insulators* (Section 10.3.4).

A well-established technological solution to the bonding of alumina insulators is a *multistage* process based on *precoating* the alumina with a paste made from powders of *molybdenum, manganese* and their *mixed oxides.* On sintering at 1500°C in moist hydrogen, a continuous, *glassy phase* is formed, which bonds strongly to the alumina, while the *sintered metal* powders form a continuous, interlocking network within the glassy phase. In the next stage, this glass/metal *interlayer* is further coated with *nickel*, either by *electroplating* or by reducing a coating of *nickel oxide*, painted onto the first interlayer. Reduction is achieved by sintering in dry hydrogen at 950°C. The *final* stage of the process is the *brazing* of the now nickel-coated assembly to a *Kovar substrate*, using a standard *silver–copper* brazing alloy.

The complexity and cost of the above process have, in some cases, led to its replacement by *reactive brazing*, in which the Kovar is *directly* brazed to the alumina using a *titanium-containing* brazing alloy, as described in the previous section. Nevertheless, the intermediate, two-phase *glass/Mo–Mn* layer developed in the multistage process provides a *transition zone* which helps to reduce the *thermoelastic mismatch* across the interface and improves the *reliability* of the join. While *reactive brazing* can provide adequate strength, and has the potential to replace the multistage process, it may still be several years before the *reliability* of the multistage Mo–Mn process is seriously challenged.

8.5.3 Ceramic–Ceramic Joints

High performance *ceramic* components have limitations of both *size* and *shape* which are inherent in the *powder processing* technologies used for their fabrication. Since the components are for the most part shaped by powder technology, the *size* must be limited to dimensions compatible with the available presses. More seriously, size-dependent *residual stresses* in the *green* (unsintered) compacts are associated with the *friction* generated between powder particles as they rearrange during *compaction*. *Drying* of the green powder compacts also generates internal stresses associated with *capillary forces*, and these too depend on the *dimensions* of the compact. Finally, *sintering* is a process of *densification* which frequently involves large changes in *linear* dimensions (commonly over 10%). The *size* of the components which can be successfully sintered is consequently limited.

The same factors restrict the *shape* of the components which can be produced. Any *reentrant angles* are sources of *stress concentration*, which enhance not only the *applied* stress, but also the *residual stresses*. Complex shapes do not lend themselves to production by standard powder technologies. Some of these restrictions can be relaxed by employing ceramic compositions which contain a molten, *glassy phase* at the sintering temperature. The *internal stresses* are then relaxed by *viscous flow* of the glass. However *liquid-phase sintering* leads to a serious reduction in the high-temperature *creep strength* of the ceramic, which can only occasionally be compensated by *post-sintering crystallization* of the glassy phase.

Reaction-bonded ceramics show commercial promise. In these materials a *chemical reaction* occurs during sintering which both compensates for the sintering shrinkage and contributes to the strength of the sintered compact. *Reaction bonding* technologies have been developed for both *silicon carbide* and *silicon nitride*, as well as, more recently, for *alumina* and *mullite* (aluminium silicate). However, although reaction bonding may reduce the *costs* of *processing*, this technology has not made much impact on the *size* and *shape* limitations.

Most high-performance ceramic components have *simple* shapes with *few* reentrant surfaces, and *cross-sections* which are typically less than 1 cm in thickness. Exceptions (for example, armour plate) are produced in *thick* sections of several inches, but only by *hot-pressing* and at considerable cost. The only viable alternative seems to be to manufacture more complex structures by the *assembly* of simple components, and this requires methods for joining *ceramics* to *ceramics*. Some attempts have been made to *weld* ceramic components directly, by first *preheating* the components into the plastic range before applying a *localized heat source* (a laser beam, for example). Success has been very limited and there is little likelihood of *commercial* applications for the *direct welding of ceramics* at the present time.

A more promising technique is the use of *glass interlayers*. The glass can be applied as a *powder* mixed into a *paste* and subsequently *liquid-phase sintered*. Although little information is available on current commercial practice, there is no doubt that such methods are *scientifically* viable. Such a *glassy interlayer* can be employed as a *transient liquid phase* (Section 8.2). The small amount of glass is either *absorbed* in the residual porosity of the ceramic components (through *capillary attraction*), or undergoes *partial* or *complete crystallization* during cooling or subsequent heat treatment.

It is also possible to assemble the components in the *green state*. *Injection moulded* parts commonly contain 20 to 40 vol% of organic additives, largely in the form of a *polymer binder*. These binders are usually *thermoplastic*, so that components can be bonded together in the green state by raising their temperature to the *softening temperature* of the polymer binder. *Ultrasonic welding* of the green components is also possible, based again on the localized softening of the polymer binder. Once the green components have been bonded it is necessary to *remove* the binder, and this *'burn-out'* can be achieved by heating to temperatures of the order of 350 to 550°C. However, rather long times are required (often several days) for all but the thinnest cross-sections. Too *rapid* burn-out results in the formation of *gas bubbles* and *bloating* of the compact. The *unsintered* assembly is extremely fragile once the binder has been removed.

Some mention should be made of *superplastic forming*, which offers a potential route to the diffusion bonding of *ceramics* without the introduction of any interlayer. *Superplasticity* is a phenomenon that has been exploited commercially in the forming of some *fine-grained metallic alloys*. Plastic elongations of several hundred percent have been observed under *tensile load* at high temperatures. The same phenomenon has now been observed in many *fine-grained ceramics*, especially when grain growth is inhibited by the presence of a second phase. The most notable example is *alumina ceramics* containing up to 20% of submicron zirconia particles distributed at the grain boundary junctions. There is no scientific barrier to the diffusion bonding of superplastic *fine-grained alumina* components, although the economic viability of such a process remains to be tested.

Most recently, many new developments have emphasized the production of *nanomaterials*. These are materials which have a stable, *ultrafine* grain size, or which contain stable particles in the *submicron* size range. Since *superplasticity* is primarily associated with *grain-boundary sliding*, it is to be expected that *nanomaterials* will offer enhanced control of superplastic behaviour, still further improving the chances of developing *diffusion bonding* as a viable process of joining simple ceramic shapes into more complex structures.

Summary

The integrity of a metal–ceramic joint requires that the high stiffness of the components and the brittle nature of the ceramic be accommodated at the join without developing localized stress concentrations. This may be accomplished by introducing a ductile or compliant interlayer, by ensuring that dimensional tolerance requirements are met, by matching the thermal expansion characteristics of the components, or, most commonly, by a combination of all these strategies. The ideal join would involve a smooth transition in properties across the interface, and this may be approximated by employing a number of interlayers of intermediate properties or by giving careful consideration to the surface profile of the join.

Thermoelastic mismatch between the components is a primary cause of failure, but good matching of the expansion coefficients of the components and interlayers is only possible over a limited range of temperature. Glasses and glass ceramics have expansion coefficients which can be controlled over a wide range and are particularly useful interlayers. While a low expansion coefficient reduces the transient stresses induced by a temperature gradient, engineering alloys (especially aluminium alloys) may have large expansion coefficients, leading to excessive thermal stresses in the region of the interface. Thus plasma-sprayed zirconia coatings have comparatively large thermal expansion coefficients, matched to that of the nickel alloy or stainless steel substrate, but the maximum thickness of coating is then limited by the residual stresses developed by the in-service temperature gradient across the coating.

Many engineering materials, metals and ceramics, are inherently anisotropic. Anisotropy may reside in the microstructural morphology (the flattened grains in a plasma coating for example) or in the partial alignment of the crystal axes in a polycrystal (crystalline texture). While some forms of anisotropy may lead to improved properties (for example, by inhibiting crack propagation across a join), anisotropy in the plane of the join is generally undesirable, leading to parasitic shear stresses across the interface.

Glass–metal seals have been around for a long time and generally fall into one of two categories: planar surfaces, such as silvered mirrors, and cylindrically symmetric joins, such as electrical feed-throughs. Provided that the dimensions of the ductile, metal component are sufficiently small, the stresses developed in the glass will be minimal, but as the dimensions of the metal increase, so the need to match the thermal expansion coefficient of the components becomes more critical. For large dimensions (as in the bonding of metal to glass tubes), several interlayers may be necessary. The final dimensions of the 'graded seal' are then of the order of the tube diameter, and this space must be allowed for in the system design.

Diffusion bonding is accomplished entirely in the solid state and involves temperature-assisted plastic flow (creep), as well as diffusion, to achieve complete physical contact of the components across the bonded interface. While creep processes control the early stages of bond formation (growth of interfacial contacts), diffusion dominates the final stages of bonding (pore shrinkage and closure). There is always some thermoelastic mismatch across the diffusion-bonded interface, and it is important to ensure that interfacial defects are too small to initiate crack propagation. Diffusion bonding requires meticulous surface preparation and is achieved in a protective or reducing atmosphere.

As in diffusion bonding, the bonding of two ceramic components is limited by the stresses developed as a result of thermoelastic mismatch and the stress concentrations associated with defects at the interface. In these brittle materials the critical defect size is generally below the limit of detection of available methods of non-destructive testing, requiring proof testing of the assembled components. Reactive brazing has been successfully developed for joining ceramic components, and relies on cation replacement in the ceramic by a component of the braze (typically titanium). The strength of a reactively brazed joint is extremely sensitive to both the initial surface preparation and the brazing parameters, but both the methods of testing the mechanical strength and the criteria for acceptability of the joint have yet to be agreed and reduced to international standards.

A most successful ceramic–metal seal is that used for alumina-based ceramics, as in the spark plug assembly of the internal combustion engine. This is again a multistage process involving the use of interlayers (one of which is ductile). Ceramic–ceramic joints have been made successfully in development projects, the more promising processes involving assembly in the green state followed by firing and sintering of the assembled components. Glass interlayers and superplastic forming have also been used to achieve a join, but both these processes appear to be some way from commercial exploitation. At the time of writing considerable interest is being shown in the use of sol–gel technology and nanometre particle sizes to arrive at improved ceramic properties, and these new technologies should be readily adaptable to the joining and assembly of ceramic components.

Further Reading

1. W. E. C. Creyke, I. E. J. Sainsbury and R. Morrell, *Design With Non-ductile Materials*, Applied Science, London, 1982.
2. P. Shewmon, *Diffusion in Solids*, Second Edition, TMS, Warrendale, Pennsylvania, 1989.
3. A. H. Carim, D. S. Schwartz and R. S. Silberglitt (eds.), *Joining and Adhesion of Advanced Inorganic Materials*, Material Research Society, Pittsburgh, 1993.
4. D. M. Mattox, J.E.E. Baglin, R. J. Gottschall and C. D. Batich (eds.), *Adhesion in Solids*, Materials Research Society, Pittsburgh, 1988.

Problems

8.1 Describe the shrink-fit process for bonding metals to ceramics. List and explain the advantages and disadvantages of the process with regard to:
   a. Geometry.
   b. Thermal mismatch.
   c. Device service temperature.
   d. Thermal shock.

8.2 What are the limitations of the different bonding processes for bonding unlike materials? Include mechanical bonding, soldering, brazing, welding and adhesives, and describe both the limitations of the process and the limitations of the joint in service.

8.3 Diffusion bonding of alumina to metals is commonly achieved by first bonding alumina to niobium, and then niobium to the metal component. Why should the alumina–Nb system be preferred for bonding alumina to metals?

8.4 Explain the transient liquid phase concept for bonding. Which parameters limit the use of this process for different materials systems?

8.5 In the design of a glass–metal graded seal, which material parameter is most likely to be the limiting design factor? Justify your answer!

8.6 Two phenomena are associated with a diffusion bonding process: creep and diffusion. Explain the role of both creep and diffusion in the diffusion bonding process and differentiate clearly between them.

8.7 Describe the functions of proof testing and non-destructive testing. Why is proof testing usually required for a diffusion-bonded joint involving a ceramic component?

8.8 What is the difference between brazing and reactive brazing? Why is reactive brazing usually regarded as necessary for joining ceramics, whereas conventional brazing is often sufficient for joining metal components?

8.9 What parameters limit the diffusion bonding of two ceramic components? Suggest some ways in which the process might be made commercially viable.

8.10 Two asymmetric alumina components are to be joined. Due to their geometry, diffusion bonding is not a viable option. Suggest two other possibilities and in each case describe the engineering limitations.

9

Adhesives

Adhesives are polymeric compounds, commonly called *glues*. As such, they are applied at room temperature to the surfaces to be bonded, and may harden either at room temperature or after some further treatment: *heating* or *irradiation.* The hardening of the adhesive bond is referred to as *curing*.

The *work of adhesion* is defined as the energy difference per unit area between the *surface energies* of the original components and the *interface energy* of the bond (Fig. 9.1a). In terms of the work required to strip the adhesive from the bonded surface under *equilibrium* (reversible) conditions: $W_a = (\gamma_{As} + \gamma_{Bs}) - \gamma_{AB}$, where W_a is the *work of adhesion*, and the subscripts As, Bs and AB refer to the energies of formation of the free surfaces of the adhesive, the matrix, and the bonded interface, respectively. The *spreading force* (Fig. 9.1b), which is the driving force for the *wetting* of the matrix by the adhesive, is: $S = \gamma_{Bs} - (\gamma_{As} + \gamma_{AB})$. If we make the gross assumption that the *surface energies* are unchanged by the curing process, then: $(Wa - S) = 2\gamma As$, so that the *work of adhesion* exceeds the *spreading force* by a factor equal to *twice* the surface energy of the adhesive.

Polymeric adhesives fail by a combination of one or another of several mechanisms. The *adhesive bond* itself may fail, leading to the recovery of the free surfaces of the adhesive and the matrix. Assuming that no other factors are involved, the *work of failure* is then just W_a (Figure 9.2a). At the other extreme, failure may propagate solely *within* the adhesive layer (Figure 9.2b). Such failures are termed *decohesion*, and if the failure is *brittle* (but reversible) the *work of failure* will be: $2\gamma_{As}$.

The *brittle failure* of adhesive bonds is common, and such bonds may have quite acceptable strengths. However, the *strength* of a brittle bond depends sensitively on the dimensions of any *defects* in the bonded area. The commonest defects are *unwetted zones*, *dirt inclusions* or regions which have *debonded* previously, often as a result of *internal stresses*. Adhesives which fail in a *ductile* manner or which give an *elastomeric* failure are far more forgiving of the presence of bonding defects.

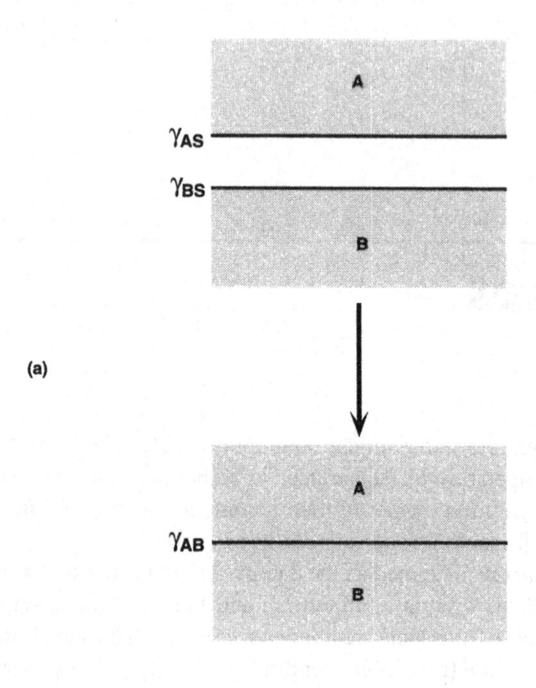

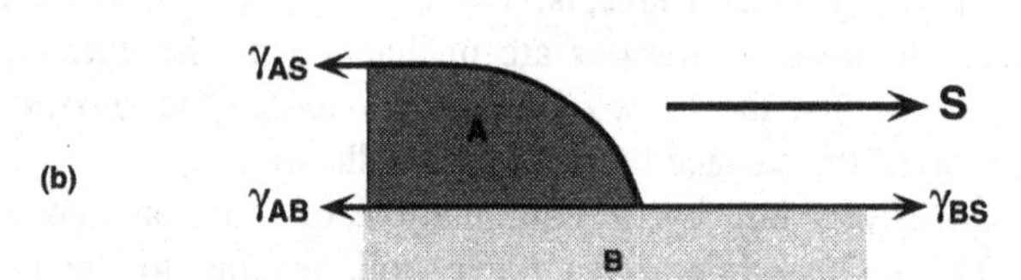

Fig. 9.1. (a) The *work of adhesion* is defined as the reduction in *surface energy* per unit area accompanying the formation of an *interface* between two mating surfaces. (b) The *spreading force* is the excess *surface tension* (force per unit length) available to drive the *wetting* of a flat solid surface A by a liquid B.

There is no sharp transition between *adhesive* and cohesive failure, and a *ductile* adhesive may show a *continuous transition* from adhesive to cohesive failure as a function of temperature or strain rate, as shown schematically in Fig. 9.3

The *fracture energy* for polymerically bonded joins is usually associated with changes in the *configurational entropy* of the giant polymer molecules. During the application of a load the *polymer chains* are forced to *extend* out of their most probable configurations, leading to a large reduction in *entropy*. It is then the $T\Delta S$ term in the *Gibbs free energy* equation, $\Delta G = \Delta H - T\Delta S$, which makes the major

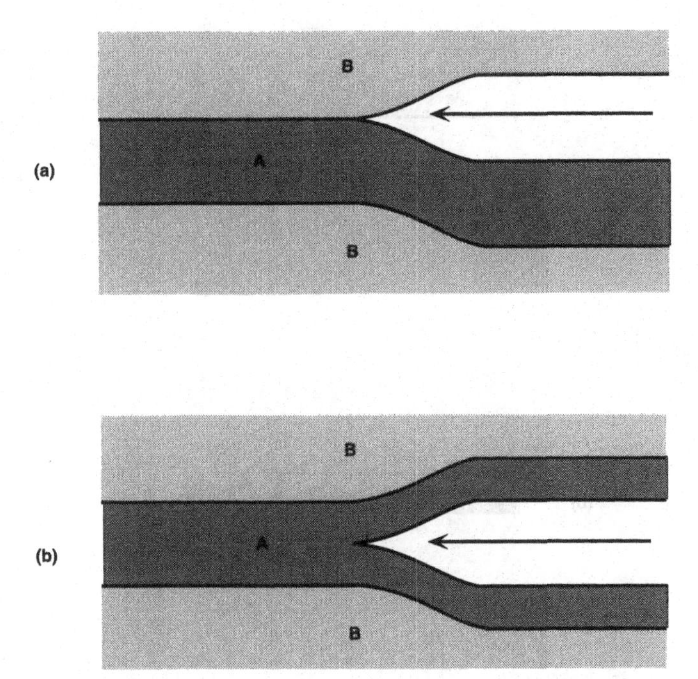

Fig. 9.2. (a) *Loss of adhesion* corresponds to the *debonding* of an adhesive layer from a substrate at the interface. (b) Loss of *cohesion* (decohesion) is the propagation of failure *within* a bonding layer, and implies that the adhesive force at the interface *exceeds* the cohesive force within the bond layer.

contribution to the *elastic energy* stored in the loaded joint. This may be compared with a *welded* or *brazed* joint, for which the change in *elastic energy* is almost entirely due to the change in *enthalpy* ΔH which accompanies elastic extension of the atomic bonds. It follows that the *bond strength* of an adhesive joint is strongly *temperature* and *strain rate* dependent. It also follows that, in general, *successful* adhesive bonds require adequate *chemical bonding* between the polymer molecules of the *adhesive* and the surface of the *matrix*. The *weakest* bonds between the polymer adhesive and the surfaces of the components are van der Waals' *polarization forces*, which in the limit may be too weak to achieve any useful bond strength. A good example is *polyethylene*, which *cannot* be bonded by simple adhesives, since the polyethylene molecule possesses no *dipole moment* and has only a very weak *polarizability*.

Ionic bonding, based on *coulombic* forces, can be established with adhesives which have *ionizable groups*, such as the *phenolic resins*, but the *strongest* chemical bonds are *covalent* in nature. In this case the active groups in the adhesive are *chemically* attached to the surface of the component. The *adhesive strength* of the bond then depends on the *surface density* of these attachment points. The density of

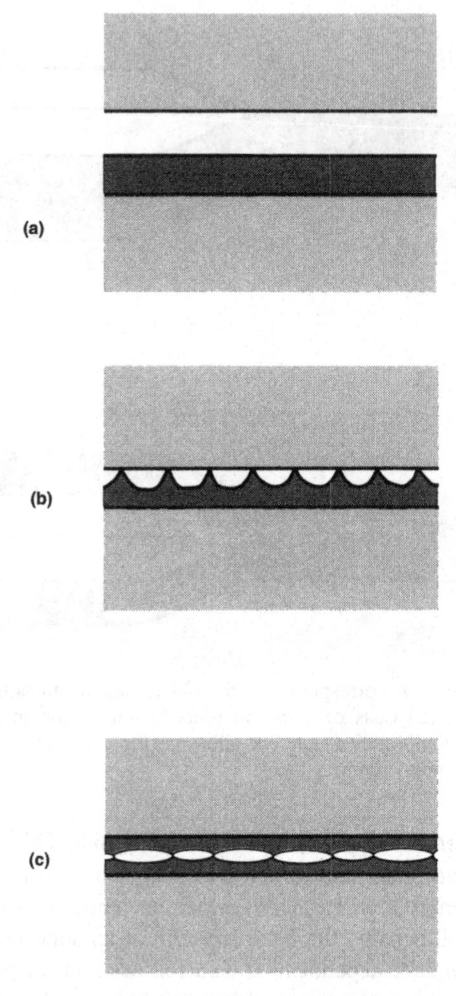

Fig. 9.3. There is a *continuum* of ductile failure morphologies at an adhesive bond: from (a); *loss of adhesion* with *no* plastic flow through (b); *ductile void formation* at the *interface* to (c), *ductile void formation* and *decohesion* within the bond with no loss of *adhesion* at the interface.

attachment points is low compared to the density of potential sites on the surface of the matrix, primarily due to *steric hindrance* associated with the bulky nature of the polymer molecules. It follows that there is now an *entropy* term associated with the number of possible attachment configurations: $\Delta S = k \log w$, where k is Bolzmann's constant and w is the number of possible *attachment configurations*. The *adhesive energy* is then the sum of the *internal energy* (the surface density of attachment points multiplied by the bond energy per point), and the *entropy* term. This is the *two-dimensional* analogue of the configurational entropy term in a *solid solution*, in which *solute* atoms replace *solvent* atoms at random. As in the case of a *regular solution*, the *internal energy* is expected to decrease *linearly* with the site occupancy of attachment points, while the *entropy* term reaches its maximum value at 50% occupancy (Fig. 9.4). However, the total number of allowed sites is limited by *steric hindrance*, and at the low site occupancies which are geometrically possible it is the entropy term which dominates the bonding energy, once more leading to a strong dependence of the adhesive bond strength on temperature and strain rate.

It follows that adhesive bonds are expected to show *transitions* in their failure modes as a function of temperature and strain rate. At low temperatures and high strain rates, *adhesive* failures are expected, the adhesive peeling away from the substrate. At high temperatures and low strain rate, failure is likely to be dominated by *decohesion*, and to occur within the adhesive layer itself.

Finally, there is a strong tendency for polymers of differing *dielectric* properties to *phase-separate* into a *polymer blend*. Most adhesives therefore contain only *one* polymer component, although they may contain particles of a *filler* phase, as well as a solvent *plasticizer* or a catalytic *hardener*, and other additives (such as colouring agents and flame retardants). The role of the *filler* is usually to increase the *elastic stiffness*, but a metal powder may be used as a filler in order to improve *electrical* or

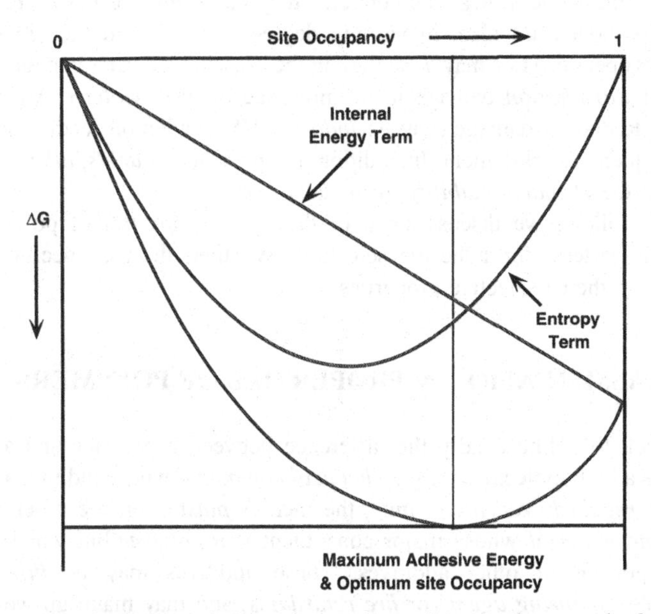

Fig. 9.4. The reduction of both *enthalpy* and *entropy* contributes to the strength of a *polymeric* adhesive bond. *Steric hindrance* commonly limits the number of available sites for polymer attachment at the interface and in consequence it is usually the *entropic* contribution which dominates the *adhesive free energy*.

thermal conductivity across the bond. The *plasticizer* is a small organic molecule whose *dielectric* properties are matched to that of the adhesive, and which is therefore *soluble* in the polymer. The plasticizer reduces the *stiffness* and, in many instances, confers improved *ductility* on the adhesive. In the event that the plasticizer has an appreciable *vapour pressure*, it will evaporate with time. A high *solvent* concentration will then allow the adhesive to be applied as a *viscous liquid*, and the bond will attain its strength as the solvent evaporates. The *hardener* is a *catalytic agent* which cross-links polymer molecules during curing. Both the *stiffness* and the *hardness* of the adhesive will then increase during curing. There will be some optimum treatment corresponding to maximum bond strength: at *short* curing times the bond will remain *soft* (and excessively ductile), but at *long* curing times the bond will become increasingly *brittle* and susceptible to flaw-induced failure.

Many adhesives develop *internal stresses* in the bonded region, not as a consequence of *thermoelastic mismatch*, but as a result of *dimensional changes* which occur either during solvent *evaporation* or during *cross-linking*. Such residual stresses can significantly *reduce* the strength of a bond, especially if the failure is *brittle*. Polymeric materials are very strain rate and temperature sensitive in their properties. Although adhesives can be found to perform over a wide temperature range (from $-110°C$ *to* $370°C$ at the time of writing), individual adhesives have a much more limited range of application. At low enough temperatures *all* adhesives become brittle, while at high temperatures they either *melt* or *pyrolyse*.

Adhesive bonds are also, in varying degrees, sensitive to the *environment* in which they operate. They may *hydrolyse* in the presence of water vapour or they may be subject to *radiation* damage in the presence of *ultra-violet* (UV) light (direct sunlight). Radiation damage may be *indirect*, UV irradiation leading to *oxidation* and subsequent embrittlement. In addition, the presence of *acids*, *alkalis* or *solvents* may affect the *chemical stability* of the adhesive.

In what follows we discuss some of the *basic properties* of polymers before considering criteria for adhesive selection. We then discuss specific *classes* of adhesive and their respective properties.

9.1 CLASSIFICATION & PROPERTIES OF POLYMERS

It is as well to define clearly the difference between a *polymer* and a *plastic*. A *polymer* is a high molecular weight *chemical compound* whose individual molecules consist of *regularly repeating units*, the mer. A *plastic*, on the other hand, is an *engineering material* whose major constituent is a *polymer* but which contains a large proportion of other additives. These additives may be *filler powders*, *plasticizers*, *colouring agents* or *fire retardants*, and may make up well over half the volume content of the material. Polymers have a molecular structure which is dominated by a *long-chain molecular backbone*. In most commercial polymers this backbone is based on a simple *carbon chain* to which are attached a variety of *halide*, *amine* and other side groups. *Stiffer* molecular chains result from polymers which include *double bonds* and/or *benzene rings* in the chain structure, and many commercial adhesives come into this category (for example the *phenolic* and *epoxy* resins). The *silicone* resins do not contain carbon in the backbone. They have a very flexible backbone consisting of an alternating sequence of *silicon* and *oxygen* atoms, with the side groups attached to the silicon. The mechanical *flexibility* of the silicone chain and the chemical *stability* of the Si–O bond give the silicone-based adhesives a very wide range of application temperature, while their exceptional *dielectric strength* frequently often makes them the material of first choice as *potting compounds* for microelectronic devices. Figure 9.5 shows schematically these common types of *polymer chain*.

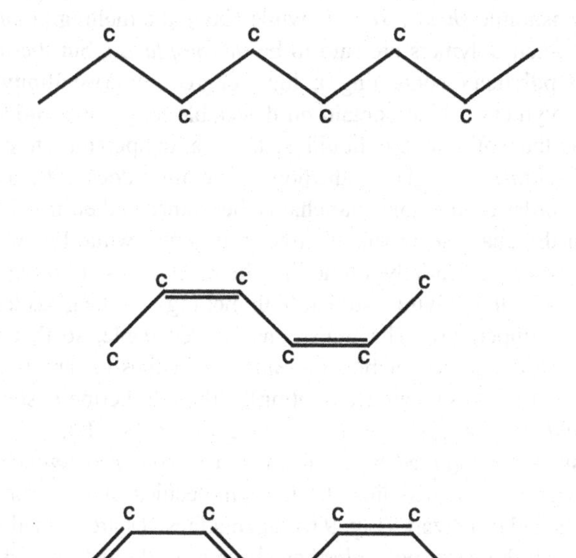

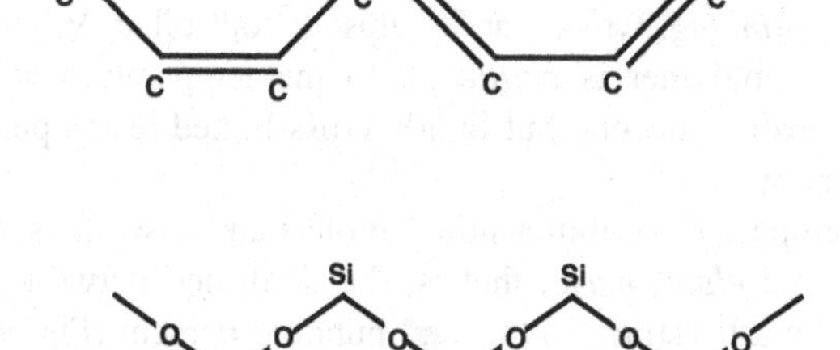

Fig. 9.5. Polymers consist of molecular chains of which the *simplest* has a backbone of *singly-bonded* carbon atoms. Stiffer chains contain carbon *double bonds* or *benzene rings* within the polymer backbone. The very flexible –Si–O–Si– repeating unit provides the backbone for the *silicone* family of polymers.

Several terms are used to classify the nature of polymers and we need to define these terms. In particular, we should distinguish the principal differences between the microstructure of polymers and that of other engineering materials. Many polymers contain some degree of crystalline order, in which the polymer chains are folded back on one another to form a *regularly repeating* structure. Such a molecular structure is indeed *crystalline*, however crystallinity in polymers is never complete, and is uncommon in adhesives, so that although the molecular arrangement must contain some *short-range* order, *no* long-range order is present. The structure is then termed *amorphous*. Were the polymer molecules to have a *crystalline structure* and a well-defined molecular weight, then a melting point would characterize a clearly defined phase transition. The melting temperature would depend on the *molecular weight*. Below the melting point the polymer would be a *solid* with a measurable *shear strength*, while above the melting point it becomes a *viscous liquid*. Such polymers are said to be *thermoplastic*, but thermoplasticity is not limited to polymers containing a high degree of crystallinity. Amorphous thermoplastic polymers, which contain no detectable long-range order, also show a transition from the solid to the liquid state at a temperature termed the *glass transition temperature*. The glass transition temperature does not mark any change in the degree of order of the molecular chains, but rather in their mobility. Above the glass transition the chain segments are free to migrate, while below the transition temperature they are essentially immobile. In most cases a *range* of molecular weights is present in the polymer, so that both melting and the glass transition occur over a range of temperature. The transitions are reversible, so that *thermoplastic* polymers are, in principle, recyclable. *Thermoplastic* adhesives are usually deposited from *solution* (that is, by solvent evaporation), although thermoplastic polymers are also readily *weldable* (as in the *vacuum sealing* of food-stuffs).

Many adhesives are supplied in the form of a *viscous resin* which reacts in the presence of a *catalyst* to cross-link the large molecules of the resin into a three-dimensional molecular network (Fig. 9.6). Such adhesives are termed *thermosetting resins* and, once all the resin molecules have been cross-linked, consist essentially of a single molecule. The mechanical properties of a *thermoset* depend sensitively on the degree of *cross-linking*, which can be closely controlled. When cross-linking is fully developed the polymer is *brittle* up to the temperature at which chemical breakdown and *pyrolysis* occurs, but lightly cross-linked resins permit considerable flexibility of the network.

If a fully developed, three-dimensional molecular network is formed, then the polymer can become *elastomeric*, that is, the deformed network may relax to its original shape, even after strains of several hundred percent (Fig. 9.7). *Elastomeric* adhesives are capable of storing large amounts of *elastic energy* before failure, in fact their *elastic energy storage* capabilities (stored energy per unit volume) are comparable to those of steel springs. In addition, the large local strains which accompany elastomeric deformation make these adhesives remarkably *insensitive* to the presence of defects in the bond plane. Elastomeric adhesives may be formed

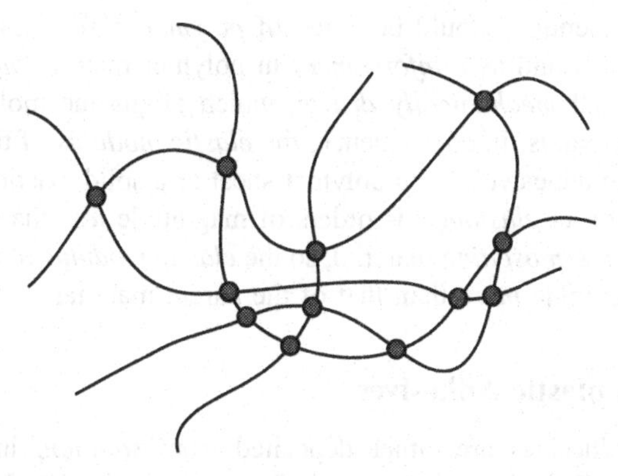

Fig. 9.6. Polymer chains can be chemically *cross-linked* by a *thermosetting* reaction, converting a low molecular weight *resin* to a flexible *elastomeric solid* whose elastic modulus depends on the *cross-link density*, and finally, at a sufficiently high density of cross-links, to a brittle, *amorphous glass*.

from *solution*, followed by solvent evaporation, or by controlled *cross-linking* during bonding, in which case the density of cross-link sites has to be carefully monitored. The *rubbery* behaviour of elastomers is *entropic* in nature: extension of the polymer *reduces* the number of available *configurations*, and it is the resistance to this *decrease* in entropy which determines the elastomeric modulus. Once the polymer chains have been extended, at a large imposed strain, the *elastic resistance* (the slope of the stress–strain curve) increases rapidly, corresponding to distortion of the *bond angle* in the carbon chains (an increase in the *internal energy*), until final failure occurs (Fig. 9.7).

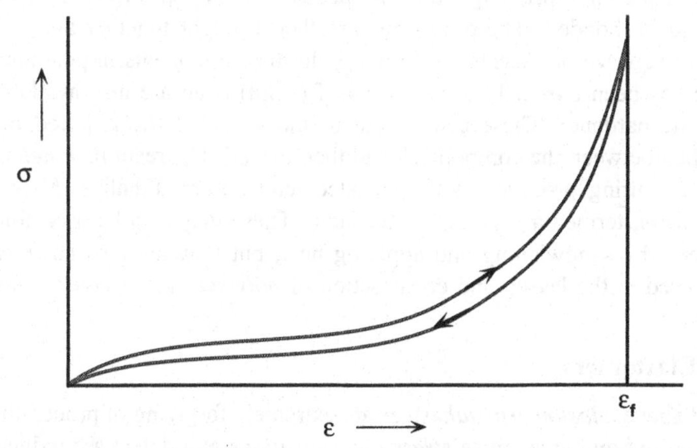

Fig. 9.7. *Elastomeric polymers* have very low elastic moduli and are capable of extremely high elastic strains—up to or exceeding 800%—with little or no *plastic* deformation ('permanent set').

Finally, some mention should be made of *polymer fibres*, primarily because of their role in *textiles* and as *reinforcement* in polymer matrix *composites*. Polymer fibres are nearly all *mechanically drawn*, which aligns the molecular backbones parallel to the fibre axis. In consequence, the *elastic modulus* of the drawn fibre far exceeds that in an adhesive film, a polymer sheet or a solid plastic body. Just as the *elastic modulus* of an *elastomer* is orders of magnitude *less* than that of a regular *thermoplastic* or *thermosetting* material, so the *elastic modulus* of a drawn *fibre* may be orders of magnitude *more* than that of the parent material.

9.1.1 Thermoplastic Adhesives

Thermoplastic adhesives are either deposited from *solution*, in which case the solvent must be allowed to *evaporate* before making the bond, or else they are applied from the *melt*. In both cases it is usually necessary to apply *heat* and/or *pressure* to the components in order to establish the bond.

Thermoplastic adhesives represent an *intermediate* range of polymer strength in that their *theoretical* energy of fracture (more exactly, the *cohesive energy density*) is intermediate between that of *elastomers* (with a *low* cohesive energy density) and that of the high-performance, *crystalline* polymers (exemplified by the *polyamides*—the nylon family).

9.1.2 Thermosetting Resins

Thermosetting resins, by comparison, must contain a *hardener* to catalyse the *cross-linking* necessary for the development of the full bond strength. This may be achieved by several, quite different, routes: in the simplest case a *viscous resin* is mixed with the appropriate quantity of *hardener* before spreading the resin on the surfaces to be bonded. These surfaces are then brought together and *pressure* is applied, to squeeze out surplus resin and hold the components in position until the adhesive has been *cured*. Flexible sheets of (solid) resin are also available, which *contain* the hardener. These can be cut to the shape of the required bond area, sandwiched between the components and then heated. The resin first *melts* and then *hardens* as curing proceeds at the elevated temperature. Finally, *fibre-reinforced resin preform*, termed a *prepreg*, is available. These may also be used for bonding components by sandwiching and applying heat, but they are also an *intermediate product* used in the lay-up and construction of *polymer matrix composites*.

9.1.3 Elastomers

As noted above, *elastomeric adhesives* are extremely forgiving of processing defects present in the bond area, since *stress concentrations* at a defect are reduced by the ability of the elastomer to retain its load-carrying capacity without failure at high strains. As a result the *integrity* of the bond may be retained, even if the bond is

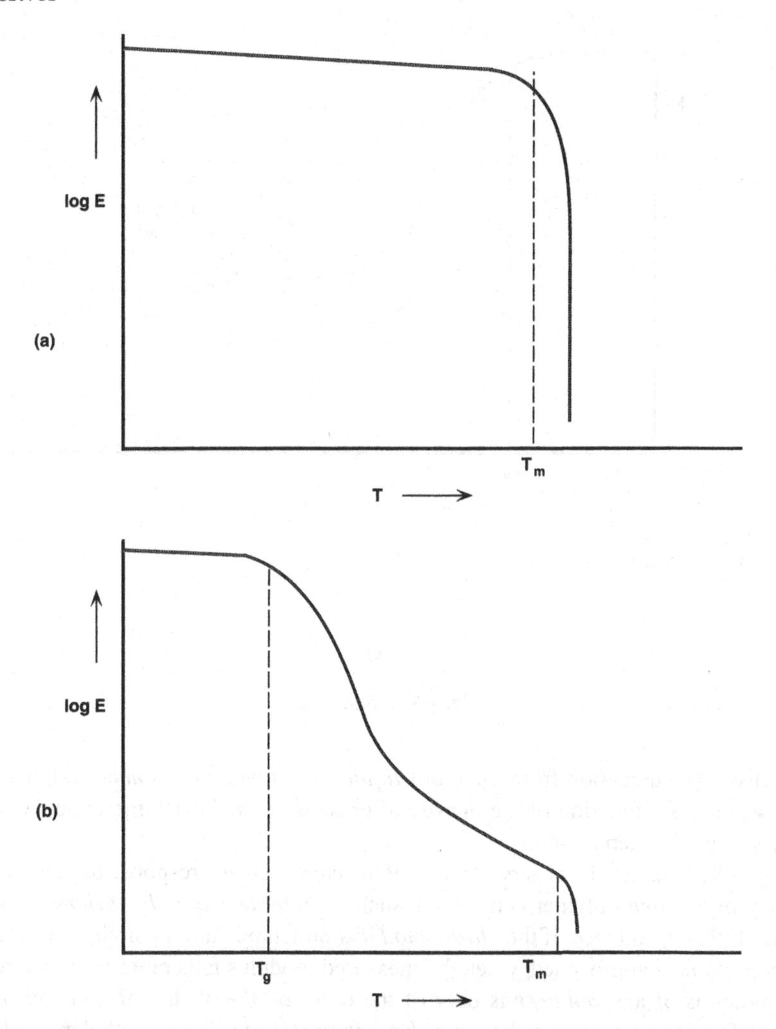

Fig. 9.8. (a) Largely *crystalline* polymers (high density polyethylene or the polyamides) show only a small reduction in *elastic modulus* with increasing temperature until the melting point T_m is reached. Because the *molecular weight* is not precisely defined, melting takes place over a small range of temperature. (b) A *glass transition temperature* T_g can be defined for *non-crystalline polymers*, at which the elastic modulus starts to fall steadily and the mechanical properties change from those of a *brittle*, glass-like material to those of a *tough*, ductile solid. Softening and melting eventually occur, with a well-defined *melting point* if the polymer is partially crystalline. (c) *Thermosetting polymers* are amorphous, *cross-linked* materials which also show a well-defined *glass transition temperature*. Above the glass transition temperature the mechanical properties depend on the density of cross-links, changing with increasing cross-link density from those of a *viscous liquid* to those of a *rubbery solid*, then increasing in toughness (and becoming *leather-like*), and eventually becoming *glassy* or brittle. Cross-linking prevents melting: at a high enough temperature the polymer will *pyrolyse* or 'burn' as the molecular structure breaks down.

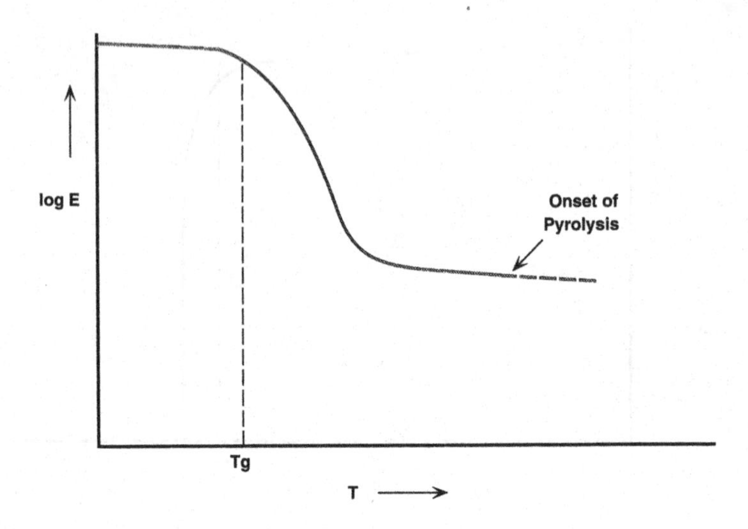

(c)

Fig. 9.8. *(continued)*

defective. The transition from a *viscous liquid* to a *rubbery elastomer* and, finally, a *brittle solid* is a function of the *density* of cross-links, and elastomeric behaviour is very sensitive to temperature.

Fig. 9.8 illustrates the observed range of the typical *elastic* response of polymers. A largely *crystalline* polymer (Fig. 9.8a), such as *high density polyethylene* (HDPE), shows little dependence of the *elastic modulus* on temperature or strain rate until the *melting point* is approached, when the measured modulus falls rapidly. (The value of the modulus of any *polymer* is only of the order of 1% of that of an *engineering alloy*.) The same polymer in the *amorphous form (low density polyethylene—LDPE)* shows a well-defined *glass transition temperature*, below which the modulus is not greatly different from that of the crystalline material, but above which it *falls*, rapidly at first and then steadily until the final, drastic drop at the *melting point* (Fig. 9.8b). In the intermediate range, between the *glass transition temperature* and the *melting point* (the ratio of the two, measured in K, is typically 2/3 for a large number of polymers), the polymer is a *ductile thermoplastic*, but may exhibit some *elastomeric* properties for specific classes of polymer having a large molecular weight and associated entanglement of the polymer chains.

Thermosetting resins exhibit a *similar* temperature dependence of the modulus, dependent on the *degree of cross-linking*. The major difference, Fig. 9.8c, is that *melting* can no longer occur, since the molecules are now chemically pinned at the cross-links. The region in which *elastomeric* behaviour is observed depends on the elastomer, but the essential condition is that the lengths of free chain between pinning points should, with the help of *thermal activation*, be able to take up a large number of *alternative configurations*. If the pinning points are too widely separated, then an imposed strain is unlikely to be recovered, and the deformation is *viscoplastic*. If the pinning points are too close together, then the number of possible configurations is severely restricted and deformation is *conventionally elastic* (that is, based *on bond distortion*, not *entropy* considerations). In the intermediate range of pinning point separations, large strains are recoverable on relaxing the stress and the material is truly elastomeric. Some *elastomeric* polymers are *thermoplastic*, and in this case the pinning points responsible for the elastomeric behaviour are assumed to be *mechanical entanglements*.

9.2 STRUCTURAL ADHESIVES

Structural adhesives are those adhesives which are intended to impart sufficient strength to a joint for it to be capable of withstanding the loads involved in structural applications. Before considering the materials used as structural adhesives we consider the basic principles involved in their *selection.*

9.2.1 Basic Considerations

Most adhesives are supplied as *viscous liquids*. The surfaces to be joined have to be *coated* with the liquid adhesive and *pressure* must then be applied to ensure complete *wetting* and establish the bond before the adhesive sets. Three separate regimes may control the *rate of wetting* of a solid surface by the adhesive, depending on the *geometry* of the system and the nature of the *external forces*. These three regimens may be described as *squeezing* (Fig. 9.9a), *spreading* (Fig. 9.9b) and *capillary adsorption*. In the *squeezing* regime a film of the viscous liquid is placed between two flat, parallel surfaces and subjected to a *normal force*. The *flow* of the liquid is a function of its *viscosity*. For a *circular plate* of radius a the time t required to reduce the thickness of the liquid film to a value h by application of a compressive force F is approximately given by:

$$t = \frac{3}{4}\pi\frac{\eta}{F}\frac{a^4}{h^2}$$

when η is the viscosity.

A more usual situation related to the formation of an adhesive joint would be the *spreading* of a circular *drop* of the liquid between the two circular plates. In this case (Fig. 9.9b) the contact area *increases* as a function of time. If the final dimensions of the joint are defined by the *radius* of the disc a and the *thickness* of the liquid film h, then solving the appropriate equations for *laminar flow* leads to a time to spread the

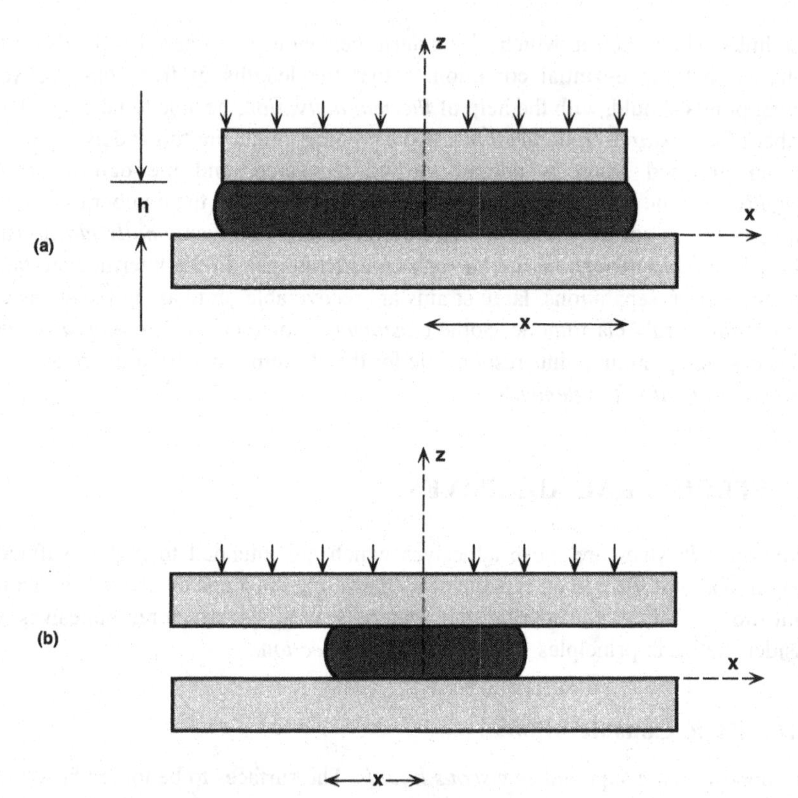

Fig. 9.9. (a) A *viscous layer* of adhesive covering the surface may be *squeezed* between two surfaces to reduce the thickness of the layer. (b) Similarly, a single *drop* of the viscous liquid may *spread* across the surface under an applied pressure.

drop $t^* = t/2$. That is, a drop will spread over the full area of the circular plate in just *half* the time required to squeeze surplus liquid from the *same* area to the *same* thickness. Of course, this presupposes a *uniformly* spreading drop with *constant* coefficient of viscosity (*Newtonian flow*), independent of both time and flow rate, which has been sited accurately at the *centre* of the plate and has the correct *volume* ($\pi a^2 h$) required to cover the join at the final, required thickness, but the rheological principle is clear!

If *capillary* forces are present, then these will *assist* spreading if the contact angle is *less* than $\pi/2$ and *inhibit* spreading if it is *greater* than $\pi/2$. In the limit, the capillary forces may promote either complete *wetting* or *dewetting*, as described in Section 2.6.1. A *porous* substrate will absorb a liquid provided that the *contact angle* is less than $\pi/2$. Assuming a *parallel plate* configuration for a capillary model, the *capillary pressure* is given by $P = 2\gamma \cos\theta/\delta$, where θ is the contact angle and δ is the plate separation. The rate of *laminar flow* in a slit u is then given by (Fig. 9.10):

$$u = -\delta^2/12\eta \cdot \Delta P/\Delta z = dz/dt$$

substituting P/z for $\Delta P/\Delta z$ and using the above relation for the *capillary pressure*:

$$z \cdot dz/dt = \delta/6\eta \cdot \gamma \cos\theta$$

After integration we obtain: $z_0^2 = t \cdot \delta \cdot \gamma \cos\theta/3\eta$, which emphasizes the dependence of *capillary penetration* z_0 on the *size* of the channels δ, the *time* t available for viscous flow and the balance of *capillary* and *viscous* forces $\gamma \cos\theta/\eta$.

Surface roughness will have a large effect on surface *spreading*. As noted in Section 7.1.2, a roughened surface can be regarded as a *network of interconnected grooves* which promote spreading of a liquid by the action of the *capillary forces*. As in the above equation for the *capillary penetration* in a slit between two parallel plates, spreading on a roughened surface will be promoted by a *low* contact angle, *coarse* groove dimensions, a *low* viscosity and a *high* value for the liquid surface tension.

A *roughened* surface will also set a lower limit to the *average separation* of the two component surfaces, and will be determined by the *amplitude* of the surface roughness. A *wetting liquid* between two *roughened* surfaces will also generate a *capillary pressure* which *resists* separation of the components. For such a system, the strength of the bond becomes *equal* to the capillary pressure, that is, the *plate separation* δ, in the above equation, is now (ignoring a constant) the effective *amplitude* of the surface roughness. This is the principle used in *adhesive tape*. The viscous adhesive is coated onto a flexible plastic backing and the peel strength of the tape is a measure of the *capillary pressure* on any surface to which the tape adheres. The peel pressure is a *maximum* when the tape adheres to a smooth surface of *minimum* roughness, such as glass. However, on a *porous* surface (paper) adhesion is

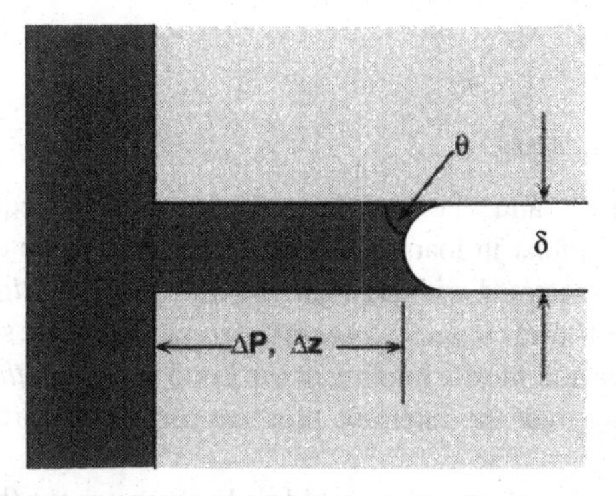

Fig. 9.10. *Capillary flow* is a dominant feature of adhesive spreading. We consider *penetration* of adhesive to a depth z in a slit between two plates separated by a distance δ.

also good, since the adhesive then wets and *penetrates* the surface fibres of the paper. On a porous surface volatile components of the adhesive can evaporate through the pores, while free ingress of *oxygen* may lead to *cross-linking* reactions which *embrittle* the adhesive. The bond between an adhesive tape and *paper* usually deteriorates quite rapidly, the tape backing eventually flaking off the paper.

The criteria for *selecting* an adhesive are for the most part straightforward common sense, but the full list may contain some surprises, as well as some criteria that require a considerable degree of experience and expertise to implement. We will discuss them under *six* headings.

9.2.1.1 SUBSTRATE

The *substrate* must be adequately prepared *before* bonding; in particular it must be adequately cleaned in order to remove all *contamination*. This includes *drying* the components, since the presence of *water vapour* is frequently a major reason for the loss of adhesive strength, which depends on the polar attachment of the adhesive to the surface. Frequently, *precoating* the surface may improve the bond strength and will be recommended by the manufacturer. Organic *coupling agents* are small, chemically active compounds that bond strongly to both the *substrate* and, subsequently, to the *adhesive*, ensuring a strong bond.

The *chemistry* of the substrate will be important in determining the nature of the *chemical bond* which can be formed with the adhesive. *Covalent* or *ionic* bonding with the adhesive is desirable, and may be achieved by *abrasion*, *irradiation* or *chemical treatment*. The *geometry* of the substrate is also a major consideration. *Roughened* surfaces increase the bond area per unit nominal area of contact and provide a degree of *mechanical interlock* which dramatically increases the *shear strength* of the bond. *Porous* substrates permit *penetration* of the adhesive into the components, improving not only the *shear strength* but also the *tensile strength* of the bond.

9.2.1.2 THE APPLICATION

The use *temperature* and the *loading conditions* are frequently the primary considerations. *Variations* in load and temperature need to be considered, and may result in problems associated with *thermal* and *mechanical fatigue*, long-term *creep* response, and short-term *impact loading*. *Stress analysis* should consider the proportions of *shear* and *tensile* loading at the bond interface (*the mode mixity*). The variations in loading over the interface area are usually important in design of the joint.

The *environment* also needs to be considered: *water vapour* (humidity) can have a disastrous affect on the long-term bond strength of some adhesives, and *industrial pollution* may be a consideration. *Organic solvents* can cause swelling and adversely affect adhesive properties. *Ultra-violet radiation* acting on adhesive bonds which are exposed to sunlight may either *depolymerize*, reducing the elastic modulus and eventually liquefying the adhesive, or increase the *degree of cross-linking*, leading ultimately to embrittlement.

9.2.1.3 Storage

Most adhesive resins have limited *chemical stability*, that is to say, they have a limited *shelf-life*. Their components are frequently *volatile*, posing a health hazard during storage and processing, as well as making them further susceptible to *degradation*. In some cases contact with the *atmosphere* is sufficient to initiate chemical changes in the resins, while in others it is the presence of *moisture* that leads to rapid loss of properties. *Refrigeration* may be required to maintain an adequate shelf-life, or the use of storage dessicants may be indicated. *Sealed packaging* may ensure retention of the chemical activity of the adhesive, but then the quantity required should be comparable with the amount packaged in order to avoid wastage.

9.2.1.4 Processing

Four stages are generally involved in preparing an adhesive bond: *mixing*, *spreading*, *assembly* and *curing*. All four stages may present *health hazards* associated with the resins, solvents and other additives. Each stage has to be adapted to the production requirements. The *size* of the components and the *chemistry* of the adhesive will dictate the equipment needed for *curing*. Large, sealed autoclaves may be required to operate at temperatures of several hundred degrees centigrade, and suitable *jigs*, employed in assembling the components, may also need to be accommodated in the autoclave.

9.2.1.5 Appearance

Appearance may not be an engineering requirement, but it is certainly a *marketing* criterion. Both the *colour* and the *finish* of an adhesive bond may be important in ensuring that the assembled product is acceptable. If the bond *'looks good'*, then it is natural for the customer to give the product the benefit of the doubt, at least until the subsequent performance of the assembly proves inadequate. Conversely, excess adhesive, staining and any appearance of slovenly assembly are likely to prejudice the potential user against the system, independent of whether or not the engineering performance is satisfactory.

9.2.1.6 COST

This again is an obvious consideration, although it is not always obvious how cost should be calculated, or how *cost effectiveness* should be measured. The cost calculation must include the costs of *raw materials*, *storage*, *processing* and *rejection* (failure to meet specifications). However, these are not immutable: *standards* can be altered by mutual agreement of the parties concerned; the cost of raw materials frequently depends on the *quantity* ordered, and both storage and processing costs can be *site-dependent*, differing dramatically from one part of the world to another.

On the other hand, it is often the case that reduced costs are generally of more concern to the customer than *improved performance*. If a new process is 20% *cheaper*, then most manufacturers will give it serious consideration, but if the new process promises a 20% *improvement in bond strength*, then the commercial response may be lukewarm, unless the customers are unsatisfied with the performance of the existing product.

9.2.2 Materials Options

The very wide range of polymeric adhesives that have been developed over the past half century present a bewildering spectrum of options for both the interested observer and the practicing engineer. In what follows we will note *some of the characteristics* of a range of *representative adhesive classes*, but without attempting to present any detailed information. In view of the speed with which new compositions are being developed and marketed, any such attempt would be obsolescent within a year or two. The order in which these examples of adhesives are discussed is deliberately random.

9.2.2.1 EPOXIES

The *epoxy resins* have proven to be one of the most successful high-strength all-purpose adhesives for the mass market. They are commonly sold as two components, the *resin* and a *hardener*, which have to be mixed before spreading on the surfaces to be bonded. The basic structure of the *epoxy* chemical chain is shown in Fig. 9.11a, and it is the epoxy end group which bonds strongly to the surface of a suitable substrate. A wide variety of grades are available, with consumer grades curing at *room temperature* over a period of hours to days. The higher strength commercial grades are cured at moderate temperatures, typically up to *200°C*. The cured adhesive is *brittle* and sensitive to *moisture*, losing strength over an extended period. Curing is accompanied by some *shrinkage*, which often generates undesirable *residual stress*. The epoxy bond degrades at temperatures exceeding 140°C. Adhesion to a metallic or plastic substrate is generally *poor* and the bonds typically fail by *loss of adhesion* rather than by decohesion. Nevertheless,

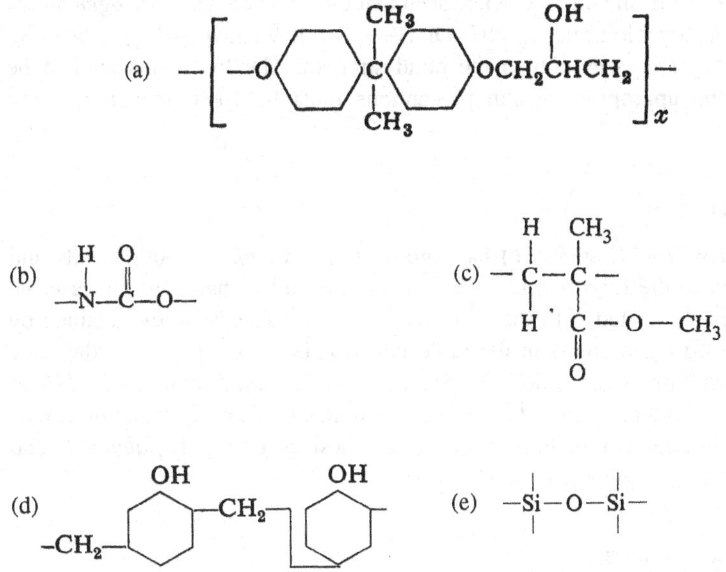

Fig. 9.11. The *repeat structure* of the polymer chains (*the mer*) for some common classes of adhesive (a) Epoxy resins. (b) Polyurethanes. (c) Methacrylates. (d) Phenolic resins. (e) Silicone compounds.

epoxy-bonded *wooden* structures generally fail in the wood rather than in the bond, and a wide range of *glass-fibre-reinforced epoxy composites* have been developed for structural applications. Rolls of flexible epoxy resin sheet *prepreg* are available which contain all the necessary additives. They can be cut to size, assembled, compressed and heated to the curing temperature, when they first *melt* and then *cross-link* to form the structural bond.

9.2.2.2 POLYURETHANE (PU)

These are also commonly sold as *two-component* systems, consisting of the resin and a second component containing a catalyst, however they may also be sold as a *single* component. Compared to the epoxy resins, they have a more limited *shelf-life* and are more sensitive to *moisture* (so that, once the container has been opened the resin should be used as soon as possible if optimum properties are to be obtained). *Curing* may be at room temperature or on heating. *Residual stresses* associated with dimensional changes on curing are not appreciable. Polyurethane adhesives generally have better low temperature strength and toughness than the epoxy resins. They are available with a wide range of *elastic response*, which includes both good stiffness and *elastomeric grades*. The general structure of the molecular chain is shown in Fig 9.11b from which it can be seen that the chain contains *nitrogen* atoms in the highly polarizable group —NHCONH—. It follows that *hydrogen bonding* will make a major contribution to the bond strength. The PU resins tend to be *highly toxic* and appropriate health precautions must be taken in handling the components.

9.2.2.3 ACRYLICS

Grafted *methacrylates* (Fig. 9.11c) have good *peel strength* for both metals and plastics, related to the high *polarizability* of the molecular chain. In this case the polarizability is associated with the —*CO·O*— group. The *flexibility* attained by the inclusion of oxygen atoms in the molecular backbone is reflected in the wide range of temperatures over which the properties are retained, typically -110 *to* $+120°C$. The polymer is resistant to *moisture* and no significant degradation can be attributed to humidity. On the other hand these adhesives are highly *flammable* and degrade over time at higher use temperatures.

9.2.2.4 CYANOACRYLICS

The characteristic chain structure of this family of adhesives includes a *cyanine* group adjacent to the *acrylate* group, increasing the *polarizability* of the polymer still further in comparison with the previous classes of adhesives. The commonly marketed '*superglue*' is of this type, notorious for its ability to stick together the fingers of the unwary! These adhesives are used for bonding *metal*, *plastics* and *rubber*, but they require a good surface finish and minimum thickness of adhesive to be effective. The surfaces to be bonded should therefore be *flat* and the bond gap minimal. *Curing* is achieved through a reaction with *moisture*. In general, the bond tends to have a poor *peel strength*, while the cost of the adhesive remains high.

9.2.2.5 ANAEROBICS

These adhesive resins only cross-link when *oxygen is excluded*, so that curing is initiated by placing the adhesive-coated components *in contact*, thus preventing ingress of air to the resin. The cure reaction is *inhibited* by surface coatings which are in a high oxidation state, such as *chromate* or *anodic* films. *Anaerobic adhesives* have good moisture and solvent resistance, and have been used to bond *metals* and *thermoset* components at temperatures up to $150°C$.

9.2.2.6 SILICONES

Silicone adhesives have limited *cohesive strength* but an exceptionally wide range of application temperatures, from -60 *to* $+370°C$. The structure of the chain is very flexible (Fig. 9.11e) and the elastomeric nature of many of the silicones gives them good *peel strength*. They an be used to bond *metal, glass, rubber* and *plastic* and have a wide range of applications in the microelectronics industry as *potting compounds*. They are, however, moderately expensive.

9.2.2.7 PHENOLICS

The *phenolics* were the first adhesives to be marketed which were not derived from natural compounds. The molecular backbone is based on *linked benzene rings* (Fig. 9.11d), with the *steric hindrance* which one would expect from this geometry. The phenolics are very brittle at low temperatures (−60°C) and require curing at elevated temperatures, but they are extremely *cost effective* and have good environmental resistance. They are the adhesives of first choice for many *laminates* and other *wood products.*

9.2.3 Process Options

In this section we discuss the *form* in which commercial adhesives are marketed, together with the options which are available for *assembling* the adhesive joint. The listing is given in an approximate order of 'popularity' for the different processes, from those most commonly used in the assembly of adhesive joints, to those less frequently met with. The objective is to present a *panorama of options*, on the understanding that this is an active area of technical development in which new ideas are continually being evaluated and applied.

9.2.3.1 CHEMICALLY REACTIVE ADHESIVES

Since *thermosetting* adhesives always involve a chemical reaction, these constitute the class of *chemically reactive adhesives*. They are commonly available either as *two components* which require mixing (the *resin* and the *hardener*), or they are supplied '*premixed*', commonly referred to as '*no-mix*'. The method of curing may be by *heating*, through a reaction with *moisture*, or by an *anaerobic* reaction (with curing *inhibited* by the presence of oxygen until the joint is sealed).

9.2.3.2 EVAPORATION/DIFFUSION ADHESIVES

A large number of *thermoplastic adhesives* are supplied either dissolved in a suitable *organic solvent*, or as an *emulsion*, dispersed in water. After spreading, the solvent is allowed to *evaporate* before assembling the joint and applying *pressure* to complete the bond. Alternatively, the solvent (water, in the case of an *emulsion*) may diffuse and be absorbed into the components, which presupposes sufficient *porosity* or *permeability* in one or other of the components to ensure that *capillary* or *osmotic* forces can separate the adhesive from the carrier.

9.2.3.3 Hot Melt Adhesives

Thermoplastic adhesives may also be supplied as thin sheet which can be cut to size and, when assembled in the joint, subjected to *heat* and *pressure* to create the join. Such processes come very close in concept to a *soldering* operation, except that the temperatures required to melt the *thermoplastics* are lower and the sealing filler is a *polymer* rather than a low melting point metallic alloy.

9.2.3.4 Delayed Tack Adhesives

The *tackiness* or *tack* of an adhesive is the force required to separate a contact immediately after it is made. It is a measure of the capillary forces or '*stickiness*' of the adhesive. In *delayed tack adhesives* a *solid* plasticizer is dissolved in the adhesive to make it non-tacky at room temperature. *Heating* the adhesive above the melting point of the *plasticizer* then results in an abrupt change in the properties: the adhesive, which is now a highly *viscous liquid*, wets the contact surfaces.

9.2.3.5 Film & Tape Adhesives

A high molecular weight polymer, blended with a cross-linking resin, can be supplied in the form of *film* or *tape*, either supported on a carrier or backing film (which must be stripped off before the bond is made) or unsupported. After assembling the adhesive in a *sandwich*, pressure and/or heat must be applied to the bond to complete the bonding process. This method ensures that the adhesive film is of *uniform thickness* and eliminates much of the wastage associated with mixing a batch of *viscous resin*.

9.2.3.6 Pressure-sensitive Adhesives

Elastomeric polymers are sensitive to *pressure*, and appropriate engineering of the polymer structure can lead to an *elastomer* that develops *tack* only under pressure. That is, the polymer is produced as a film or tape which undergoes a sharp transition in its mechanical properties *only* when it is squeezed within the bond gap.

9.2.3.7 Conductive Adhesives

Finally, it may be desirable to have an *electrically* or *thermally conducting* adhesive, for example to prevent *electrostatic charging*. The introduction of a metallic filler powder will achieve this, and *silver*, *copper* and *aluminium* have all been used for this purpose. Solvent-based, silver-filled *acrylates* are commonly used to mount specimens in the scanning electron microscope, where the requirements include adequate bond strength, low vapour pressure, and the absence of *electrostatic charging* in the vacuum environment. *Thermal conductivity* combined with *electrical insulation* is commonly achieved with an *alumina* filler powder.

The volume fraction of filler required to ensure that the *percolation threshold* for conductivity has been reached (so that the powder particles contact each other to form a continuous conducting path) is of the order of 30%. Bearing in mind that the *density* of the powders used is higher than that of the adhesive (by a factor of the order of three or more), this translates into a *weight fraction* of filler of 0.75 to 0.85. The *size distribution* of the filter powder has a major effect on the packing efficiency of the particles and has to be controlled.

9.3 JOINT DESIGN

The most efficient *design* for an adhesive joint is the *tongue and groove*, lap-joint construction, which has been familiar to carpenters, joiners and cabinet makers for close to a thousand years (Fig. 9.12). In this joint a *tensile stress* σ applied to one member is transferred to the other member by *shear stresses* across the interface. The tensile strength of the regions of the joint *perpendicular* to the applied load make a negligible contribution to the joint strength.

We assume that the tongue has a width b equal to twice the width of the arms of the groove and that there is no bond across the segments of the interface perpendicular to the applied load. As a result the stress in the tongue falls to *zero* at the end, as does that at the ends of the arms. In an excellent first approximation (the *shear-lag model*) the *shear stress* τ is then uniform along the bond interfaces, *parallel* to the applied *tensile stress*. If w is the dimension of the bond in the *transverse* direction, then: $d\sigma \cdot bw = 2\tau w \cdot dl$, where dl is an increment of length along the interface, and the factor two arises because two *symmetrically placed* interfaces are involved. Integrating over the length of the joint gives: $\sigma \cdot b = 2\tau l$,

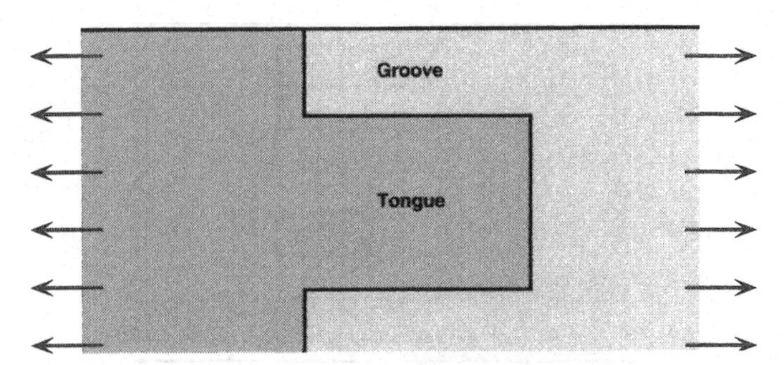

Fig. 9.12. A *tongue and groove* design relies on the *shear strength* of the adhesive joint to transmit a *tensile stress* acting on one of the components to the other component over the length of the join.

which implies that *optimizing* the tongue and groove joint requires making the ratio of the *length* of the tongue to its *width* equal to the ratio between the *tensile strength* of the tongue and *twice* the *shear strength* of the joint. Note that the *total* width of the joint is $2b$, so that the strength of the joint is reduced to one *half* that of a monolithic bar.

It is not necessary for the length l to be all in the same plane, and either space restrictions or a limited shear strength may sometimes make it advantageous to use a *multiple tongue and groove* design. In this case the *shear-lag analysis* yields: $\sigma \cdot b = 2n\tau l$, where n is the number of tongues (Fig.9.13).

A better strength may be attained using a *tenon and mortice joint* (Fig. 9.14a), which places the adhesive in *mixed* compression and shear, and uses *mechanical locking* to improve the joint reliability. The use of *sideplates* will allow a joint to be designed *equal* in strength to the monolithic bar, although at the cost of the increased dimensions associated with the sideplates (Fig 9.14b).

It is common to use an *asymmetrical* assembly in many adhesive joints but this introduces a *bending moment* into the loading geometry. In the simplest situation the joint is *rotated* out of the axis of loading and a *shear stress concentration* develops at the ends of the joint, together with a *tensile* component (Fig. 9.15a). This shear stress initiates a mode II dominated crack and breakaway starting at the end of the joint, a failure mode known as *peeling*, as shown in Fig. 9.15b. Many other joint geometries are possible, but the general principles ought now to be clear: *good* adhesive joints are *symmetrically* designed for *shear loading*, with *mechanical interlocking* as an additional option for improving the strength.

Summary

Viscous polymeric adhesives are applied at room temperature and subsequently harden, sometimes requiring heat or irradiation to reach their full strength (curing).

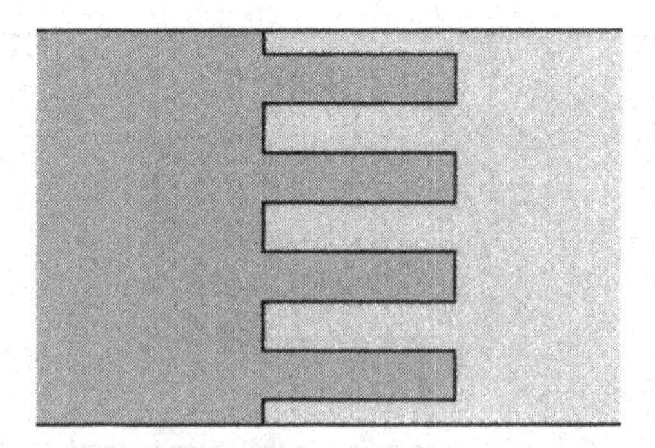

Fig. 9.13. In carpentry, it is common to make use of a *multiple* tongue and groove to reduce *stress concentration* and improve *reliability*.

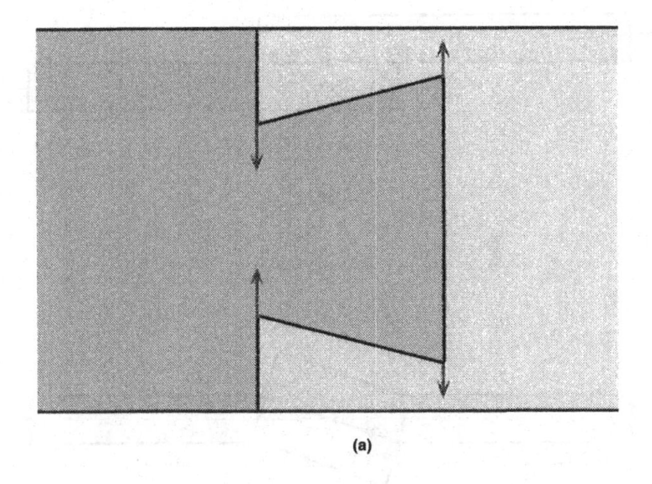
(a)

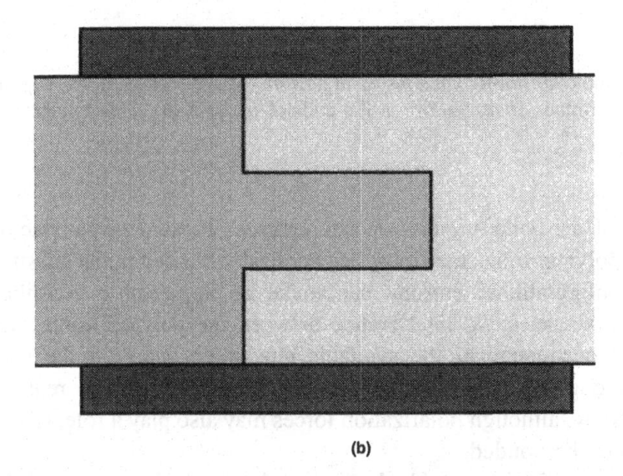
(b)

Fig. 9.14. (a) In a *tenon joint* the mechanical *constraint* prevents shear failure at the bonded interface, but introduces a *stress concentration* at the reentrant angles. The 'waists' (arrowed) now become the weakest points. (b) In ply construction (composite joins) a tongue and groove can be reinforced by *side plates*, but at the cost of an increased cross-section.

Failure of an adhesive bond may occur at the interface (adhesive failure), or within the adhesive layer (cohesive failure), or by a combination of the two morphologies. The failure of the bond may be brittle or it may involve considerable plastic flow, accompanied by the nucleation, growth and coalescence of cavitation. Many adhesives are elastomeric, exhibiting a very large and reversible elastic compliance. The large configurational entropy associated with the long chain polymer molecules contributes an appreciable fraction of the cohesive failure energy, leading to a strong

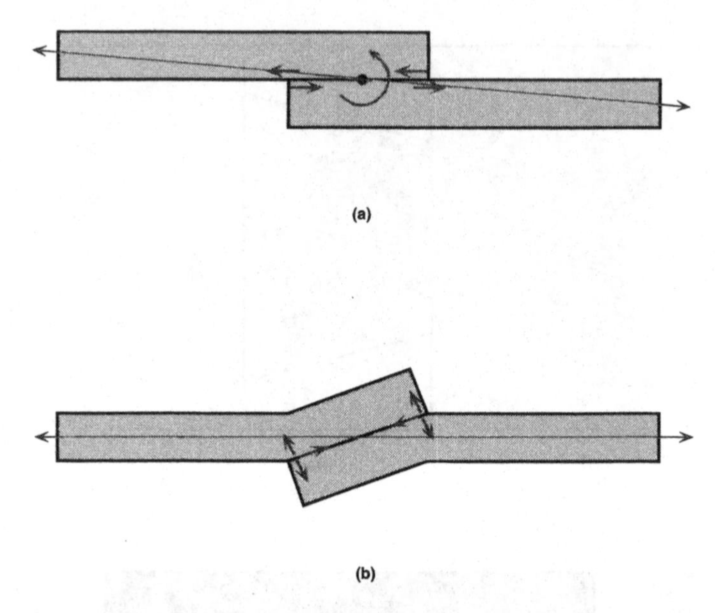

Fig. 9.15. A simple *lap-joint*, when tested in *tension* (a) experiences a *bending moment* which initiates a mixed-mode *shear failure* at the ends of the join (b). This failure mode is termed *peeling*.

dependence of the bond strength on temperature. To achieve satisfactory adhesive strength the polymer molecules must be chemically bonded to the substrate, but once again the configurational entropy can make an appreciable contribution to the strength. In practice, steric interference between the polymer molecules results in only a small occupancy of the available attachment sites on the surface of the substrate. Strong chemical bonding to the surface generally requires ionic or covalent bonding, although polarization forces may also play a role. Chemically inert surfaces cannot be bonded.

In polymeric adhesives residual stresses develop as a result of the chemical reactions occurring during curing, but these are not usually cause for concern. Brittle adhesives are, of course, sensitive to the presence of bonding defects (dirt, debonded regions, gas bubbles and sometimes microcracks), but ductile and elastomeric adhesives can be very defect tolerant. Additions of a filler powder to the adhesive may be used to reduce the compliance of the joint or even, in the case of a metallic powder filler, to confer thermal and electrical conductivity across the interface. Plasticizers increase the compliance. Many adhesives contain a solvent, which evaporates after application, or a hardener, which catalyses a cross-linking reaction between the polymer molecules. In many cases the adhesive bond is sensitive to the environment, especially to the presence of moisture and sunlight. Nevertheless, polymer adhesives can be found to cover a wide temperature range, from −110 to 370°C.

Polymers are generally classified according to their mechanical behaviour. Thermoplastics have well-defined glass transition temperatures and melting points, and have a range of ductility at intermediate temperatures, above which they are viscous liquids. Thermosetting compounds undergo an irreversible cross-linking reaction, remaining solid at higher temperatures until chemical breakdown occurs (pyrolysis). In elastomeric compositions the polymer chains are linked to one another at widely separated points, leading to a large number of alternative configurations. Under stress the material deforms reversibly to very large strains. Thermoplastic and elastomeric adhesives are commonly deposited from solution, but the thermoplastics can also be welded and bonded by the application of heat. Thermosetting compounds undergo a chemical reaction during curing to develop the required toughness and strength.

Most structural adhesives are supplied as viscous liquids and the joint must be formed by clamping the components to be joined. Adequate time must be allowed for the adhesive to wet the surfaces of the components and fill the space between them. The required pressure depends on the joint dimensions and the extent to which capillary forces assist wetting. The latter are largely determined by the surface finish, that is the roughness of the two surfaces. Porous media (such as paper) constitute a special case in which the adhesive flows into the pores and forms a mechanically locked bond.

Most adhesives have a limited shelf life and some may require refrigeration or storage in sealed containers with a dessicant. As a general rule the adhesive must be mixed and applied to the surfaces, which are then assembled for curing. An important consideration is the size of the components which, together with the chemistry of the adhesive, will dictate the curing equipment. Both the appearance of the finished bond and the cost of the adhesive may be important considerations.

A bewildering range of adhesives is now commercially available. The epoxy resins are among the most successful thermosetting adhesives, but they are degraded by moisture and have poor adhesion to metals and plastics. Polyurethanes are available in a wide range of grades of varying compliance. They are extremely versatile and their low temperature strength and toughness are generally better than that of epoxy, although their precursors may be extremely toxic. Acrylics and cyanoacrylics are effective over a wide range of temperature, and can be used for bonding metals, plastics and rubber. Other compositions include the anaerobics, silicones and phenolics.

While most adhesives are applied as liquids, thermoplastics may also be supplied as a sheet preform which is clamped between the surfaces and subsequently heated. A sheet preform may also be used for thermosetting adhesives where the resin is below its softening point at room temperature and the cross-linking reactions occur on heating above the softening point. Pressure sensitive and delayed tack adhesives also involve a solid/liquid transition, the former induced by an applied pressure and the latter by the application of heat.

The design of an adhesive joint is generally aimed at avoiding tensile stresses and therefore emphasizes a lap-joint construction in which it is the shear strength of the joint which is critical. In this type of joint the tensile stress in one component is transferred to the other component by shear stresses across the joint. It follows that there is an exact relationship between the shear strength of the joint, the tensile strength of the two components, and the required dimensions of the joint itself.

Further Reading

1. F. W. Billmeyer, Jr., *Textbook of Polymer Science*, Second Edition, John Wiley & Sons, Chichester, 1971.
2. S. Wu, *Polymer Interface and Adhesion*, Marcel Dekker, 1982.
3. W. J. Feast and H. S. Munro (eds.), *Polymer Surfaces and Interfaces*, John Wiley & Sons, 1987.
4. L.-H. Lee, *Adhesion and Adsorption of Polymers* (2 Vols.), Plenum Press, 1979.
5. B. W. Cherry, *Polymer Surfaces*, Cambridge University Press, 1981.
6. I. M. Ward, *Mechanical Properties of Solid Polymers*, Wiley Interscience, Chichester, 1971.

Problems

9.1 Explain the difference between adhesion and cohesion.

9.2 Describe the phenomenological difference between adhesive failure and cohesive failure. Why is this issue especially important for glues and adhesives?

9.3 Why does the work of adhesion, $w_{ad} = (\gamma_{AA} + \gamma_{BS}) - \gamma_{AB}$, usually differ from the work of failure? Under what circumstances is this also true of a brazed joint as well one which has been glued?

9.4 Explain why entropy plays a major role in the change in free energy which accompanies the failure of polymer-based materials.

9.5 What is the role of entropy in the bonding energy of an adhesive polymer?

9.6 Which types of polymer are expected to melt and which types of polymer are expected to pyrolyse on heating? What is the structural difference responsible for the difference in response to heat of these two types of polymer?

9.7 Define the difference between a plastic and a polymer. Try to be both concise and accurate!

9.8 How does an elastomeric adhesive differ from other types of adhesive? Explain the molecular origin of these differences.

9.9 What class of polymer is 'epoxy'? Give two applications for an epoxy bonded system and describe two engineering limitations of such a system.

9.10 What surface preparation is required for glueing? When would a rough surface finish be better than a polished surface and why?

Part III

APPLICATIONS OF JOINING

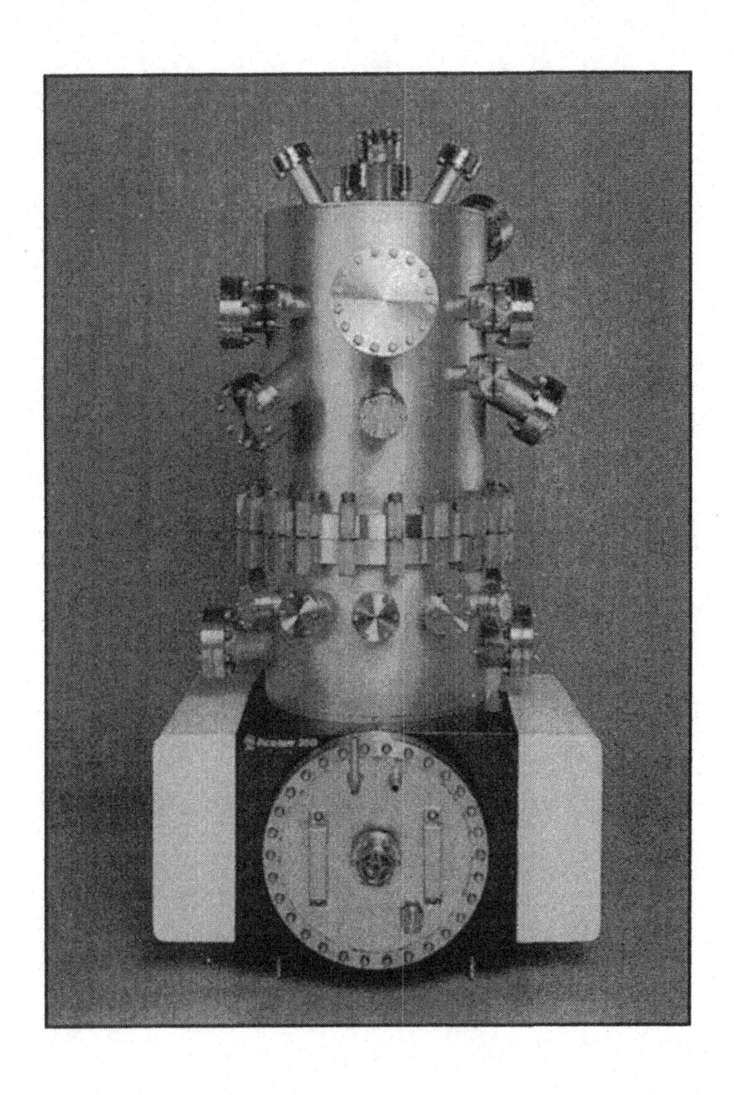

10

Vacuum Seals

Vacuum seals constitute an area of application for a broad range of joining processes. *Vacuum technologies* have developed rapidly over the past fifty years, in response to the demands of some very diverse industries: *food technology*, *microelectronics* and *specialty alloys*. However, the principles employed in the design and operation of *vacuum seals* have changed remarkably little. The combination of the technological importance of *vacuum processing* with the proven reliability of *vacuum sealing* methods makes the subject of this chapter a particularly useful peg on which to hang our first attempt to unify and coordinate the topics covered in previous chapters.

Vacuum seals fulfil a wide range of functions, in addition to the obvious one of providing a *gas-tight joint*. *Ease of assembly* and *disassembly* are often needed, but very often seals are also required to accommodate *dimensional* limitations and the *misalignment* of components. The vacuum seal may be a means for *transmitting movement*, often involving more than one degree of freedom. Figure 10.1 shows a simple scheme for a *bellows seal*. Expansion or contraction of the bellows provides movement in the *z-direction*, which may be combined with angular displacements to obtain movement in the *x–y plane*. The bellows thus allows for *outside* geometrical control of the an assembly positioned *inside* the vacuum system.

An *electrical lead-through* is a common requirement, either to *transmit power* into the vacuum chamber, for heating a substrate or controlling a magnetic field, or to *detect a signal*, as in a wide range of calibrated vacuum gauges. In general, lead-throughs may be of *four* types: electrical lead-throughs required to *detect* or *transmit* a signal, *low-power* connections, operating at much less than a kilowatt (as in an electric light bulb), *high power* transmission seals, generally operating at *low* voltages and *high* currents, and those seals designed for *high voltage* but *low* current connections (as in an electron microscope gun chamber). The *first* are typical of *all* vacuum systems, the *second* of vacuum furnace requirements, and the *third* of

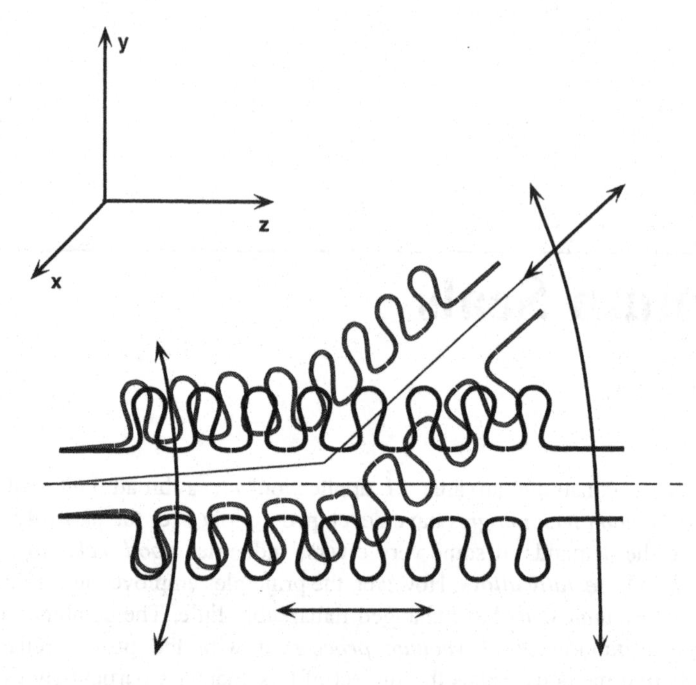

Fig. 10.1. The combination of a displacement *parallel* to the z axis of a metal bellows with *rotation* about the two axes x and y, which are perpendicular to z, permits positioning anywhere within a limited volume *inside* a vacuum system to be controlled from the *outside*.

specialized *ion beam* and electron deposition (*sputtering*) systems, while the *fourth* is characteristic of the requirements of transmission or scanning electron microscopes.

10.1 VACUUM CHARACTERISTICS

From the design engineer's point of view, the difference between a *vacuum chamber* and a *pressure vessel* is that the pressure vessel may have to withstand pressures of hundreds or thousands of atmospheres, while a vacuum system never has to withstand a pressure greater than *one atmosphere*. It follows that satisfying the *mechanical* requirements needed to prevent implosion of a vacuum system is not particularly difficult.

On the other hand, the *pressure ratio* across the wall of a *pressure vessel*, P_1/P_2, is usually less than 10^3, while that across a *vacuum seal* is unlikely to be less than 10^4, and is more likely to be in the range 10^8 *to* 10^{10}. It is this enormous *gradient of gas concentration* across the seal that presents the major obstacle to the design, construction, operation and maintenance of *vacuum seals* – if gas *can* get in to the system from outside, *it will!*

10.1.1 Pressure & Temperature

At sufficiently low pressures *Boyle's law*, $PV = RT$, is obeyed by all low molecular weight gas species, since the *mean free path* between molecular collisions is then orders of magnitude greater than the molecular dimensions. It follows that, for a given system of *fixed* volume, *pressure* is a linear function of *temperature*. Since the *pressure* on the wall of the system is due to the *transfer of momentum*, caused by molecules striking the surface and rebounding, the pressure is also proportional to the *number* and *molecular mass* of the gas species, and to their *velocity*.

The standard international unit used to measure pressure is the *Pascal* (1 Pa = 1 Nm^{-2}), but a commonly used unit in vacuum technology is the *Torr*, named after the Italian scientist *Torricelli*. Atmospheric pressure is 760 Torr. A good *high vacuum system* will operate at 10^{-6} Torr (1 μTorr) and thus has a pressure ratio which approaches 10^9 across the vacuum seals. *Vacuum melting* of specialty alloys is typically performed under a pressure of the order of 10^{-2} to 10^{-4} Torr, while *microelectronic processing* frequently requires that the pressure be below 10^{-9} Torr.

For a *given* temperature the molecular *momentum* is a constant, so that the molecular *velocity* is inversely proportional to the molecular *mass*. From this it follows that the *lighter* gaseous species travel faster than the *heavier* species and, in a *pressure gradient*, will diffuse *faster* down the gradient. The *lightest* gas species (*hydrogen and helium*) are therefore very much more mobile than the heavier species. They are pumped more rapidly *out* of the system *and* leak more readily *into* the system.

10.1.2 Leak Rates & Pumping Speeds

At *constant* temperature the *quantity* of gas that is transferred through a pumping system is proportional to the product of the *volume* transferred multiplied by the *pressure* at which it was transferred: $Q = PV$. In a leaky seal, a *small* volume of gas leaking into the system is *amplified* by the pressure ratio into a *large* volume of gas in the system. Both *leak rates* and *pumping speeds* are measured in units of litres per second (l s^{-1}), and refer to the *volume* of gas transferred per second *from* the low pressure region (*pumping speed*) or *into* the low pressure region (*leak rates*). In a pumping system the gas is extracted in a *series* of pumping stages. A high vacuum, vapour diffusion pump, extracting gas from a system at 10^{-6} *Torr*, may be backed by a mechanical rotary pump operating at, perhaps, 10^{-3} *Torr*. It follows that the *pumping speed* required for the *backing pump* is only 10^{-3} of that required for the *high vacuum diffusion pump*. However, the diffusion pump can only operate efficiently if it is *adequately* backed. For example, a 500 l s^{-1} diffusion pump backed by a 10 l s^{-1} rotary pump will enable the diffusion pump to retain its *full* pumping speed unless the pressure in the system should rise to over $10/500 \times 10^{-3}$ Torr, that is 2×10^{-5} Torr. Above that pressure the effective speed of the diffusion pump falls linearly as the pressure rises, limited by the speed of the rotary pump.

The *mean free path* of gas species is the average distance between intermolecular collisions in the gas phase, given by:

$$\lambda = 1/\sqrt{2} \cdot T/(p\sigma)$$

where λ is in nm, T is the absolute *temperature*, p is the *pressure* in Torr, and σ is the *molecular cross-section* in nm^2. At moderate pressures, below atmospheric, most momentum transfer occurs by collisions in the *gas phase*. Mass transfer in a pressure gradient, in this regime, can be described in terms of the *dimensions* of the system and the *viscosity* of the gas. For a small hole (or gas leak!) with a large pressure differential, we can assume that *all* the molecules impinging on the hole are transmitted to the low pressure side. The number transmitted per unit cross-section per unit time is given by:

$$N = 1/4 \cdot P/(kT) \cdot 2/\sqrt{\pi} \cdot (2RT/M)^{1/2} \ \mathrm{m^{-2}s^{-1}}$$

As noted previously, a low molecular weight species will leak faster than a high molecular weight species, and this is used for leak testing: a helium jet applied to the external site of the leak will increase the pressure inside the system, since the helium is transported faster than the heavier oxygen or nitrogen.

At sufficiently *low pressures* collisions with the walls of the system will dominate *momentum transfer*. That is, the molecules of the gas phase now interact primarily with the *walls of the system* rather than with each other. This *low pressure regime* of gaseous diffusion is termed *Knudsen flow* and the gas now obeys *different* laws of mass transfer in a pressure gradient. For a cylindrical tube of length l and diameter d the *rate of flow*, Q in l s^{-1}, is given by:

$$Q = \sqrt{2\pi}/6 \cdot [d^3 \Delta p / l\sqrt{\rho}]$$

For a circular hole of cross-section A the *rate of flow* is given by:

$$Q = A\Delta p/\sqrt{2\pi\rho}$$

In any vacuum system a primary objective is to ensure that the available pumping speed of the vacuum pumps is used *effectively*. This requires that the *dimensions* of the system should be sufficient to permit adequate throughput of the gas and ensure a uniform pressure in the *working chamber*, close to that at the *pump aperture*. In terms of *vacuum seals*, this means that *all* vacuum valves should operate in such a way as to cause *minimal* obstruction to the gas flow when they are open, while adequately fulfilling their *sealing function* when closed. Some successful *designs* for vacuum valves are shown in Fig. 10.2. The *butterfly valve* (Fig. 10.2a) clearly presents the *least* obstruction, but is difficult to design to close without leaking. The

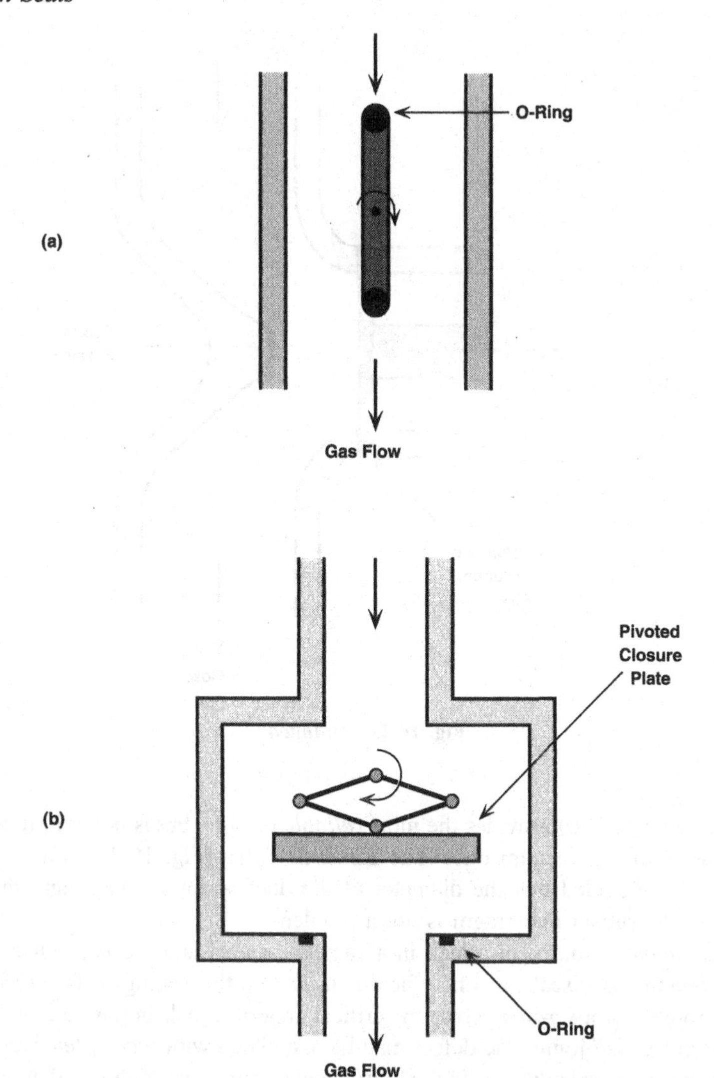

Fig. 10.2. Several strategies can be used to design a *vacuum valve* so as to minimize the loss of *pumping speed* when the valve is open. (a;) A *'butterfly' valve* may not reduce the pumping rate when *open* but is difficult to seal when *closed.* (b) The *'gate' valve* gives excellent closure but has to be bulky if the pumping speed is not to be restricted in the 'open' position. (c) The *'Edward's' valve* is relatively compact and efficient, but relies for its operation on the flexibility of a *rubber membrane* with the associated *outgassing* problems.

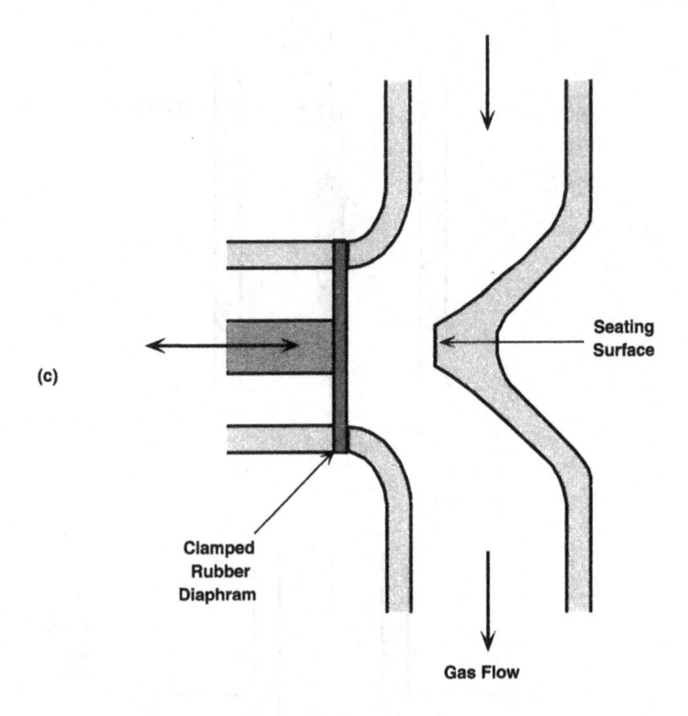

Fig. 10.2. (*continued*)

gate valve (Fig. 10.2b) provides the most *reliable* closure, but is bulky in relation to the diameter of the vacuum duct. The *Edwards valve* (Fig. 10.2c) is an effective compromise, provided that the diameter of the duct is not too large and that *out-gassing* of the rubber diaphragm is not a problem.

The *commonest* source of a leak in a vacuum system is a *sealing defect*. In the case of *demountable* seals, this may be due to *dirt* on the sealing surface (grit in an O-ring groove), or a *scratch* across the critical area of a seal. In the case of *brazed*, *soldered* or *welded* joints, the defect may be associated with *incomplete wetting* of the joint, *porosity* or *inclusions* in the filler metal or *microcracking* in either the *filler metal* weld bead or the heat-affected zone. The *size* of the defect which can give rise to a significant leak is easily estimated. In the high pressure regime, the rate of flow is dictated by the viscosity of the gas, and the appropriate equation is due to Poiseuille:

$$Q = \Delta p/l \cdot (\pi d^4/128\eta) \cdot p_m$$

where d is the diameter of the pinhole, l is the leak path length through the wall of the vacuum vessel, Δp is the pressure differential, η is the gas viscosity and p_m is the mean pressure in the leak.

If the *pumping speed* of the system is inadequate to cope with the leak, then the pressure will rise. For a *500 l s*$^{-1}$ pumping speed, a vacuum of 10^{-6} Torr and a 2 mm leak path, the *maximum* allowable pinhole has a *radius* of less than 10 μm. Since the *leak rate*, in l s^{-1}, depends on the pressure ratio, it will *increase* rapidly as the pressure *decreases*, and the equilibrium pressure in a system which is *entirely* limited by a leak rate will be that at which the pumping speed is *equal* to the leak rate. The *leak rate* can always be measured by sealing off the pumping system and then measuring the *pressure rise* as a function of *time*. For a *constant* rate of mass transfer through the leak, the rate of change of pressure should be *inversely* proportional to the pressure: $dP/dt = a \cdot V/P$, where a is a constant and V is the *volume* of the system. In many cases the rate of increase of pressure obtained on shutting down the pumping system does *not* obey this law. If the pressure tends to saturate at some limiting value, then we may conclude that *degassing* is limiting the attainable vacuum, the ultimate pressure being the effective vapour pressure of the degassing species in the system.

10.1.3 Vapour Pressure & Degassing

At *low pressures* it is usually the rate at which gas is *evolved*, from either a liquid or solid phase, which limits the ultimate vacuum. This process of *gas evolution* is termed *degassing*, and the *rate* at which materials *degas* is a strong function of *temperature* and *pressure*. *Degassing* begins when the *partial pressure* of the degassing species in the system falls *below* the *equilibrium vapour pressure* for the same species. The *equilibrium vapour pressure* depends on the *heat of sublimation* and the ambient *temperature*, and if the evaporating species is present in *solution* then the appropriate *phase transition* can be written as: [A] = (A), where the *square* brackets indicate the *solid solution* and the round brackets the *vapour*. The *equilibrium constant* for the degassing reaction is then: $K = p_A/c_A$, where we assume that the ratio of the *activities* in the gas and solid phases can be replaced by the ratio of the *vapour pressure* in the gas to the *concentration* in the solid. The *equilibrium constant* is related to the *free energy change* which accompanies this reaction, and this free energy change includes *both* the *heat of sublimation* of the pure component *and* its *heat of solution* in the solid.

The *volatile species* which are found in vacuum systems include some unusual cases, and a few examples may be helpful:

1. *Low molecular weight* species, present as residual, unreacted fractions in high molecular weight polymers, will contaminate a vacuum system with *carbonaceous species*. These break down under *irradiation*, depositing a layer of *carbon* on any exposed free surface. *Vacuum grease* and elastomeric *O-ring seals* are both potential sources of *carbon contamination*.
2. Some *metals* have unusually *high* vapour pressures in relation to their melting points. *Vacuum melting* of aluminium alloys will result in severe loss of volatile

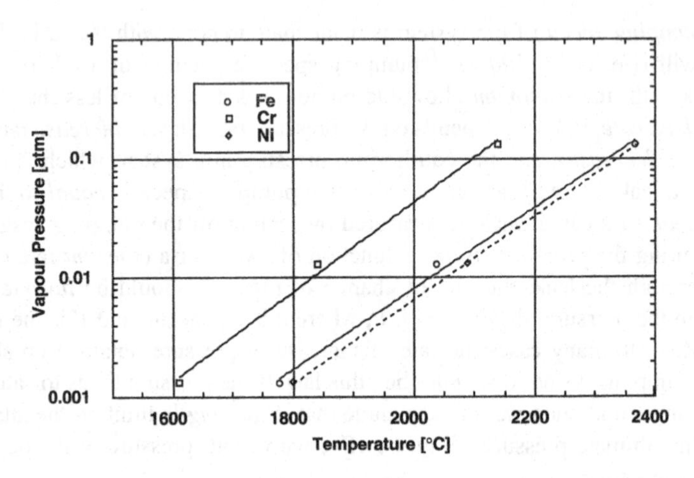

Fig. 10.3. In high temperature *vacuum metallurgy* alloy losses are frequently associated with *evaporation* of one or other of the components. In the vacuum casting of alloy steels and nickel alloys, *chromium* is often the alloying element with the highest vapour pressure.

magnesium and *zinc* additions, and *metal vapour* will deposit on the walls of the system by *vacuum distillation.* Similarly, *vacuum melting* of alloy steels is often accompanied by evaporation of *chromium*, which has a much higher vapour pressure than its neighbouring transition metals in the periodic table (Fig. 10.3).

3. *Silicon* forms a volatile *sub-oxide*, SiO, in addition to the chemically stable form of silica (SiO_2). The formation of the sub-oxide is a frequent intermediate stage in the degradation of silicon nitride components during oxidation, providing a mechanism for *vapour-phase* transport of silica and preventing the formation of a protective oxide film.

4. Many *refractory* metals (tungsten, molybdenum, tantalum and niobium) form *volatile oxides*, rather than the more common protective or semi-protective films. Quite *low* oxygen partial pressures are then sufficient to promote *mass transfer* in the gas phase from the heated refractory metal component, *via* the volatile oxide which is then deposited on a cool surface. The thinning of *tungsten* light-bulb filaments is by *oxide evaporation,* not by evaporation of tungsten. *Molybdenum oxide* is volatile at temperatures as low as *900°C,* requiring that molybdenum heat shields in high temperature furnaces be protected, either by a *reducing atmosphere* or by adequate *cooling*.

5. *Hydrogen* is quite soluble in many engineering alloys and, since it is an *interstitial* impurity, can migrate very rapidly to a free surface. *Degassing* of hydrogen from the walls of a *stainless steel* system is usually the major contributor to the residual pressure in *ultra-high vacuum* systems operating in the 10^{-9} to 10^{-10} Torr range.

Much of the *degassing* in vacuum systems originates from the *desorption* of surface species, rather than by diffusion from the bulk to the free surface. Such contaminants can be either *chemically* or *physically absorbed* on the surface, physical absorption being by *van der Waals' forces* and chemical absorption by *covalent* or *ionic bonding.* The commonest physically absorbed species are *water vapour* and *carbon dioxide,* with carbonaceous species (such as *methane*) also appearing in the mass spectrum.

The pressure in the system will reach a *steady state* when the *rate of degassing* is equal to the *pumping speed,* minus the *leak rate.* In the *absence* of a leak rate, the time dependence of the pressure on sealing the system is quite characteristic, since degassing will *cease* when the pressure in the system reaches the *equilibrium vapour pressure* of the degassing species. On measuring the rate of increase in pressure with pressure the *initial* behaviour is similar to that observed with a leak, but as the equilibrium vapour pressure is approached, the rate of increase of pressure tends to *zero.*

It is often important to know the *composition* of the residual gas in the vacuum system, but the *range of pressure* is often a good indication of the most probable species. At pressures in the *low vacuum* region, 10^{-2} to 10^{-4} Torr, the residual gas is mainly air (*oxygen* and *nitrogen*) with some physically desorbed *water vapour* and *carbon dioxide,* together with *hydrocarbon species* (from residual oil and grease). In a *high vacuum* system (10^{-5} to 10^{-7} Torr) *carbon monoxide* and *water vapour* commonly dominate, *desorbed* from the internal surfaces where they are present in either the chemically or physically absorbed state. In *ultra-high vacuum* systems (10^{-9} to 10^{-11} Torr) *hydrogen* is the dominant species, *desorbing* after diffusing to the surface from solution in the bulk solid. At sufficiently low pressures *helium* diffuses through the walls of the system from the *atmosphere* and is the commonest residual gas species.

As the system is *heated,* the composition of the residual gas will also tend to change. *Water vapour* and *carbon dioxide* are evolved by *desorption* at temperatures below 150°C, *carbon monoxide* desorbs at intermediate temperatures, while *hydrogen* diffuses rapidly out of stainless steel at 350°C.

10.2 SEALING AT MODERATE VACUUM

A *moderate vacuum* is here defined as *10^{-2} to 10^{-4} Torr,* and is frequently sufficient for the vacuum requirements in *chemical engineering* and *metallurgical* plants. *Sealing* in these systems must be on a large scale, if large quantities of gas are to be pumped at low pressures. This can only be achieved if the system dimensions are adequate. Tube diameters of *500 mm* are not unusual.

In what follows we look at some common solutions to *vacuum sealing* problems and comment on their design advantages and limitations. An important point to remember is that the vacuum system operates under a *uniform pressure* of about one atmosphere, and that this provides a nearly constant closure force for *all* vacuum sealing systems. Any forces applied by *other* means, such as a bolted flange, are then *in addition* to atmospheric pressure.

10.2.1 Glass Joints

The *cup and cone* ground glass joint has already been used as an example of a simple mechanical joint (Fig. 1.3), and we have noted that sealing is accomplished by a *film of grease* between the two ground surfaces. For *vacuum* applications, the *vapour pressure* of all constituents of the grease must be sufficiently low to meet the vacuum requirements, while the *viscosity* of the grease must be in the appropriate range: too *high* a viscosity and air bubbles will be trapped in the seal; too *low* a viscosity and the grease film may not be maintained, allowing the ground glass surfaces of the cup and cone to come into direct contact, so that *continuity* of the grease film is lost and disassembly becomes difficult.

Cup and cone seals are used in diameters up to about 50 mm. A number of alternative designs are suitable for larger diameter, *demountable* glass joints. One possibility, used in *chemical engineering plant*, is shown in Fig. 10.4. This is a ground *butt-joint*, the glass tube ending in a *conical flange*. The flange may be *either* mated to a similar component (Fig. 10.4a) *or* sealed to a metal component (Fig. 10.4b). The seal is made with an *O-ring*, either set in an O-ring groove in the *metal* component or mounted on a metal *gasket ring*. In both cases the dimensions of the O-ring mounting are designed to prevent excessive pressure being placed on the O-ring (which would result in it acquiring a *permanent set*). Metal flanges whose internal diameters exceed the maximum outer diameter of the glass flange are padded with a thick *fibrous insert*, used to locate the components and apply uniform pressure to the O-ring seal. The one atmosphere sealing pressure of the vacuum is the *dominant* force on the O-ring. The additional *closure force* on the flange is only applied to guarantee the integrity of the seal before the system is pumped.

10.2.2 Rubber Gaskets

A key component in the assembly of most *demountable* systems is the flexible gasket used to make the seal. The commonest cross-section is the *O-ring*, which is available in a wide range of materials and dimensions (Fig. 10.5). The diameter of the O-ring *cross-section* can be selected *independent* of the diameter of the O-ring, and will depend on the type of seal.

Other gasket cross-sections are available for particular sealing geometries. The *L-section*, used to seal a *vacuum bell-jar* (Fig. 10.6) is readily located by the bell-jar geometry and can be seated on a centre-ground metal base-plate. To be effective the seal must be *clean* and the base-plate free of *damage*, especially radial scratches. As with other elastomeric seals, a thin coating of a *vacuum grease* assures that the seal is *leak-tight* and can be readily *demounted.* In large vacuum evaporation and physical vapour deposition systems L-section seals have diameters of up to 0.5 m. For a bell-jar of *radius* R and *wall thickness* w, the *compressive stress* acting at the base of the L-gasket is just: $\pi R^2 P/(2\pi R w)$, or $\sigma_c = P(R/2w)$. That is, atmospheric pressure acting on the cross-section of the bell-jar is balanced by the compressive stress in the

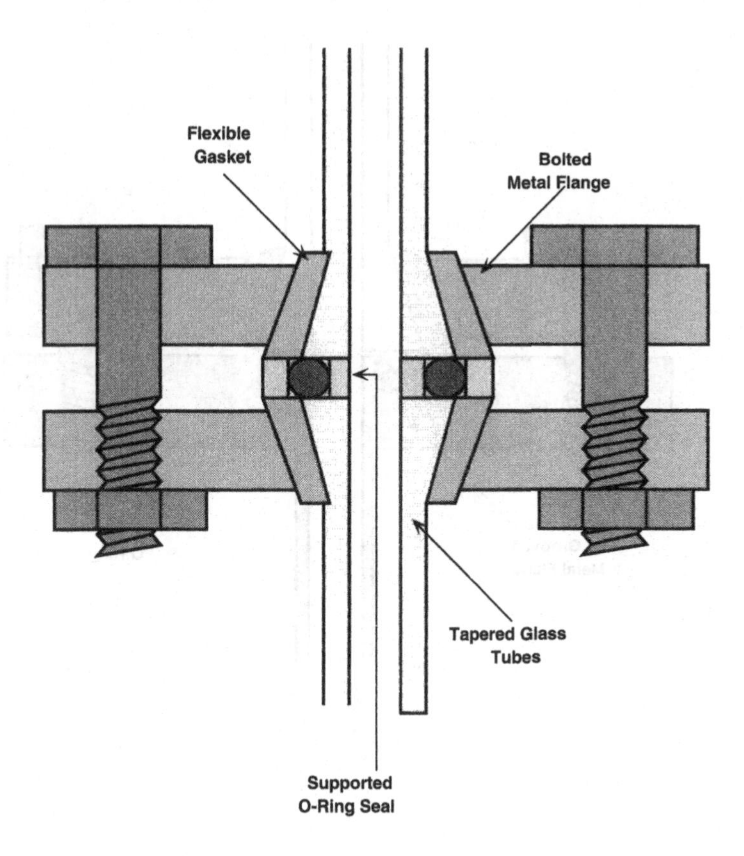

Fig. 10.4. Tapered glass tubing whose *butt ends* are ground flat can be vacuum-sealed effectively up to very large diameters. (a) Two tubes are clamped together using an *O-ring seal*. The clamping pressure is applied via a *flexible gasket* to avoid uneven loading of the heavy-gauge glass tube. (b) A *butt-ended* glass tube can be clamped to a grooved *metal* component using an *O-ring seal*.

walls of the jar. For a 150 mm radius bell-jar with a wall thickness of 5 mm and operating under atmospheric pressure, this stress comes to about 150 MPa – quite sufficient to compress even a hard rubber gasket to *half* its original thickness.

More complex cross-sections are also used, as well as *rectangular*, *square* or other gasket shapes. In general, these are more difficult to clean, less reliable and less readily available as standard items. *Reliable* seal geometries are not easy to design once the simple O-ring has been abandoned, nor is *maintenance* trivial. It is easy to

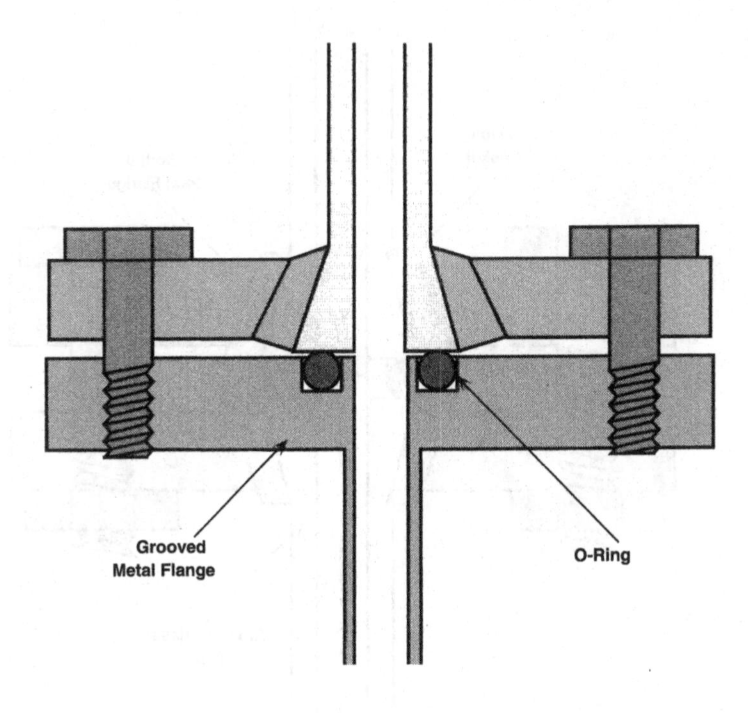

Fig. 10.4. (*continued*)

scratch a mating surface during the loading of a system, and remarkably difficult to remove the scratch once formed.

Materials selection for elastomeric vacuum seals is based on the following three engineering requirements:

1. *Elastomeric response* to the sealing stress, with minimum plastic deformation (*permanent set*) for the duration and temperature of the operating life.
2. *Ease of maintenance*, including cleaning and degreasing with an organic solvent.
3. Minimal vacuum *degassing* after pumping down the system.

In general natural and SBS synthetic rubbers are *not* suitable, and a *nitrile rubber* with a carbon-black filler is the preferred choice. Attempts have been made to reduce

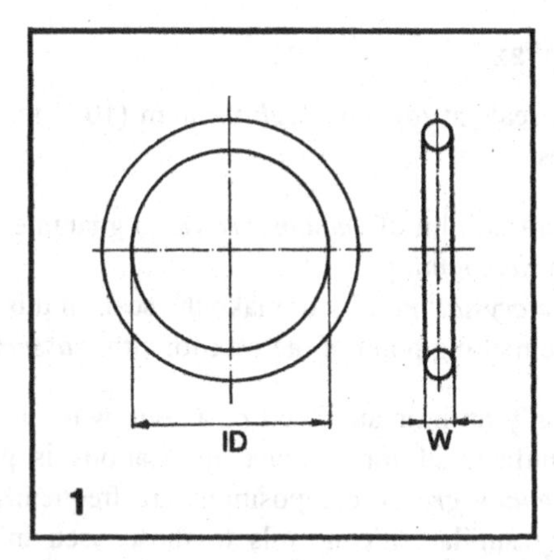

Fig. 10.5. A very wide range of *O-ring sizes* is commercially available for vacuum applications, typically defined in terms of their *internal* diameters (ID) and by the O-ring cross-section (W).

degassing by reducing the solubility of contaminants in the seal material, using O-rings manufactured from a *crystalline* polymer. Materials such as *polyamide* (Nylon) or *polytetrafluorethylene* (PTFE), have been used as vacuum seals, but have not been very successful. These materials creep under load, and relax appreciably during the lifetime of the seal. This increases the *contact area* between the seal and the component, until the *compressive stress* is no longer sufficient to maintain the integrity of the seal.

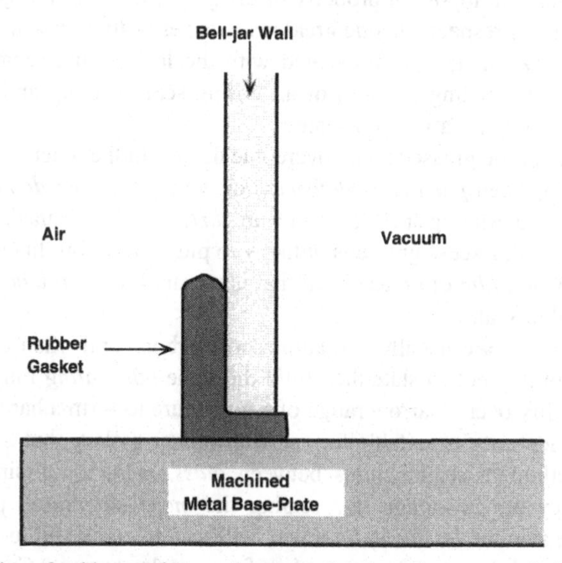

Fig.10.6. The *L-shaped* rubber gasket used to seal the rim of a vacuum bell-jar relies on *atmospheric pressure* to make the seal.

10.2.3 Grease & Wax

Demountable vacuum seals at *low* and *high* vacuum (10^{-2} to 10^{-6} Torr) are generally one of two types:

1. Those that rely on a thin film of *vacuum grease* to guarantee the integrity of the seal and yet permit *demounting.*
2. Those that employ a *crystalline wax* to make the seal. In this case, the joint must be heated *above* the melting point of the wax for either *assembly* or *disassembly.*

A *grease* is commonly an intimate blend of a *soap* with an *oil*, but the composition of greases manufactured for *vacuum* applications is generally a guarded commercial secret. *Vacuum grease* compositions are frequently based on *silicone* chemistry, and employ similar silicone oils to those used in diffusion pumping systems. These have sharply defined boiling points, excellent chemical stability and remarkably low *vapour pressures* at room temperature (below 10^{-8} Torr). *Silicone greases* are commonly used at pressures down to 10^{-6} Torr, and occasionally in systems operating in the 10^{-8} Torr range. Manufacturers' specifications guarantee that the *vapour pressure* of these high vacuum greases does not exceed 10^{-8} Torr.

The grease fulfils *two* functions. Its low shear strength allows the grease to act as a *lubricant*, ensuring that the elastomer has full freedom to relax *parallel* to the plane of the sealing surface, so that the compressive stress is *uniform* over the contact area of the seal. However, the grease also forms a *boundary layer* at the contact surface, and acts as a more effective *barrier* to diffusive mass transfer than a direct metal/rubber contact. An important property of the grease is its *solubility* for contaminant gases, and in this respect silicone greases are superior to hydrocarbon formulations. Ease of *demounting* is also associated with the low shear strength of the grease, which allows the mating surfaces of an O-ring seal or a cup and cone joint to be separated by applying a gentle torque.

The virtues of the grease cannot overcome defects in the other components. An O-ring which has been *plastically deformed* or which has been *damaged* by gouging must be replaced. Mating surfaces which are *dirty* must be cleaned. *Radial scratches* must be removed. Excess grease is unlikely to plug a leak, but likely to contribute to degassing. A *thin film* of grease is all that is required to form a *boundary layer* and effect a reliable seal.

Vacuum waxes are usually crystalline polycarbons and often contain a carbon-black filler. In the molten state they fulfil the same *lubricating* function as a grease, but they solidify over a narrow range of temperature to form a hard, rigid join. They are brittle, and hence unsuitable for use with rubber gaskets, but they can be a very effective method of sealing joins between *glass, metal* or dissimilar *glass/metal* couples. They are unsuitable for sealing the *more compliant*, polymer (plastic) components, sine the *less compliant* wax will tend to crack and lose adhesion. Both the *adhesive* and the *cohesive* strengths of wax seals are poor. *Cup and cone* joints are probably the commonest configurations for wax seals, but metal tubing can also

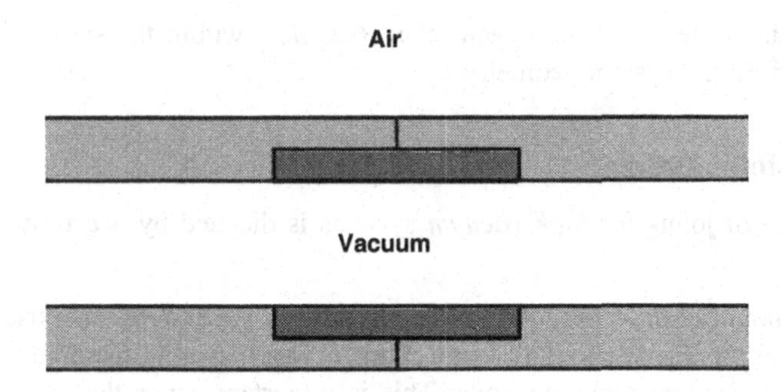

Fig. 10.7. Two lengths of metal tubing can be joined using a *sleeve*, but an *internal* sleeve may be easier to guarantee free of *leaks*, and should have significantly fewer *degassing* problems than an external sleeve.

be joined by using a *metal sleeve* to ensure that any *tensile* stresses on the tube are transferred in *shear* at the joint (Fig. 10.7). Similarly, in a cup and cone metal to glass join, the *glass* is made the cup, to ensure that it is subject to *compressive* rather than *tensile* stresses. The *melting point* of the vacuum wax is low (typically 80 to 120°C) so that the need to *heat* the components to make the seal is not a major problem. In *metal* systems waxed joints are frequently used for temporary assemblies, where more permanent construction would require *brazed* or *soldered* joints.

10.2.4 Soldered Joints

Soldered joints are perfectly satisfactory for *moderate* vacuum requirements (such as the backing lines for a high vacuum system), when these are constructed using *copper* tubing with *brass* accessories. Care must be taken to ensure that soldered joints are *completely wetted*, with no recessed, unwetted areas exposed to the vacuum system. It is also extremely important to ensure that all *flux resin* and *dross* have been removed from the soldered joints *before* assembly of the system. This is not easy when long lengths of tubing are involved.

The standard classes of *solder* are acceptable for vacuum applications, including some *low melting point* solders which may be used for special applications, such as sealing to glass and the attachment of electrical lead-throughs.

10.3 HIGH VACUUM JOINTS

In the pressure range from 10^{-5} *to* 10^{-8} *Torr* more thought needs to be given to the design and assembly of joints. We are now entirely in the *Knudsen* range of pressures, so that gaseous species are individually pumped and interact *only* with the walls of the system, not with each other. *Gas flow* within the system is therefore dominated by the system geometry.

10.3.1 Joint Design

The design of joints for *high vacuum* systems is dictated by *two* primary requirements:

1. *Minimum obstruction of the pumping path in the system.* This usually means maintaining the diameters of all connections *at least* equal to the nominal diameter of the high vacuum pumping units. This is important, since the rate of gas flow depends on a power of the diameter (Section 10.1.2), and *any* constriction will seriously reduce the rate of flow.
2. *Minimum surface area within the system.* Minimizing the ratio of the internal surface area to the volume of the system helps to minimize *degassing*, which is linearly proportional to the *total surface area*. Reducing geometrical complexity and ensuring that the *minimum* surface area of rubber gaskets and O-rings is exposed to the vacuum *also* reduces degassing. Placing an O-ring close to the *edge* of a flange, avoiding a deep recess, is *good* design, and so is ensuring that the O-ring contact exposes the *minimum free surface* of O-ring (Fig. 10.8a and b).

The same criteria should be applied to *brazed* and *welded* joints, which should never be designed in such a way as to leave a *recessed groove* exposed to the vacuum. Such grooves are impossible to clean and provide a near-inexhaustible reservoir for contamination during degassing. Figure 10.9 shows some examples of *good* and *bad* design for brazed and welded joints.

10.3.2 Brazed & Welded Joints

Soldering is not particularly useful for joining components in *high vacuum* systems, in part because *brass* components are unacceptable due to the high vapour pressure of *zinc*, the major alloying addition in brass, at even moderate temperatures. *Copper* tubing is also seldom used for high vacuum applications, since it is slowly oxidized when not under vacuum, and the copper oxide film is a major source of contamination from degassing.

On the other hand, *indium* can be used to seal optical windows (*borosilicate* glass rather than plate glass) into a metal vacuum system, since the very soft indium has a *negligible* vapour pressure and can be *cold-welded* to form a good bond between the glass and the thin chromium oxide film formed on stainless steels. A loop of indium wire is placed on the glass and *pressure* is applied through a bolted flange and a rubber gasket (which does not see the vacuum, Fig. 10.10).

Most high vacuum systems are constructed from *stainless steel*, and the preferred method of assembly of the components is by *welding.* It follows that the stainless steel must be a *weldable* grade, not subject to *weld decay* (corrosion and cracking in

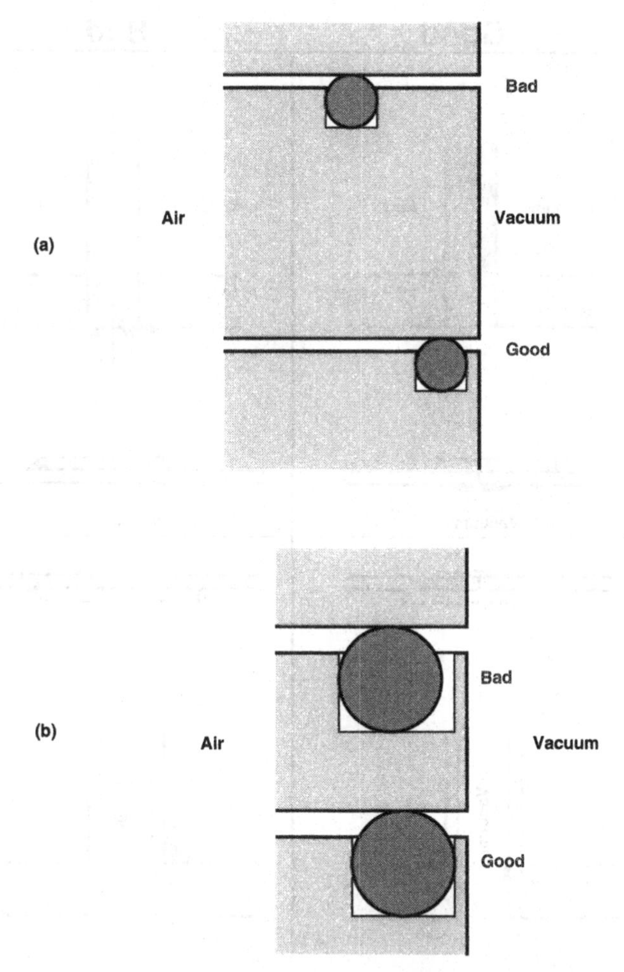

Fig. 10.8. (a) *O-Rings* should be sited in the near *proximity* of the main pumping path and *not* recessed far from the internal diameter of the sealing assembly. (b) The *O-ring* should fit snugly into its groove so that the compressed seal exposes the *minimum* area of the O-ring to *degassing*.

the heat-affected zone). *Brazing* is used if the necessary welding expertise is not available, or where *design* dictates a preference for brazing. An example would be the brazing of a *flange* onto a *thin-walled* tube (Fig. 10.11). For high vacuum assembly, the large area of this *brazed* join guarantees support for the thin walled tube, while a *welded* assembly would be more difficult to design.

The preferred *brazing alloys* for vacuum assemblies are usually members of the *Ag–Cu* family. Zinc alloy additions are undesirable, because of the low sublimation

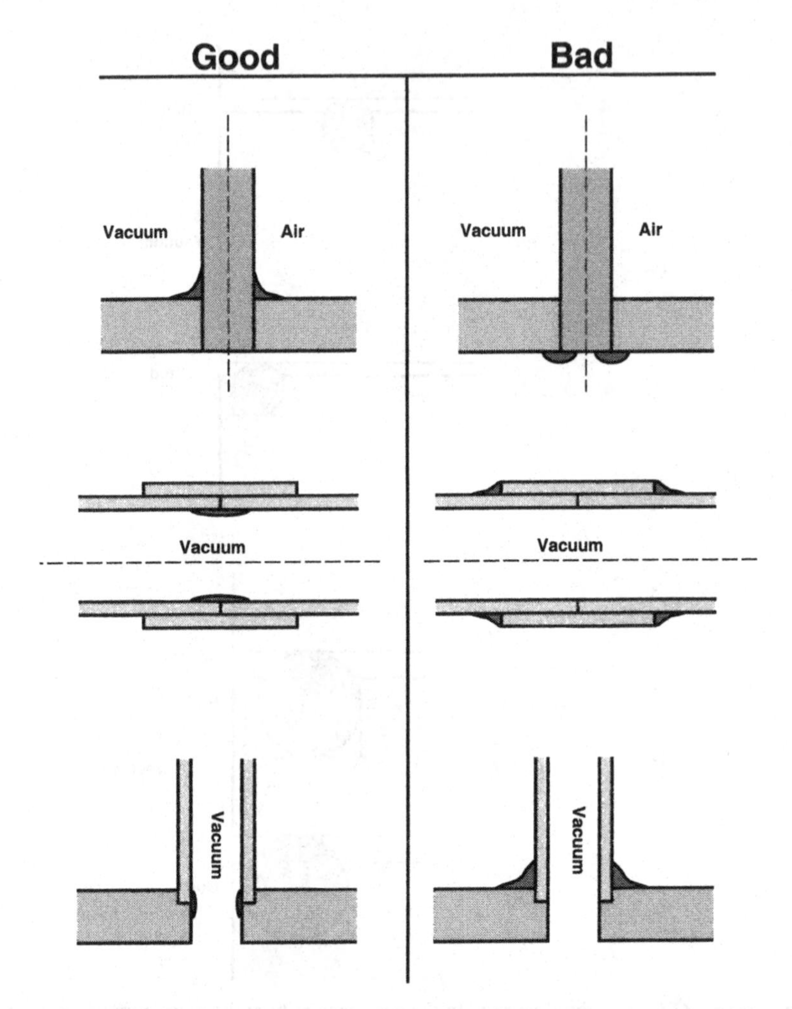

Fig. 10.9. Some examples of *good* and *bad* designs for *brazed* vacuum assemblies.

energy of zinc. Any zinc evolved *during* brazing is liable to be deposited on other parts of the system and contaminate them permanently.

Weldable stainless steels contain an alloying element (typically niobium) to getter *residual carbon* and prevent precipitation of chromium carbide in the heat-affected zone during welding. Alternatively, a *low carbon* stainless steel may be specified, in which the carbon content is reduced below the level at which carbide precipitates. Thus a *316* austenitic stainless steel is available in a *316L* low carbon designation.

Arc-welded joints for high vacuum systems must fully penetrate the cross-section, eliminating reservoirs of contamination and degassing on the internal surfaces of the system. All finished welds must be *scrupulously* clean, and residual flux and slag

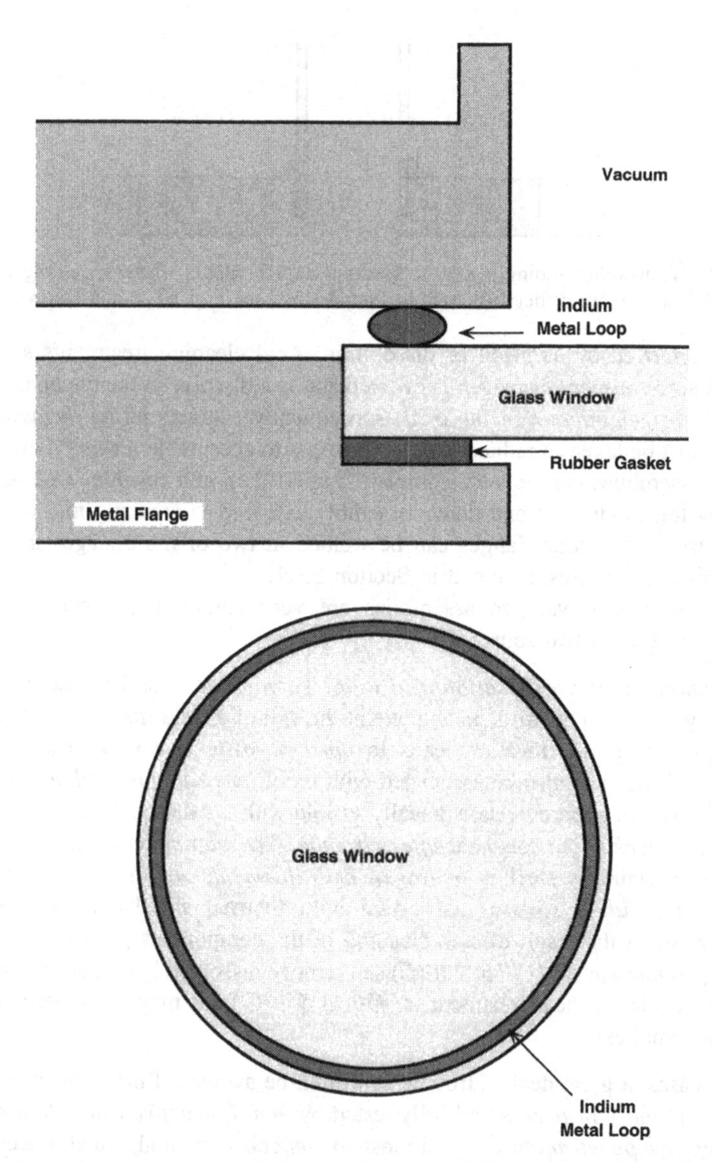

Fig. 10.10. A glass *window port* can be sealed to a *metal flange* using a loop of soft *indium* wire. This is particularly useful when the window port is coated with a layer of *electrically-conducting* tin oxide to prevent *electrostatic charging* and the indium wire serves to earth the assembly.

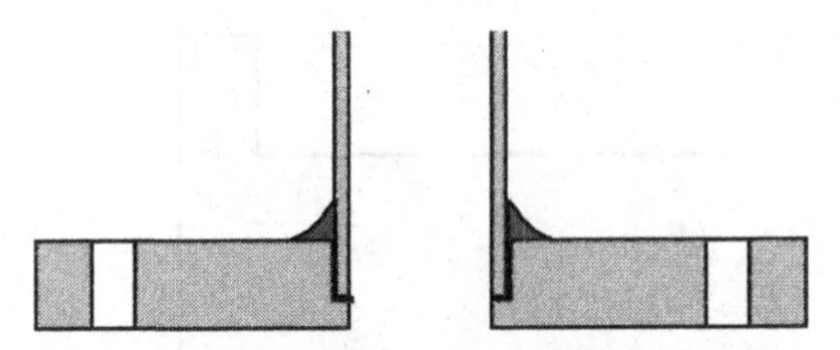

Fig. 10.11. Thin-walled tubing is easy to *braze* to a thick flange, whereas *welding* would be technically much more difficult (though not impossible, see Fig. 10.20 and below).

must be removed by abrasion or dissolution. Acid cleaning treatments should be *avoided*, since *nascent hydrogen* is evolved and can dissolve in the metal, and either lead to *hydrogen microcracking* or be subsequently evolved during *degassing*.

Care must be taken to reduce *residual stresses* to acceptable levels. Many *welded* vacuum assemblies can be very complex (Fig. 10.12), and *residual stresses* introduced by the welding of one flange assembly can lead to cracking in a *previously* welded assembly. Large flanges can be welded in two or more stages in order to reduce residual stresses, as noted in Section 5.1.1.

Leaks in welded vacuum assemblies are very common and can usually be attributed to one of two sources:

1. *Pinholes at the termination of a weld.* These often arise because melting of previously-laid weld metal is incomplete at the end of a weld track (Fig. 10.13) and *shrinkage* during solidification leads to *open porosity* or *microcracking* in this region. *Pinholes* may also be associated with *inclusions* and *gas bubbles* within the weld. The site of such defects is usually visible with a hand-held lens.
2. *Microcracking in the heat-affected zone.* The commonest cause of *microcracking* in stainless steel is *hydrogen embrittlement*, but the two contributing factors are *residual stresses* associated with thermal shrinkage and *dissolved hydrogen*, often the result of acid-cleaning of the components prior to welding. A low temperature anneal (*150 to 200°C*) can remove dissolved hydrogen. *Stress relief* typically requires a heat treatment at *450 to 550°C* (and may be unnecessary for simple assemblies).

In all cases, a great deal of trouble can often be avoided if *all* welds and brazed joints are *visually inspected* and fully tested by *non-destructive evaluation* prior to assembly. *Dye penetrant* testing is a most *cost-effective* method, but all traces of the solutions and particle dispersions used for testing *must* be removed before assembly, preferably by immersion in an *ultrasonic bath*.

10.3.3 Demountable Joints

The use of O-ring seals and vacuum grease is quite acceptable in *demountable* joints for high vacuum systems, provided that the appropriate grades of low vapour-

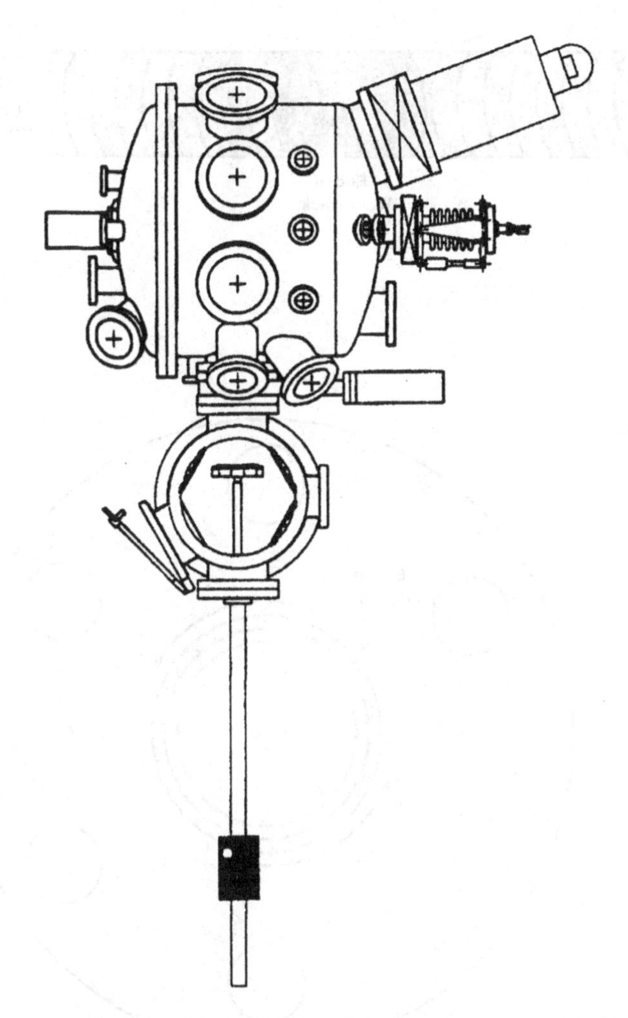

Fig. 10.12. The *stainless steel vacuum chamber* of an Auger spectrometer is designed to hold a vacuum of better than 10^{-9} *Torr* and may contain upwards of 30 individual welds, with over 15 ultra-high vacuum flanges, some of which may exceed 0.5 m in diameter.

pressure grease and vacuum O-ring are employed. The *minimum* quantity of grease should be used and the *minimum* area of O-ring exposed to the vacuum, as previously discussed.

Cup and cone joints are *not* acceptable for high vacuum, primarily because the tubing diameters necessary to ensure adequate gas flow are usually too large for the cup and cone design. This leaves the *bolted flange* as the only easy solution to a readily demountable seal. This is not a serious problem, since the *sealing pressure* is

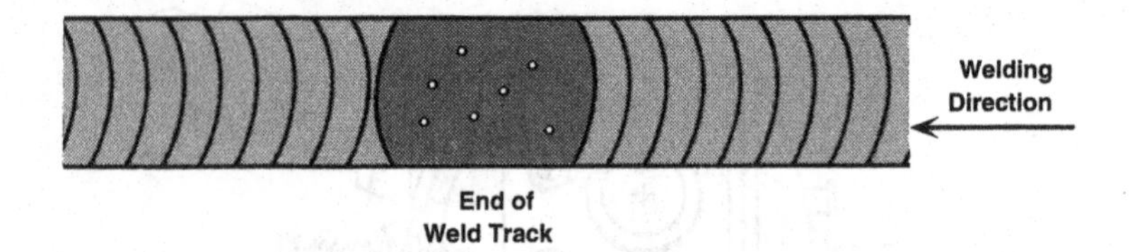

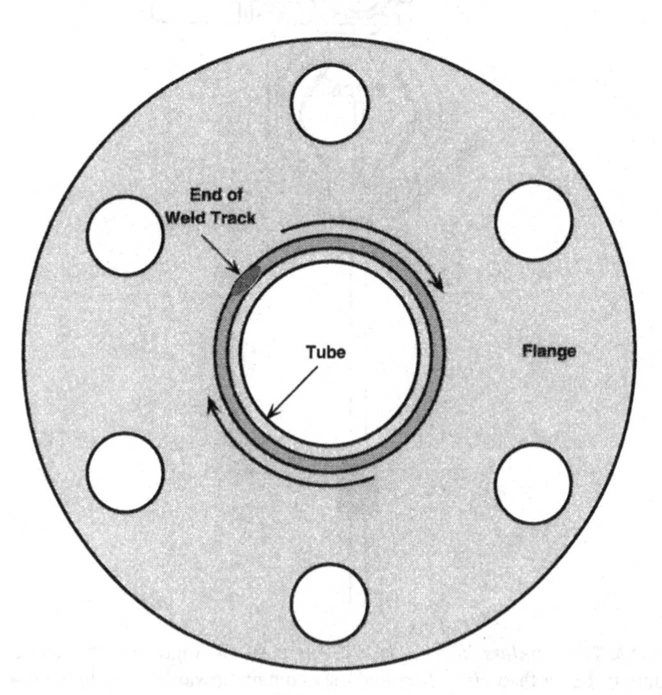

Fig. 10.13. *Pin-hole porosity* and *microcracking* providing a leak path is most likely to be found at the *end* of a weld track, where microshrinkage is often a maximum.

provided by atmospheric pressure, and the bolts can be loosened before releasing the vacuum. An *alternative*, mechanically elegant, solution is the *wedge clamp* (Fig. 10.14). In this design a single thumb screw serves to clamp the flanges of a demountable O-ring seal, using a hinged and tapered bracket. Such a system is very satisfactory for small systems, but a bolted flange is necessary at larger diameters to ensure that the seal is secured uniformly before pump-out.

Fig. 10.14. An *O-ring joint* held by a *tapered* clamp can be quickly and easily assembled or disassembled using a single thumb-screw and without the need for a bolted flange. From Vacuum Generators UHV components catalogue.

Waxed joints can be used for *high vacuum* assemblies, but they are liable to crack and are only a temporary solution. If they are to be demounted and reused *complete* cleaning is necessary, with extensive subsequent degassing. Attempts to reuse a wax seal *without* solvent cleaning, to avoid excessive degassing, are commonly frustrated by dirt in the seal and incipient pyrolysis of the wax.

10.3.4 Electrical Lead-throughs

A wide range of *vacuum lead-throughs* are available commercially for mounting in both metal and glass vacuum assemblies. The electrical conductor may be mounted in a glass *insulator* sealed in a metal sleeve and ready for soldering or brazing into a metal vacuum system (Fig. 10.15). Alternatively, a wide range of glass *pinch seals* are available and ready for assembly by a glass blower into the wall of a borosilicate glass vessel. *Both* types of assembly are suitable for *high vacuum* applications, although those designed for metal systems are often more suitable for *soldering* to brass or copper, and may not be so easily *brazed* to stainless steel.

Pinch-seal electrical lead-throughs are designed for *low voltage, low power* applications. *High current* lead-throughs are usually copper, to accommodate the *high current density*, and the electrodes may have diameters of several milli-

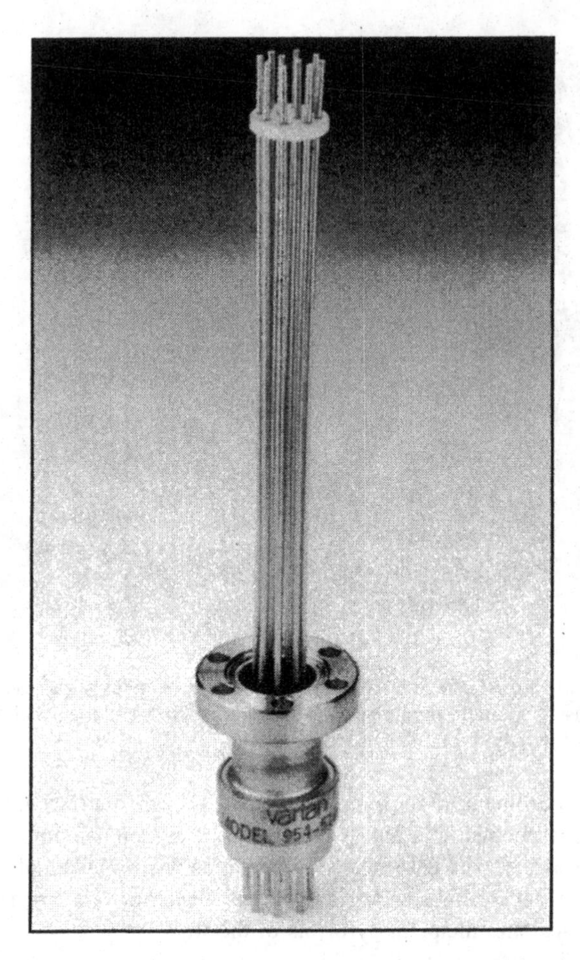

Fig. 10.15. A multiple-electrode, *high-current electrical feed-through* bonded directly to an *alumina* insulator and flange. This unit can be brazed into a stainless steel vacuum system. From Varian Vacuum Products catalogue 1991/92.

metres. Accommodation of the large *thermal expansion* mismatch between copper and the alumina-based insulator usually relies on a *molybdenum manganese/nickel* transition region (Section 8.5.2). The insulator also has to be bonded to a Kovar sleeve. This may then be directly *brazed* or *arc-welded* to the vacuum assembly, or else to a demountable flange. A typical assembly is shown in Fig. 10.16.

Even more stringent *electrical* requirements need to be fulfilled for *high voltage* applications, especially in the range above 10 kV, even though the power requirements may be modest (with currents in the milliamp range). A typical application would be for an *electron microscope* operating at *100 kV*, and such an assembly is shown in Fig. 10.17. The major requirement is for the prevention of *high voltage*

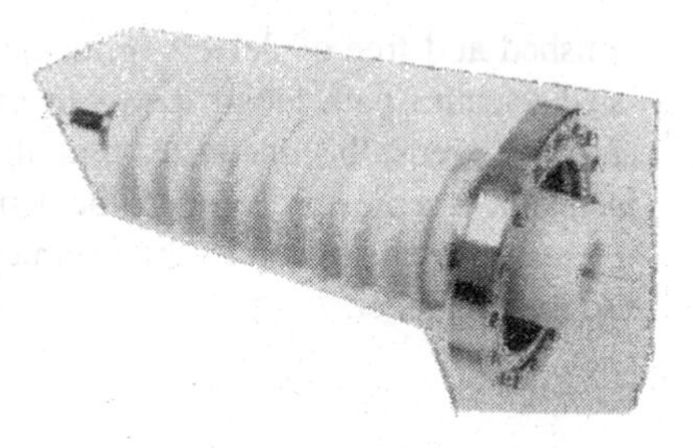

Fig. 10.16. A *high-voltage* electrical lead-through must have a sufficiently large tracking path to prevent *electrical breakdown* at both ends of the electrode that is, both *within* and *outside* the vacuum system. From Vacuum Generators catalogue UHV components.

breakdown. Outside the vacuum chamber this is achieved by conventional means, but inside the chamber some special precautions are needed.

Roughened surfaces provide sources of electric *field enhancement* which lead to breakdown (analogous to mechanical stress concentration at a notch), so that the internal surfaces of the *insulator* and surrounding *potential shield* need to be highly

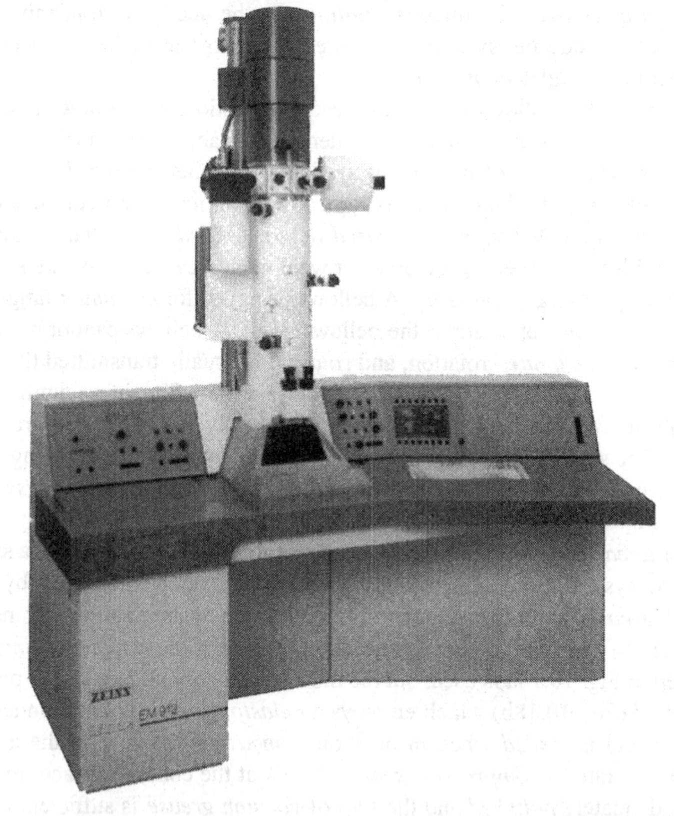

Fig. 10.17. The high-voltage *transmission electron microscope* is one of the most demanding applications for a *high-voltage lead-through*, both in terms of the electrical stability (tracking resistance) and in terms of the vacuum requirements. From Vacuum Generators catalogue UHV components..

polished and free of defects (principally *scratches, pores* or *inclusions*). The insulated vacuum path length between the *earthed* components and the high voltage *electrode* must be adequate, and this may be accomplished by contouring the insulator. A major problem can be deposition of a layer of *carbon contamination* on the insulator as a result of the chemical breakdown of residual hydrocarbons in the vacuum chamber.

10.3.5 Transmitting Motion

Vacuum systems employed in industrial processing often require that *displacement* within the system should be controlled from outside. Casting molten metal in vacuum requires *rotation* of the crucible through an angle of the order of 90°. Positioning a sample in a vacuum assembly may require displacement in *three orthogonal directions*. Continuous *rotation* may be needed to transmit *mechanical power* from outside the system. What are the options available for accomplishing these operations under *high vacuum*?

A simple *bellows* allows for a wide variety of motion, and bellows assemblies are available which permit a sample holder to be accurately positioned in three dimensions. The *range* of movement available is limited by the *dimensions* of the bellows, with allowed displacements typically being no more than 20% of the bellows length. The bellows can be *brazed* or *arc-welded* into a flange, and the join should be fully penetrated by the braze or weld metal, leaving no pockets of potential contamination through *degassing*. A bellows designed for *maximum* fatigue strength will have *minimum* curvature in the bellows' wall. A bellows cannot be relied upon to transmit even a *limited* rotation, and *rotation* is usually transmitted through an *O-ring seal*. A single O-ring is unreliable, since a thin film of vacuum grease will eventually *dewet* after repeated operation, especially if *moisture* is present on the outside of the seal. A more reliable seal is obtained by backing an O-ring seal with a packing which maintains a reservoir of *vacuum grease* on the *high pressure side* of the seal (Fig. 10.18a).

Larger *linear* displacements, such as are required for the insertion of a sample into a vacuum system from a *prepumping chamber*, can be achieved by sliding a *polished, greased shaft* through an O-ring, but some contamination is transmitted on the surface of the shaft. Better results are obtained if the *double O-ring* assembly illustrated in Fig. 10.18a is used, but the most satisfactory sliding seal is probably the *Wilson seal* (Fig. 10.18b) which employs an *elastomer disc* with an undersize inner hole. The seal is *dished* when in position, *concave* side out, and the atmospheric pressure generates a *compressive sealing stress* at the contact surface. Provided the shaft is adequately *polished* and the film of *vacuum grease* is sufficient to maintain the seal, the Wilson seal can allow for large linear displacements with no appreciable degradation of the vacuum, at least in the 10^{-5} Torr range. *Continuous* rotation presents a real problem which is not readily solved. At *low* mechanical loads

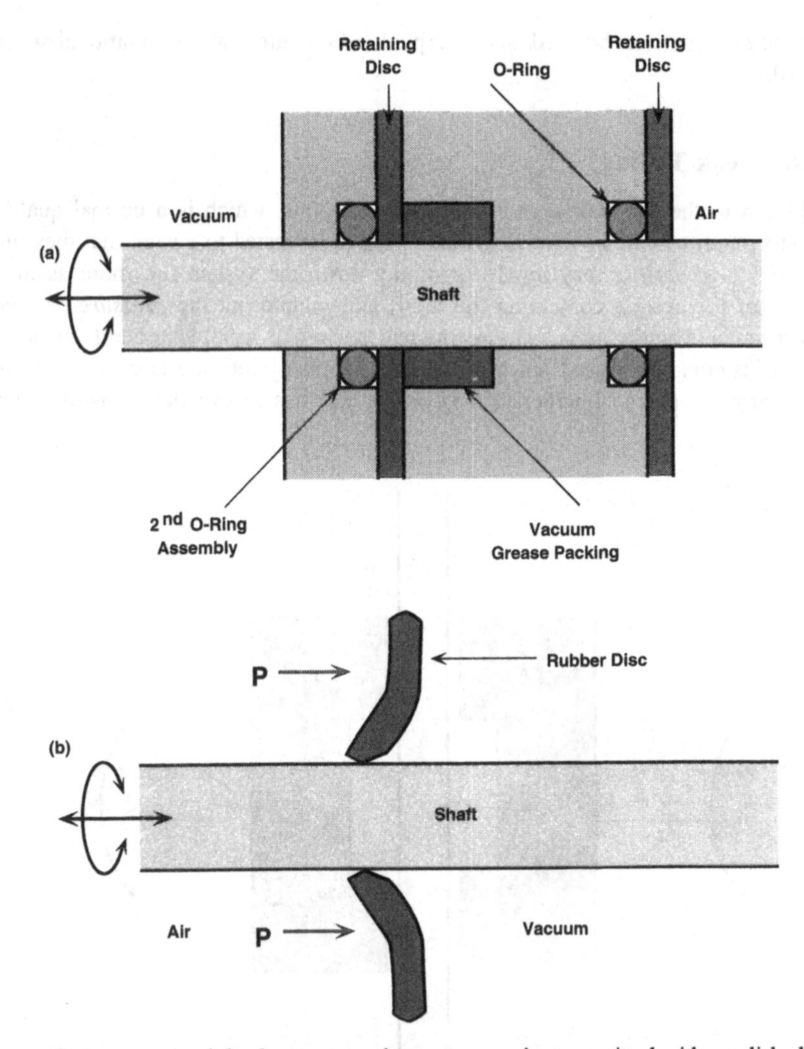

Fig. 10.18. Large *axial displacements* and *rotations* can be transmitted with a polished shaft. (a) Twin O-rings ensure that *parasitic bending* is avoided, while packing the space between the O-rings with *vacuum grease* minimizes ingress of contamination on the sliding shaft. (b) In the *Wilson seal* O-rings are replaced by *rubber disc seals* whose internal diameter is less than that of the shaft. The sense of the disc displacement is such that the *external pressure* acts to *compress* the seal onto the shaft.

magnetic coupling can be used, and is capable of operating at *any* rotational speed (Fig. 10.19).

10.3.6 Leak Testing

In addition to the *non-destructive evaluation* of joints, which is a normal quality assurance requirement, *vacuum joints* also have to be tested to ensure that they are *leak-tight*. *Leak testing* may involve pumping down the system (or, more usually, that part of the system containing the joint) and monitoring the *pressure change* when a small quantity of a *volatile* organic solvent is painted around the join. *Methanol* is commonly used, since it readily *wets* most joints and is sucked rapidly through any pinholes or microcracks. *Methanol* also has an excellent sensitivity for

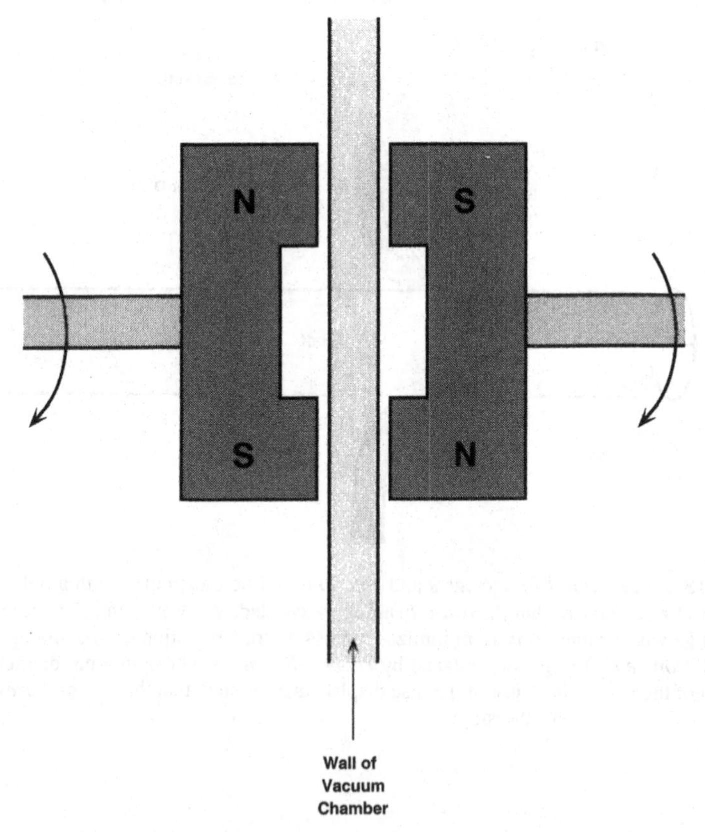

Fig. 10.19. A *magnetic coupling* can be used to transmit *rotation* with no specific limitation on *angular velocity* (and no leak path), but the *torques* that can be transmitted are rather limited.

most pressure-measuring devices (the methanol vapour gives a much *larger* vacuum gauge response than the *same* pressure of oxygen or nitrogen).

More sophisticated leak testing may be done using a *mass spectrometer* attached to the system. This provides information on the *composition* of the residual gas, which in itself is useful in diagnosing the *sources* of contamination. When a jet of *helium* gas is played over the surface of the joint, rapid diffusion (mass transfer) of the low atomic-weight helium down any leak path results in a rapid response on the *mass spectrometer*. The *helium leak-tester* is a standard quality control tool in the assembly of high vacuum systems.

All *leak detection* is plagued by the limited *spatial resolution* of the available techniques. In practice the application of methanol with a fine camelhair brush can locate a leak to within a few millimetres, but no better than that, and only provided that the seal is readily accessible. It is also important to realize that all leak detection starts with the detection of *large* leaks, and then, as the pressure is reduced, proceeds to the detection of the *smaller* leaks. Very often the repair of a seal containing a large leak at a moderate pressure will lead to the discovery that the *same* seal then contains a *small* leak, which is only detected at *high vacuum*. As noted earlier, it is usually possible to distinguish between a *leak* and *degassing* by monitoring the *time dependence* of the *pressure* when the pumping system is shut off. However, such a test says nothing about the *source* of the problem.

10.4 ULTRA-HIGH VACUUM SEALS

Below 10^{-8} *Torr*, in the *ultra-high vacuum* range, it is no longer possible to use O-rings and vacuum grease. Soldered joints are also unacceptable, and brazed joints are only used for special requirements. *Stainless steel* is the standard material of construction, and a *fully welded* structure is the norm. *Glass components*, such as vacuum gauges and viewing ports, are attached to stainless steel by *arc-welded, graded Kovar seals*. Thin-walled stainless steel tube is also *arc-welded*, and not brazed, to a bolted flange, and the *design* of the weld is important (Fig. 10.20a). The tube is welded to a *lip* on the inside of the flange, complying with the requirement that the *minimum surface area* should be exposed to the vacuum, with no reentrant angles or grooves.

Sealing gaskets for bolted flanges are made from soft *electrolytic copper*, the seal being made by a *knife-edged wedge* machined into the stainless steel flange (Fig. 10.20b). The copper gasket is cleaned by chemical polishing, and the knife-edge cuts into the copper in *shear*, to form a metal/metal *cold-welded* bond. Since the seal is not *elastomeric*, the flanges must be bolted together as *uniformly* as possible, using a *torquemeter* and tightening diagonally opposite bolts in sequence. For a *six-bolt* flange in which the bolts are numbered clockwise from 1 to 6, this tightening sequence would be: *1, 4, 2, 5, 3, 6*. A sealed copper-gasket joint is *demountable*, but

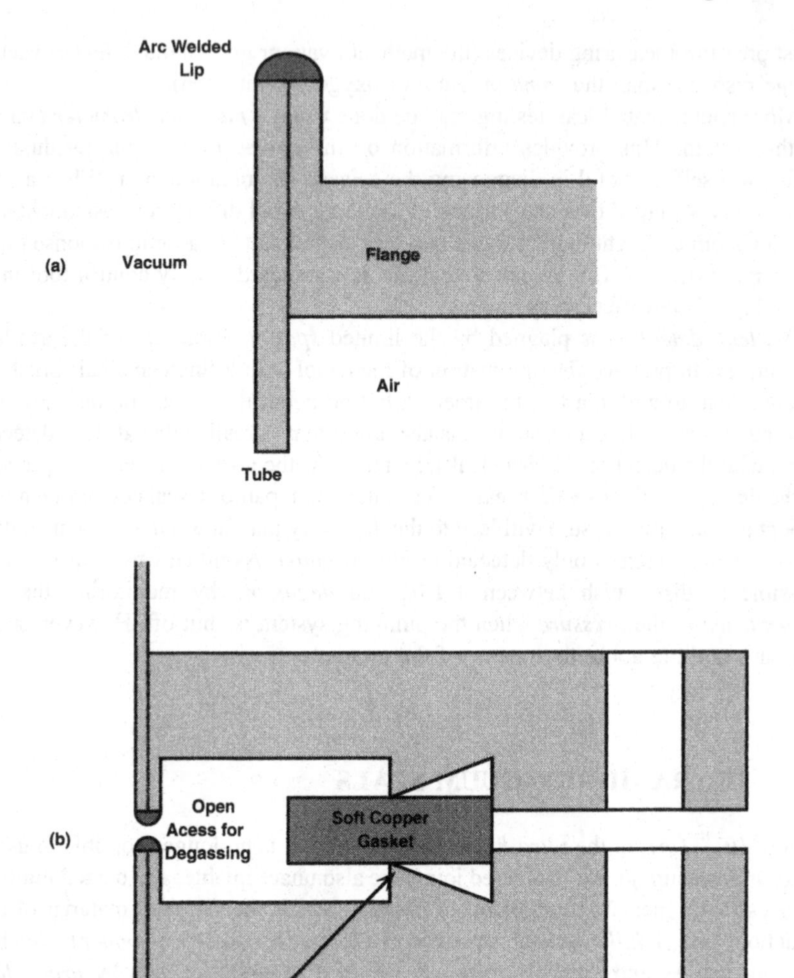

Fig. 10.20. (a) *Ultra-high vacuum* flanges are assembled by TIG welding to *thin-walled* stainless steel tubing, using a narrow lip on the internal diameter of the flange. (b) The *demountable vacuum seal* is made with a soft *copper gasket* compressed between *knife edges* machined into the flange. The larger flange thickness in the external region of the bolt-holes serves to protect both the *knife-edge* and the *weld* from accidental damage during handling.

this also must be done by exerting a *uniform* tensile load, applied through *demounting* bolts located in screw threads on one of the flanges. The *securing* bolts are first removed and then the *demounting* bolts are inserted and tightened to exert a *compressive stress* on the second flange and break the seal.

If the power requirements are small, *electrical lead-throughs* for ultra-high vacuum are made using *glass pinches. High voltage* and *high power* electrodes are identical to those used for high vacuum, except that the electrode units are now sealed into the system by *arc-welding,* rather than brazing, the Kovar mounting into a stainless steel flange assembly.

10.4.1 Comparing Options

The elegant design of the *copper gasket* and *knife-edge seal* for ultra-high vacuum assemblies was developed over a period of time, and it is instructive to look at some of the alternative options which have been discarded. It was early recognized that a plastically deformable, *soft metal* was the best sealing option for a *bakeable* vacuum system intended to operate at 10^{-10} *Torr*. In addition to *copper*, both *gold* and *aluminium* were considered to have possible advantages. (*Indium*, which has a very low vapour pressure and was used to seal *glass* in high vacuum systems, melts at too low a temperature for a *bakeable*, degassed system.)

Pure *aluminium* is extremely soft and has a melting point (660°C) well above any practicable degassing temperature. It has a very low *vapour pressure* and the *oxide* is stoichiometric and stable. A loop of 1–2 mm diameter aluminium wire placed on a centre-ground, bolted stainless steel flange can be *cold-welded* by compression to yield a vacuum tight seal. Unfortunately, the expansion coefficient of aluminium (20×10^{-6}°C^{-1}) is nearly *twice* that of stainless steel, and *leaks* tend to develop as a result of *differential shrinkage* following bakeout. The situation is not helped by two other features. *Oxidation* of aluminium is accompanied by a large *increase* in volume and this expansion also generates *residual stresses* which led to leakage. Also, aluminium diffuses readily into stainless steel, forming a high strength *diffusion bond.* On *demounting* the aluminium seals, the surface of the flanges is damaged by pull-out of diffusion bonded regions of the flange, which makes reuse difficult.

Loops of thin, 0.2 mm diameter, *gold* wire were also used for *ultra-high vacuum seals*, with considerably more success. However, *oxidation* of the external surfaces of the flanges during *bakeout* made it difficult to *relocate* successive seals after *demounting,* while the *cost* of recovery of the gold wire for *recycling* proved to be a further practical difficulty. Comparison of the gold wire seal with the geometry of the copper gasket and knife-edge seal shows that the knife-edge is essentially *unoxidized* during bakeout and *undamaged* during demounting, while the collection of the used copper gaskets for scrap presents no problem.

10.4.2 Degassing Requirements

Stainless steel systems for the *UHV* range require *degassing*, and this, of course, includes the vacuum seals. To reach the *lowest* pressures, below 10^{-10} *Torr*, the system must be baked out at a temperature of the order of *450°C* for sufficient time to enable *adsorbed* and *dissolved* gas, primarily *hydrogen*, to diffuse to, and escape from, the surface. Most vacuum processing requirements are satisfied by degassing the system for a prolonged period at *300°C*, at which temperature most *chemisorbed* surface contamination is released. If the system is only required to operate in the *low* 10^{-8} *to high* 10^{-9} *Torr* range, then it is quite sufficient to degas at a temperature of *120 to 150°C*, and remove *physisorbed* species (predominantly CO_2 and H_2O). This is often done by winding *heater tape* around the system. All UHV vacuum seals must be able to withstand bakeout, and in many cases *hybrid* systems include *both* bakeable UHV sections, which are differentially pumped, *and* high vacuum sections, which contain O-ring seals and other *non-bakeable* components.

Most systems must be let down to atmospheric pressure, if only for *maintenance*, and it is important that this should be done with *minimum* contamination. The time required to repump the system to ultimate vacuum can be greatly reduced by admitting *dry nitrogen* when the vacuum is broken.

Summary

Vacuum seals employ almost every known type of join and cover a very wide range of applications, from food technology to microelectronics and outer space. Vacuum seals may be demountable or permanent. They may be required to transmit a signal (optical or electrical) or movement (linear or rotational). The pressure gradient across the seal may correspond to low vacuum (four orders of magnitude!) or ultra-high vacuum (up to 10^{10}), and these enormous pressure gradients present the major challenge to the ingenuity of the designer.

At the low pressures present in vacuum systems Boyle's law is always obeyed. The mass of gas transferred through the system is determined by the dimensions of the system, the sources of the gas and the pumping speed of the vacuum pumps. The sources of gas are either leaks (controlled or accidental) or out-gassing from the components or contents of the vacuum system. At sufficiently low pressures collisions between gas molecules are infrequent and flow is determined by collisions with the walls of the system. The commonest sources of leaks are sealing defects, often associated with dirt and scratches (which can be very difficult to detect, let alone eliminate).

Degassing occurs in several stages. Physically absorbed, low molecular weight species are evolved from the walls at low temperatures. Water vapour is a common example. Chemically absorbed species, such as carbon monoxide, are degassed at higher temperatures, while gas dissolved in the walls will only diffuse to the surface and escape at elevated temperatures. Volatile oxide species are commonly observed in vacuum systems when components are heated (silicon monoxide, for example), and some metals, such as magnesium and zinc, have unusually high vapour pressures at quite moderate temperatures. Many polymer formulations give off hydrocarbon species when exposed to vacuum.

Vacuum systems used in metallurgical processing may have massive dimensions, requiring welded seals and rubber O-rings and gaskets exceeding 500 mm in diameter. In the chemical and petrochemical industries it is not uncommon to have demountable glass seals of similar dimensions. A key component in such seals is the rubber gasket or O-ring. To ensure that such elastomeric seals are both leak-tight and have a reasonable service life careful attention must be paid to the design of the joint. The elastomer must also be resistant to outgassing. Many demountable seals rely on vacuum grease or wax to prevent leaks, and must have a very low solubility for gaseous species if they are to succeed. Soldered joints are frequently employed to fulfil low vacuum requirements, and can be used not only to join metal components but also glass.

For high vacuum requirements joints must be designed to avoid obstruction of the pumping path and minimize the ratio of surface area to volume in the vacuum system. This applies equally to demountable O-ring seals and gaskets, as well as to brazed and welded joints. The preferred brazing alloys belong to the Ag–Cu system. Zinc-containing compositions should be avoided because of the high vapour pressure of zinc. Stainless steel components should be of weldable grade and arc welding is the preferred process. Leaks are very common and can often be attributed to pinhole porosity associated with inclusions of dross and slag. Microcracking is also common and is generally due to the presence of residual stresses and atomic hydrogen in the HAZ. Visual inspection and non-destructive evaluation before assembly both help to reduce the costs of rejected components.

Electrical lead-throughs and the transmission of motion both present the designer with specific problems which have generally been solved by a combination of pragmatic common sense and long experience. In many cases a bellows can be employed to avoid the ingress of gas associated with a sliding seal. In ultra-high vacuum systems sliding seals are generally unacceptable, although they may be differentially pumped. Demountable seals generally rely on the plastic shear of a metal gasket, usually soft copper sitting on a stainless steel knife edge. To achieve ultra-high vacuum the whole system must generally be baked, typically to about 350°C. This automatically eliminates all possibility of using polymer components for ultra-high vacuum seals.

Further Reading

1. M. H. Hablanian, *High-Vacuum Technology*, Marcel Dekker, Inc., New York, 1990.
2. J. F. O'Hanlon, *A User's Guide to Vacuum Technology*, John Wiley & Sons, Chichester, 1980.

3. A. Chambers, R. K. Fitch and B. S. Halliday, *Basic Vacuum Technology*, Adam Hilger, New York, 1989.

Problems

10.1 What differences in engineering requirements account for the differences in the design of vacuum and pressure seals?
10.2 Explain briefly the principles of operation of both rotary and diffusion pumps.
10.3 What would you suggest to be the most common cause of a leak in a vacuum system? Justify your answer.
10.4 In a UHV system operating at minimum pressure, hydrogen is usually the dominant species found in the residual gas. Why should this be?
10.5 Assume that you are heat treating a low vapour pressure metal in UHV. Would you expect the metal surface to remain chemically passive? Explain the reasons for your answer.
10.6 What is meant by the 'permanent set' of a vacuum gasket?
10.7 Describe two functions of vacuum grease employed in a demountable seal.
10.8 In the 10^{-5}–10^{-8} Torr range of vacuum the geometry of a vacuum joint may have a large effect on the pumping speed of the system. Explain why.
10.9 Discuss the limitations of an aluminium gasket as a seal for UHV applications. Why is copper a better choice?

11

Microelectronic Packaging

The *electronics industry* is a key factor in industrial development all over the world, and the technology associated with *electronic devices* controls progress in all sectors of *commerce, industry* and *society*, both now and for the foreseeable future. Applications range from agriculture to food processing, from heavy industry to pollution control, and from accounting and banking to education and communications. The revolution that began over a century ago with the discovery of *electricity* and the rapid electrification of our society was a revolution in the *sources of power* and the *means of communication*. The invention of the *computer* and the discovery of the *transistor*, both now half-a-century old, transformed the communications industry: in *data acquisition* media, in the capacity for *data processing* and *storage*, and in the speed of *data transmission*.

A key feature of the electronics industry is the drive for *faster* data processing, and *more efficient* data acquisition and transmission, and the three main areas of current hardware development are in *sensor technologies* for data acquisition, *microminiaturization* for data processing and *fibre optics* for data transmission. In the present chapter we will examine a few of the *joining* and *sealing* requirements specific to the electronics industry, and in particular, the impact of *miniaturization* on the *selection* of joining methods.

11.1 MINIATURIZATION

The rapid development of electronics has presented the *materials engineer* with a series of technical challenges which show no sign of approaching a limit. Consider the following examples: the control of semiconductor *purity* at the *ppm* (parts per million) level; the growth of accurately oriented semiconductor *single crystals* (*silicon*, the workhorse of the semiconductor industry, and a wide range of other semiconductors); the drawing of low signal-loss *optical fibres* for long-range data transmission; the assembly of *multilayer ceramic packages* with tolerances in the micron range; the replacement of *rigid* printed circuit (PC) assemblies by *flexible* boards.

This list is long and the successes impressive, but perhaps the major challenge is the drive to *smaller dimensions* for the active device elements. Semiconductor processing based on *photolithographic imaging* is restricted by the *wavelength* of the radiation used to form the image. *Optical* imaging with glass lenses in the visible range (usually green light at 0.5 μm) has a minimum 'point' image diameter of about *0.3* μm, and the various processing steps in device manufacture increase this to the order of *0.7* μm. It follows that this is the current *minimum dimension* for an active element on a semiconductor chip. The *speed* with which data can be processed is primarily limited by the *distance* the signal has to travel, so that *reducing* the distance between active elements is the most effective method of improving the *rate* of data processing.

The *resolution* available in *optical* processing could be improved by reducing the *wavelength* used for photolithography to the *ultra-violet* range (down to a wavelength of the order of *0.3* μm), and this can be done by using special lenses, but the gain is not large. *Electron-beam writing* is used for experimental microcircuitry with a minimum dimension approaching *20 nm*, but the process is slow and unsuitable for large scale integration and mass production. *X-ray lithography* is a possible alternative, and *soft X-ray lithography* is one more area of challenge in materials science. The question is: how will this drive to *miniaturize* the scale of the components affect their *microstructural stability*, especially the *interfacial stability* between the components.

11.11.1 Microstructural Stability

The *equilibrium structure* of a surface was discussed in Chapter 2. As the dimensions of the active components in an engineering system are reduced, the *stability* of the surfaces and interfaces is increasingly affected by 'end effects'. The *interface* can no longer be treated as a *two-dimensional* transition region between two grains of different composition, structure or orientation. Even different *orientations* for the interface between two misoriented grains will have different interface structures (Fig. 11.1). If a *low energy* boundary orientation exists, any *general* orientation may *reduce* its energy by developing a low energy, *facetted* surface. If the grain size is *small* the grain boundary junctions will *constrain* the grain boundary plane to be *curved*. For the *isotropic* case (surface energy independent of orientation) the minimum energy configuration will then be a *curved surface* for which the curvature ($1/r_1 + 1/r_2$) is the *same* at every point. For an *anisotropic* polycrystal, the boundaries at *large* grain sizes will be dominated by a facetted morphology, but at sufficiently *small* grain sizes they will form curved boundaries. At some *intermediate* grain size a transition is to be expected, when boundaries facetted on *specific crystallographic planes* first make their appearance.

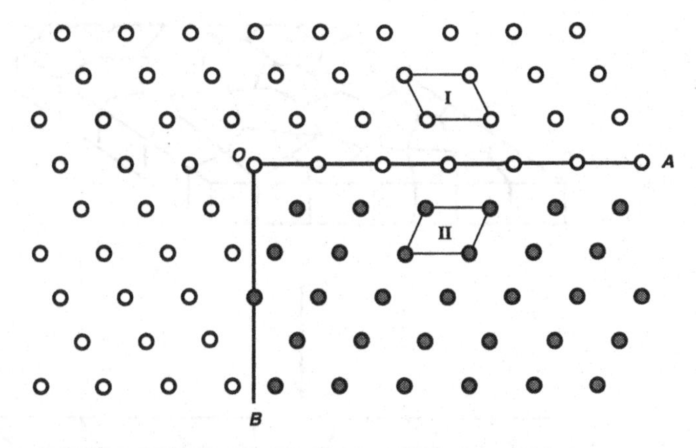

Fig. 11.1. The point lattices of the two grains I and II lie in a *twin orientation* and the segment of boundary OA is a *coherent twin boundary* which would be expected to have a *low interface energy* and hence be *facetted*. The segment OB, however, shows *lattice misfit* distributed over a boundary width of the order of the lattice periodicity. For this segment *boundary curvature* is possible, dictated by separation of the *grain boundary junctions* and hence the *grain size*.

When the size of the *active element* in a microelectronic circuit approaches the *microstructural dimensions*, similar morphological transitions will occur. A *conducting strip* in a microelectronic device is a metal polycrystal which has to carry extremely high *current densities*. It is not easy to stabilize the grain structure of the strip, and *mass transfer*, associated with the current flow, leads to the formation of *holes* at grain boundary triple junctions at one end of the strip, and the growth of *hummocks* at the triple junctions at the other end. This process, termed *electromigration*, is one of the factors limiting the useful *operational life* of microcircuits. What happens when the dimensions of the conducting strip are changed? There are *two topological transitions* (Fig. 11.2). The *first* occurs as the *thickness* of the strip approaches the *grain size*, when most of the boundaries intersect both the top and bottom surfaces of the strip (Fig. 11.2a). In the *isotropic* case the *curvature* is then reduced to $1/r_1$, since the *equilibrium* boundary is always *normal* to the surface of the strip. This is the common feature of *conducting strips* used in most circuits – the grains run perpendicular, from top to bottom of the film. The *second* transition occurs when the *width* of the conducting strip approaches the grain size. *Isotropic* boundaries in *equilibrium* now run *across* the film, *normal* to the axis of the strip and forming *cusps* in what is termed a *bamboo structure* (Fig. 11.2b).

Which of the three structures is the *most* desirable? If the grain size is much *less* than the strip dimensions, then the total boundary area per unit volume of conductor is *large*, and electromigration is *easiest*. On the other hand, a *bamboo* structure can very easily enlarge the edge cusps by *electromigration*, producing a *waist* in the cross-section which *increases* the local current density, *accelerating* the process

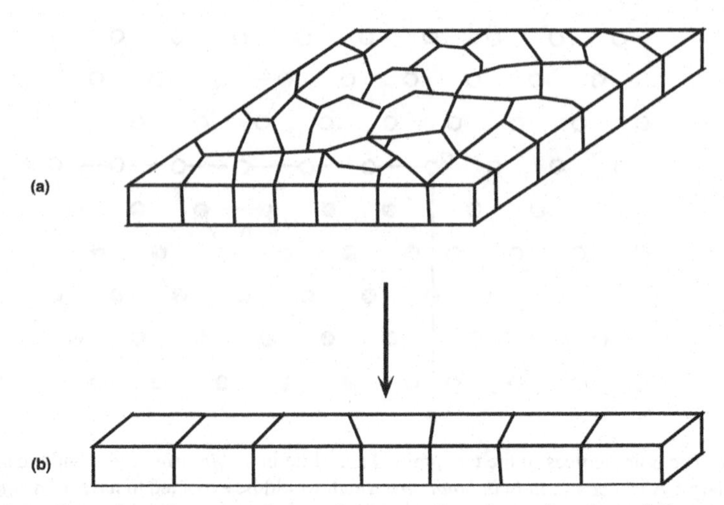

Fig. 11.2. *Grain boundary morphology* is expected to be a function of the component dimensions. (a) The *near-random* orientations of the *boundary normals* in the bulk material tend to a morphology in which *all* the boundary normals lie *in the plane* of a thin film as the *film thickness* approaches the *grain size.* (b) For a *thin wire* or connector strip, the *boundary normals* will align themselves *parallel* to the wire axis, either as the *diameter* is reduced to the *grain size* or as the grains *grow* to the diameter of the wire.

until electrical contact is broken. The *optimum* grain size, which ensures the *longest* working life, is actually that represented by the *intermediate* case (Fig. 11.2a), in which the grain size *exceeds* the film thickness but is *less* than the film width.

This example indicates that *microstructural stability* depends not just on the *microstructural* factors and the *operating conditions* (temperature and current density), but also on *component dimensions*. In general it is the free energy of the *system* which determines *microstructural stability*, so that *fine-grained* materials are *less* stable than *coarse-grained* materials simply because they have a *larger* grain boundary interface area per unit volume, and hence a *higher* total internal energy.

A major effort is being expended on the development of *nanomaterials*, materials having a grain size in the *submicron* range (typically *10 nm to 0.1 μm*). These materials are inherently *less* stable than those with grain dimensions in the *micron* range, but may have inherent advantages for some applications. *Two-phase* nanomaterials should be more stable than *single-phase* materials, since a stable, *discontinuous* second phase inhibits grain boundary migration, while two *continuous, interlocking* phases can only undergo grain growth if mass transfer can take place freely. (That is, *diffusion* of one or more constituents over distances comparable to the grain size.) So far *nanomaterials* have made no impact on the design of *microelectronic devices*, although such materials would certainly seem to have advantages as *substrates* and *multilayer capacitor* elements.

Many *thin films* which have been grown on an oriented *single crystal* substrate have an exact *orientation relationship* with the substrate, and are said to be *epitaxial*. Epitaxially grown layers are themselves *single crystal*, and hence may be microstructurally *more* stable than their polycrystalline counterparts. Such *epitaxial films* are an increasingly common component of microelectronic devices.

Two factors determine the *microstructural stability* of a component:

1. The effective *diffusion distance* of the constituents over the working life of the component. This is determined by the parameter $2\sqrt{Dt}$, the normalizing parameter for the diffusion distance (Section 5.2.2).
2. The *mobility* of the microstructural features per unit driving force. For example the *velocity* of grain boundaries during grain growth, driven by changes in *curvature* (surface energy per unit volume).

Clearly, the *smaller* the dimensions of the active element, the *less time* will be required before changes in microstructure affect device performance.

11.2 PACKAGING REQUIREMENTS

Microelectronic components have to be *packaged* for assembly into engineering systems, and the *packaging* has to fulfil certain engineering requirements, just as would be expected for any *macroscopic* component of an engineering system. These *engineering requirements* for electronic packaging may be summarized as follows:

1. *Electrical contacts*. There has to be a means by which *electrical signals* are transmitted and collected between the *device* and the *outside world*. It is a curious feature of most devices that their *external* size is determined by the limited manual dexterity of the assembly worker, and not by the size of the very small active elements in the package. This is particularly true of the size of the *package* in relation to the *chip*, on which the circuit is mounted. The dimensions of the *electrical connections* are similarly massive in comparison to the *voltage* and *current* requirements of the chip.
2. *Heat dissipation*. In general, the *smaller* the device the *more* heat generated per unit volume. Provision for *heat dissipation* is a major packaging requirement. Adequate *heat paths* often depend on the thermal properties of the substrates used for chip mounting. *Alumina* has excellent thermal conductivity, but *beryllia* is better. However beryllia powders are highly toxic, and the manufacture of beryllia substrates is a hazardous process. *Aluminium nitride* also has excellent thermal conductivity, and has replaced alumina for some applications. Some sealed devices are *helium gas-cooled*, since the low molecular weight of helium gives it excellent heat transfer characteristics.
3. *Environmental sealing*. Microelectronic circuits contain components with drastically different *chemical* properties in electrical contact; they are therefore

susceptible to *galvanic corrosion* in the presence of *humidity*, as well as to attack by a variety of industrial *pollutants*. Adequate sealing from the environment is necessary to protect the circuit and its connections from *oxidation* and *corrosion*.

4. *Mechanical fragility*. The *small scale* of microelectronic components makes them susceptible to *mechanical damage*, especially *vibration*. It is one of the tasks of the packaging to protect against such damage. A range of industrial test procedures has been developed to evaluate the resistance of the package to *mechanical shock*, *vibration*, and *thermal cycling* at both cryogenic and elevated temperatures.

Two basic strategies are used for packaging, and these are illustrated schematically in Fig. 11.3. In Fig. 11.3a the *microchip circuit* is protected from the environment by a *hermetically sealed can*. The *chip* is bonded to a *header* and the *contact wires* are welded to *Kovar pins*. The header, the pins and the can are all assembled on a *glass seal*. In Fig. 11.3b environmental protection is provided by a *potting compound*. *Electrical contacts* are made to a *lead frame* from the *chip microcircuit* which is again mounted on a *header*. The thermosetting *potting compound* is then cast to completely enclose both the microcircuit and the contact wires, leaving only the *lead frame electrodes* exposed. The *dimensions* of both these assemblies are selected for *ease of handling* and bear no relation to the size of the *microcircuit*. The chip dimensions, on the other hand, are made *as small as possible* in order to reduce the *signal transmission times* within the circuit.

11.2.1 Electrical Contacts

The *contact wires* in the microcircuits illustrated in Fig. 11.3 are typically only *20 μm* in diameter or even less. These wires have to be *welded* to the tabs deposited on the chip, which are only a few microns thick, and then to the electrode pins at the other, which are a few tenths of a millimetre in diameter.

These bonds are made by *ultrasonic welding*. The *wire* is fed through an *alumina capillary* (Fig. 11.4) and an *electrical discharge* is used to form a *bead*. This is bonded to the *contact pad* on the microcircuit by transmitting ultrasonic energy from a *piezoelectric transducer* down the alumina capillary. The wire is drawn *through* the capillary to the *electrode pin* and a second ultrasonic weld is made to the top of the pin. Finally, a *spark discharge* through the pin *melts* the wire, leaving a solid bead ready for the next contact pad weld.

When the chip is mounted on a *ceramic substrate* the electrical conducting paths are deposited *directly* onto the ceramic, and electrical contact with the system is made *either* via electrode pins bonded *through* the ceramic *or* at the edge of the substrate via a *leadless* contact. Both options are illustrated in Fig. 11.5. With the rapid improvements in processing has come increased *reliability* for these *leadless* ceramic contacts.

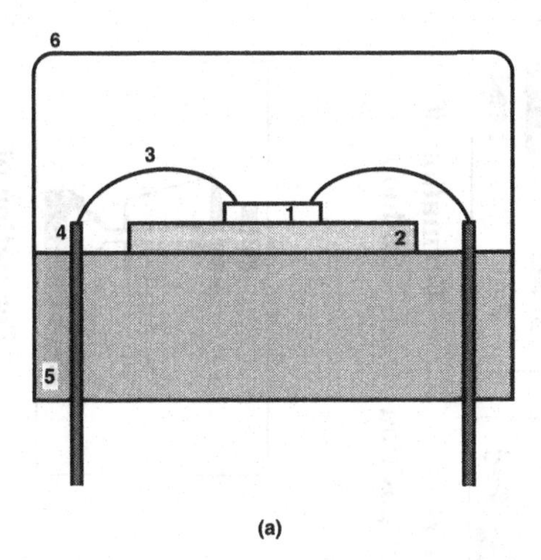

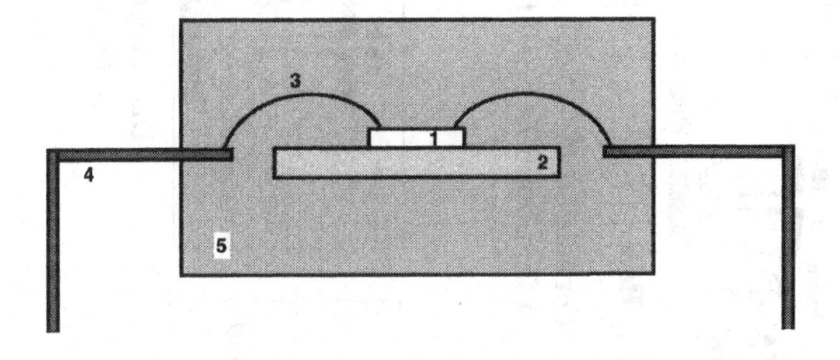

Fig. 11.3. The silicon *chip* is much smaller than the component *microcircuit,* which has to be capable of being handled and *mounted* in a system. *Two* strategies are commonly used. (a) The *chip* (1) is mounted on a *backer* (2) and the electrical *connections* are made with fine wires (3) attached to electrode *pins* (4) which are set into a glass *mount* (5), the whole assembly being hermetically sealed into a metal *can* (6). (b) The *chip* (1), *backer* (2) and wire *leads* (3) are mounted as before, but the leads are connected to a *lead frame* (4) and the assembly is set into a thermosetting *potting compound* (5).

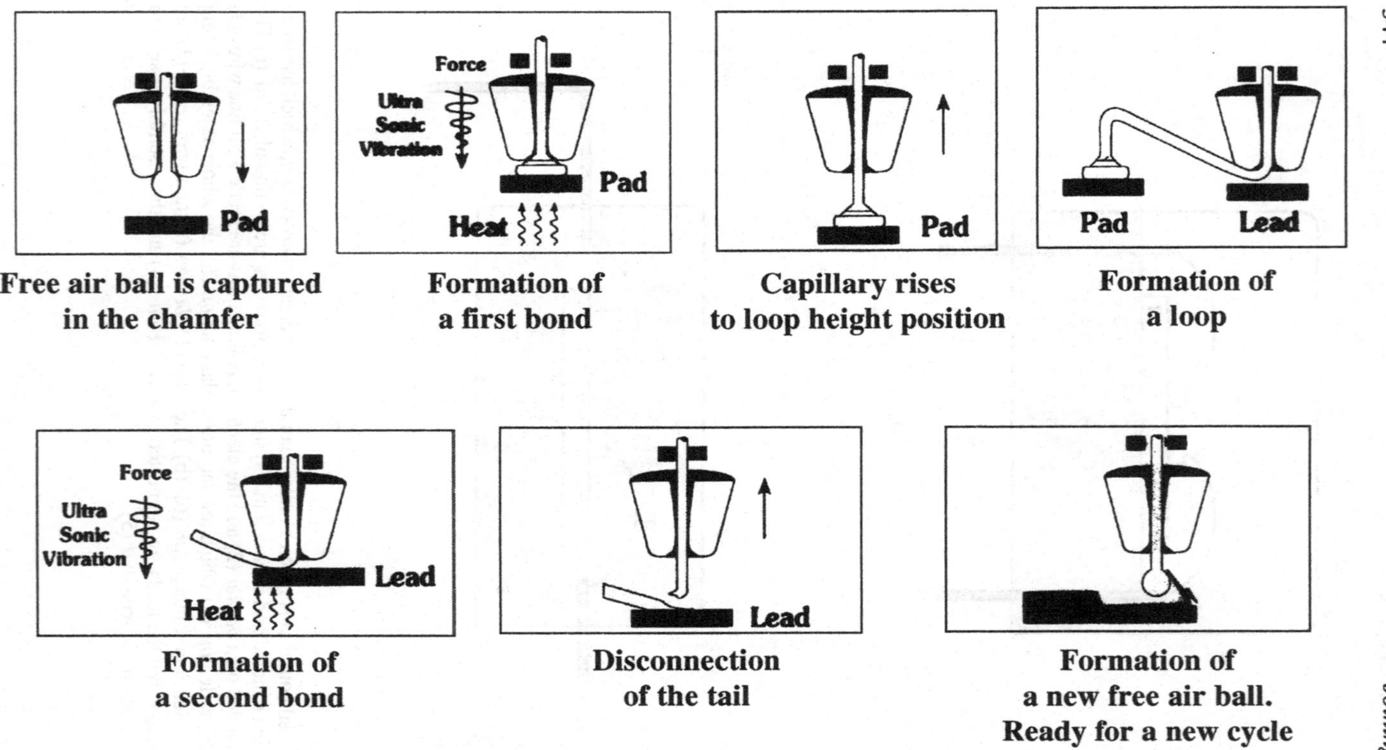

Fig. 11.4. *Ultrasonic welding* is the preferred method for attaching the *leads* from the *chip* to the *electrodes* which make contact with the external world. A bead is formed at the end of the connecting wire. The *bead* is bonded to the *contact* on the chip, the ultrasonic energy being transmitted from a *piezoelectric transducer* down the ceramic capillary *wire guide.* The *wire* is drawn through the *capillary* and a second *ultrasonic bond* is made to the external *electrode*. An *electric discharge* passed through the *electrode* melts the *wire*, terminating the connection and forming the *bead* ready for the next weld.

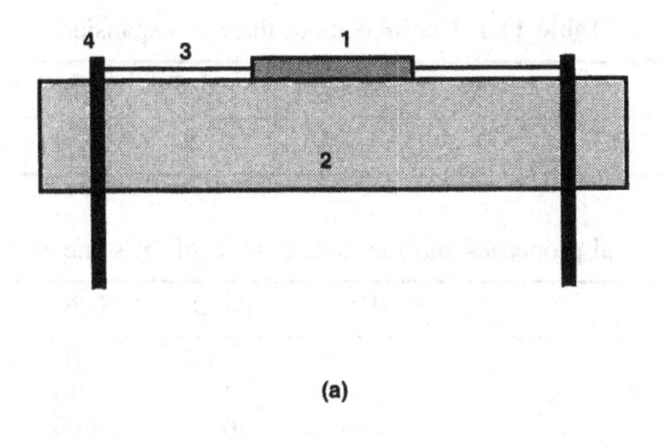

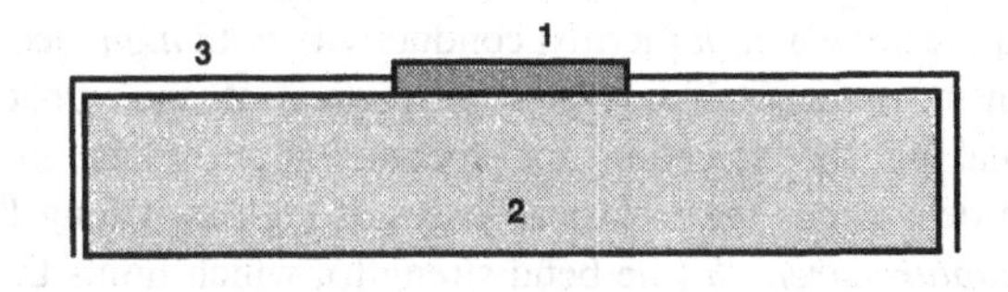

Fig. 11.5. *Two* techniques have been developed for making the electrical connections to a *chip* (1) mounted on a *ceramic substrate* (usually alumina) (2). (a) Electrical *leads* (3) are printed and fired onto the *ceramic* and *diffusion-bonded* to *pins* (4) set into holes in the ceramic to form a *pin grid*. (b) The electrical *leads* (3) are printed *over the edge* of the ceramic substrate, where they make *direct* contact with the rest of the system in a *leadless* assembly.

11.2.2 Ceramic Substrates

The *mounting* for a microchip on a substrate may be based on one or other of *three* technologies:

1. *Headers and cans*
2. *Lead frames*
3. *Ceramic sheet*

Good *adhesion* between a silicon chip and a substrate requires that the *thermal expansion mismatch* should be less than $10^{-4}\ °C^{-1}$. Table 11.1 lists the coefficients of thermal expansion for *silicon, Kovar* (for the range 100 to 500°C) and some candidate substrate ceramics: Al_2O_3, *BeO*, *AlN*, and *MAS* (magnesium aluminium

Table 11.1. Coefficients of thermal expansion

Material	Kovar	Silicon	Al_2O_3	BeO	AlN	MAS
Exp. coeff. $°C^{-1} \times 10^{-6}$	4	2.5	8	9	4.2	2.0

Table 11.2. Physical properties and mechanical strength of some electronic ceramics

Material	BeO	Al_2O_3	Si_3N_4	MAS	AlN
Dielectric constant	6.5	9.6	>10	5.0	10
Thermal conductivity ($W\ m^{-1}\ K^{-1}$)	250	26	30	5	150
TRS (MPa)	200	400	600	120	300

silicate – the mineral cordierite). *Glass substrates* are chosen to have an expansion coefficient equal to *Kovar*, about $4 \times 10^{-6}\ °C^{-1}$.

Heat dissipation and *electrical insulation* requirements demand that substrate materials should combine a *high* thermal conductivity with *high* electrical resistance. Some compromise is inevitable. Other properties of the substrate which are important in making the selection for specific applications are the *dielectric constant*, which determines the electrical losses at high switching frequencies, and the *transverse rupture strength* (the bend strength), which limits the dimensions of the substrate which can be used without exceeding an acceptable *probability of failure* under service conditions. Some typical values for the *dielectric constant*, the *thermal conductivity* and the *transverse rupture strength* (TRS) are listed in Table 11.2. *Alumina* and *silicon nitride* are less suitable for *high frequency* applications because of their *high dielectric constant*, while MAS substrates are a poor choice for applications requiring *high power dissipation*, since MAS has a very *low thermal conductivity*.

11.3 ELECTRICAL INSULATION

An important component of microelectronic devices is the *electrical insulation* of the elements of the microcircuit. All electrically insulating films for this purpose have in common the requirement for a *high breakdown voltage*, together with the need for *uniform deposition* (at thicknesses which are typically in the *submicron* range).

Silica films are grown on silicon chips by *controlled oxidation*. Ceramic films are grown by *chemical vapour deposition* (CVD) while polymer films are used for both *photoresist layers* and *protective coatings*.

Each of these classes of *insulating film* has its own range of application, for example:

1. *Silica* is an interlayer dielectric and gate oxide in solid state devices.
2. *Silicon nitride* is a capacitor dielectric and a diffusion barrier.
3. *Polyimide films* are used as protective coatings in *gallium arsenide* technology.

11.3.1 Silica Insulating Films

Thermal oxidation of silicon to form an insulating, amorphous silica film on a silicon chip takes place in two stages. In the *first* stage, up to about *100 nm*, the rate of oxidation is limited by a combination of the rate of *gas arrival* and the rate of *surface reaction*, and the growth is *linear*. For *thicker* films *diffusion through the oxide* is the rate limiting process, so that growth is then *parabolic* in time. Three factors affects the growth rate in addition to the *partial pressure* of oxygen and the *temperature* of oxidation:

1. *Surface orientation.* The oxide film grows *faster* on {111}-oriented silicon surfaces than on {100}-oriented surfaces.
2. *Dopant species. Boron* and *phosphorus,* present as dopants in the silicon chip, *inhibit* the oxidation process.
3. *Gaseous impurities.* The presence of *water vapour* and *HCl* affects the rate of oxidation.

11.3.2 Chemical Vapour Deposition

Chemical vapour deposition (CVD) can replace *thermal oxidation* in the formation of silica insulating films. The temperature over which CVD occurs varies widely:

$$SiH_4 + O_2 \rightarrow SiO_2 + 2H_2 \text{ at temperatures of the order of } 400°C$$

$$SiCl_2H_2 + 2N_2O \rightarrow SiO_2 + 2HCl + 2N_2 \text{ at about } 900°C$$

The temperature range of the latter process lends itself to *controlled doping*, for example by additions of PH_3 to the gas phase in order to introduce *phosphorus.*

CVD can also be used to deposit uniform thin films of other *insulating ceramics*, for example:

$$SiH_4 + 4NH_3 \rightarrow Si_3N_4 + 12H_2 \text{ at a temperature of } 700°C.$$

11.3.3 Photoresist Polymers

Several polymers have been used as *photoresists*, all of which operate in much the same way. The *polymer* is deposited from solution as a *thin film* (usually by centrifugal spreading). It is then *irradiated* through an *optical mask* to activate the polymer. After irradiation the polymer is *developed*, either to remove *unexposed polymer* or to remove the *exposed polymer*. The developed polymer film should have

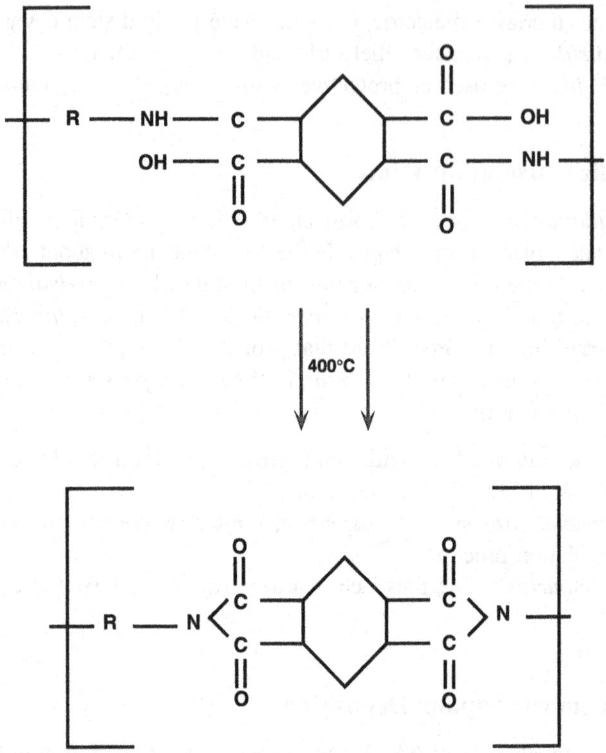

Fig. 11.6. *Polyimide* can be printed as a *photoresist* and then fired, when the polymer loses *water* to form a durable and reliable *dielectric film*.

similar dielectric properties to *silica*, and act to *electrically isolate* the chip in the regions selected by the imaging process.

Polyimide films are common photoresists used in microelectronic fabrication. After developing the resist, the polymer can be *fixed* by heating to *400°C*, when *water vapour* is evolved and the chemical stability enhanced (Fig. 11.6). Alternatives to polyimide are *polymethylmethacrylate, epoxy resins* and some elastomers, such as *isoprene* and *butene*.

11.4 JOINING PROCESSES IN MICROELECTRONICS

The *range* of processes used in joining the components of a microelectronic circuit covers most of the methods which have been treated in this text:

1. *Connecting wires* from the microcircuit are soldered to the *pins* which connect to the external system.
2. *Microchips* are bonded to the *substrate* which supports and insulates them from their surroundings.
3. *Conductive strips* must adhere to their *substrate* to ensure the integrity of the circuit.

Many of these processes, used on the *microscale*, are based on *liquid-phase* phenomena, and there are many *hybrid* processes. The *ultrasonic welding* technology, outlined above (Fig. 11.4), is a *solid state* process, involving *vibration-induced plasticity* beneath the alumina capillary tip, but one stage in the process involves *melting* of the connector wire, to detach the welding head from the second weld and form a metal *bead*, ready for the next weld.

Polymerization of a photoresist resin or polymeric protective film converts the *viscous precursor* into a *flexible polymer*. *Soldering* a connection requires a *melting* and *solidification* heat cycle. The bonding of a *ceramic substrate* to *glass* is accomplished by heating the assembly *above* the softening temperature of the glass (above the glass transition temperature).

The factors to be considered in *microjoining* do not differ in principle from those factors which separate success from failure at the *macroscopic* level, for example:

1. *Surface roughness* needs to be at an acceptable level. *Microsurfaces* are *not* atomically smooth, and *grooving* at grain boundaries is often apparent, as are *surface facets* associated with *anisotropy* of the *surface energy*. *Undercutting* at the edge of a connector strip is *unacceptable*, and edge transitions should be reasonably smooth. Fig. 11.7 shows two examples of the *surface topologies* observed in microcircuits.
2. *Contamination* is a major concern in *all* microelectronics processing, and extreme precautions are taken in all production lines to ensure that *particulates* are removed from the atmosphere and that personnel in the production area do not introduce foreign substances on their clothing or person. The same concern applies to the *joining technology*. It is important to appreciate the reasons for this concern. The *semiconductor properties* of the active components on a microchip are determined by *trace additions* or *dopants*. These are present in quantities which have to be controlled at the level of *parts per million* (ppm); any *external* source of contamination can drastically affect these properties.
3. *Chemical reactivity* of the surface is also important. *Silica* is a stable oxide, but will still react with *some* components. For example, it is reduced by *titanium*:

$$Ti + SiO_2 \rightarrow TiO_2 + Si$$

4. *Mutual solubility* leads to *interdiffusion* of the constituents, depending on the *temperature*, resulting in changes in the *electrical* and other properties of the active elements.

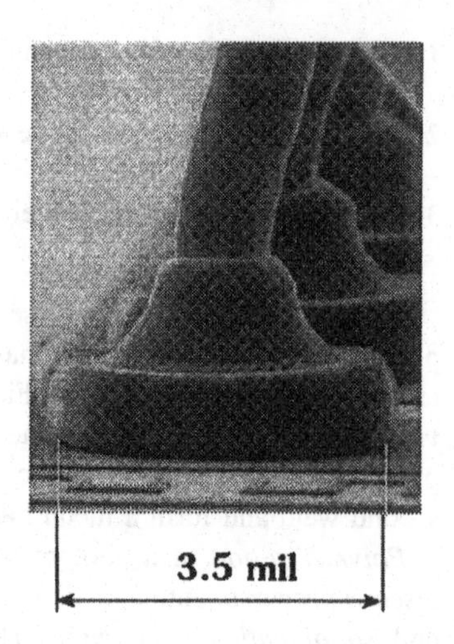

(a)

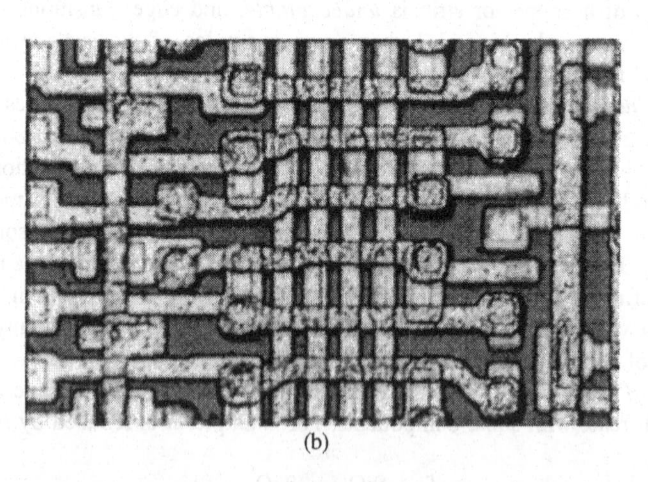

(b)

Fig. 11.7. In *microelectronic circuitry* the topological scale of the microstructure and the geometrical dimensions of the components are similar. (a) *Ultrasonic welds*. (b) *Connector strips*. (b) Reproduced with permission of IOP Publishing Ltd., from Grovenor: *Microelectronic Materials (1989).*

11.4.1 Chip Bonding Processes

Several distinct technologies have been developed to bond a *chip* to its *substrate* in a microelectronic package. In this section we describe *three* of these technologies and some of their *variants*. These processes are *complex*, involving several production stages, and each has specific areas of application, which are continuously developing and changing.

11.4.1.1 BACK-BONDING

A layer of *solder*, about *40 μm* thick, wets the *contact pads* as these are lowered, inverted, onto a bath of molten solder and then removed. Among the *solder compositions* used for this process are:

1. *Gold/silicon* eutectic alloy (3.6% Si) with a melting point of 370°C.
2. *Gold/tin* eutectic alloy (20% Sn) with a melting point of 280°C.
3. *Epoxy resin* filled with silver powder (thermosetting at 150°C).

11.4.1.2 WIRE BONDING

Ultrasonic welding is used to make the connections (Fig. 11.4), as described in Section 11.2.1.

11.4.1.3 TAPE BONDING

An *array* of contacts is made simultaneously (with a minimum separation of about *50 μm*). The assembly is shown schematically in Fig. 11.8. The *gold-coated copper strip*, bonded to *polyimide tape*, makes contact with a *gold bump*, bonded to an *aluminium alloy* strip, through a *chromium interlayer*. The aluminium alloy strip is metallized on the microchip and the tape-bonded assembly is potted directly into the final package.

11.4.1.4 THE FLIP-CHIP PROCESS

This process was developed a quarter of a century ago. It is illustrated in Fig. 11.9. The *solder layer* wets a *chromium interlayer* deposited on the *aluminium alloy* connector strip. *Inverting* the assembly over the contact pads mounted on a carrier film allows the *molten solder* to form a pendulous drop, which makes contact and wets the *contact pad*.

11.4.1.5 BEAM LEADS

If *gold leads* are deposited *directly* on the chip, which is then etched to leave the ends of the gold contacts protruding, then these *microbeam leads* can be *soldered* directly to an external circuit.

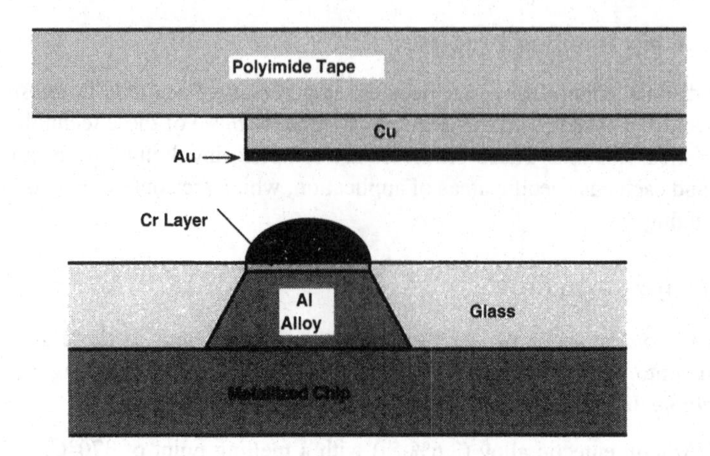

Fig. 11.8. The *tape-bonding* process in which a gold 'bump' is *diffusion bonded* to a thin gold layer on a copper connector in a *directly potted* assembly.

11.4.2 Potting Compounds

Potting the microcircuit is a common method of protecting it from the environment. The polymer compositions used for this purpose include the *epoxy resins*, *silicones*, *polyurethanes* and *polysulphides*. Each group has specific advantages and disadvantages.

Epoxy resins are inherently *brittle* and sensitive to moisture. This makes them a poor choice for demanding applications (for example, service in tropical climates). The *silicone rubbers* are stable over a wide range of temperature and are moisture-resistant, but they have *poor adhesive strength*. Breakdown of *adhesion* will allow *moisture* to penetrate at the interface. *Polyurethanes* have *highly toxic* precursors. *Polysulphides* have high precursor *viscosities*, which may inhibit full *penetration* of the potting compound, leaving *gas bubbles* and *voids* in the assembly.

11.4.3 Printed Circuit Boards

Although not strictly relevant for a discussion of the implications of miniaturization, some consideration should be given to the *mounting* of microelectronic circuits. This is commonly on a *printed circuit board* (PCB), which is usually a rigid, glass-fibre-reinforced epoxy laminated with a 30 μm thick copper sheet on which the circuit connections are printed using photoresist technology. Surplus copper is removed with an acid etch, the circuit leads being protected from etching by the developed photoresist pattern.

Double-sided PCBs are also used, as are *multilayer* PCB assemblies, in which contacts between layers are run through holes in the PCB. More recently, increasing

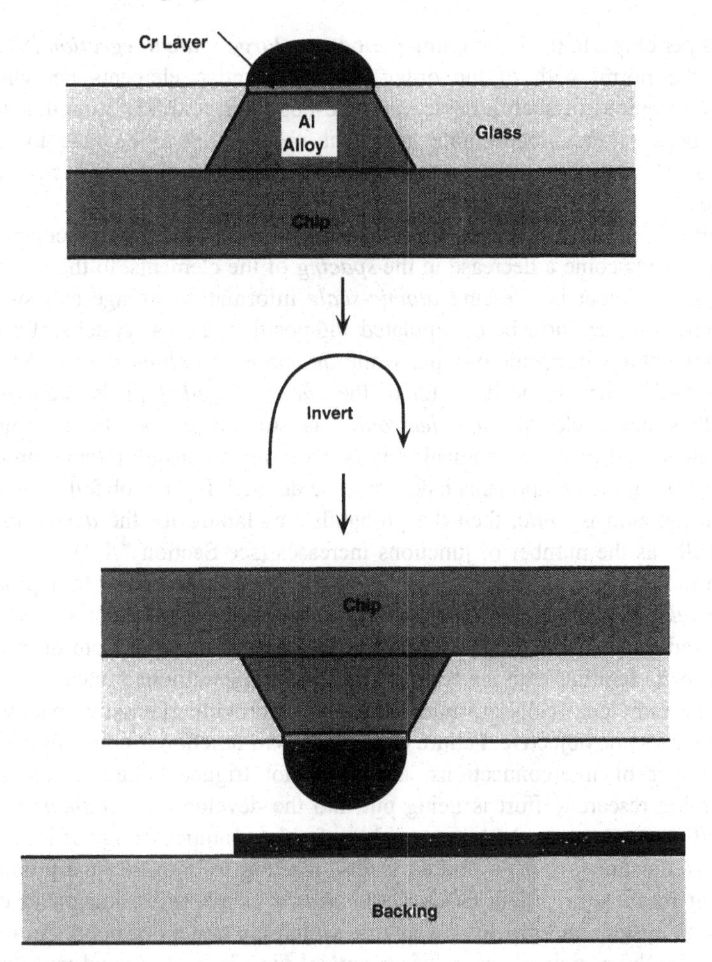

Fig. 11.9. The *flip-chip* process developed in the early 1970s. A *solder alloy* replaces the gold bump and the chip assembly is *inverted* so that the pendulant drop of molten solder makes contact with the electrical connection.

use is being made of *flexible* PCBs which can be more easily accommodated to the spatial limitations of an engineering system.

11.5 RELIABILITY & SIZE

A quarter of a century ago there were many experts who doubted that it would be possible to increase the number of active elements in an *integrated circuit* (IC) much beyond the then limit of *large scale integration* (LSI, containing perhaps 10 active elements per chip). In the intervening period *very large scale integration* (VLSI) has become the norm, with of the order of 10 000 active elements per chip. The *reliability* required of such a device is mind-boggling, and VLSI circuits are only feasible because they incorporate sufficient *redundancy* to be able to function satisfactorily, even when a proportion of the *interconnects* and *switches* are defective.

In parallel with the increase in the *number* of active elements comprising a microcircuit has come a decrease in the *spacing* of the elements, to the point where current development is stressing *atomic-scale* information storage and switching. *Individual atoms* can now be manipulated and positioned on a crystal surface, using the scanning tunnelling microscope, while *quantum tunnelling devices* have been demonstrated in which the diameter of the *quantum well* is in the *subnanometre* range. This new field of *nanoelectronics* is still some way from commercial development, and the technological criteria necessary to ensure reliable production and assembly of the components have yet to be defined. If the probability of a single defective junction is *finite*, then the probability of failure for the *device* increases dramatically as the number of junctions increases (see Section 7.1.3).

When the density at which the *active elements* are packed starts to approach the *atomic packing densities*, then the whole concept of *product reliability* needs to be re-evaluated. In particular, the engineer may have much to learn from the biologist, who is already familiar with the concept of *redundancy* in living systems. In a living system the individual cells are interconnected to provide alternative pathways for achieving the same objective. Failure of some cells to function is compensated by the large number of interconnections, and does not trigger failure of the system. Considerable research effort is being put into the development of *smart* materials and *intelligent* systems, which are capable of either compensating for local failure (the *smart* function), or, to a limited extent, reacting to eliminate the problem (an *intelligent* response). In both cases, active *sensor* elements are integrated into the engineering system and provide a signal to which the system respond. An example should make the principle clearer: if an *optical fibre* is incorporated into the fibre-reinforcement of a composite, then a change in stress level will affect the optical properties, and a laser signal transmitted through the fibre can be used to monitor the load on the component.

It is not science fiction to imagine an engineering system which is capable of rapid adjustment to external conditions by the incorporation of sensor devices whose electro-optical signals are converted into an appropriate response by a computer operating through adaptable nanoelectronic devices. This extension of *robotics* will certainly depend on successful interfacing of the sophisticated components, that is, on the *joining processes* which are employed.

In this final section we restrict ourselves to *classical* failure modes. The *scale* may be drastically reduced in a microelectronic circuit and some failure modes may be new, but nevertheless many of the *modes* of failure are still those familiar to the materials engineer in aircraft, ships, vehicles and bridges!

11.5.1 Failure of Electrical Contacts

The destruction of a metallic interconnect strip through *electromigration* (Section 11.1.1) constitutes a completely *new* failure mode, unique to *microcircuitry*, but many common failures are simply the result of *mechanical* or *thermal fatigue*. *Soldered* joints, in particular, experience cycles of mechanical and thermal stress. The (largely unsupported) connecting wires may be subject to *vibration*, especially if the microcircuit is mounted in a *powered system*, such as a car or aeroplane. The *frequency of vibration* may be high, and soldered joints, unlike steels, have no *fatigue limit*.

Since the fine contact wires are mounted on (comparatively) massive supporting chip assemblies and electrode pins, the dominant mechanical stresses are *shear* at the interface, and a *mixed-mode* fatigue failure is the expected result.

11.5.2 Packaging Failures

Fig. 11.10 illustrates *three* possible failure modes associated with the assembly of a *microchip package*:

1. A *shrinkage crack* associated with *thermal expansion mismatch* may break an electrical contact in a metallic interconnecting strip.
2. *Mechanical vibration* may lead to a *shear dominated* fatigue failure between an *interconnect* and a *bonded lead* at a welded or soldered joint, as described above.
3. *Poor adhesion* due to thermal mismatch and inadequate bonding may result in a *delamination* failure between a *microchip* and its *substrate*.

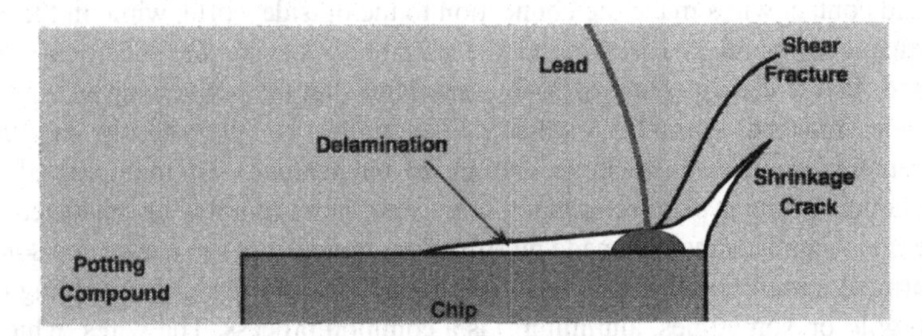

Fig. 11.10. *Potted assemblies* frequently fail. (a) *Shrinkage cracks* in the potting compound are associated with the thermosetting process and can provide paths for ingress of water vapour. (b) Mechanical vibration, especially at high frequencies, may be inadequately damped and lead to *fatigue failure* in shear of the electrical connections. (c) Poor adhesion of the potting compound can lead to *delamination* at the chip surface, again providing a path for environmental corrosion.

Summary

The continuing drive towards further miniaturization of microelectronic components, with functional components approaching the submicron level, is accompanied by a continuing demand for improved tolerances in assembly and a scaling down of the dimensions of the interfaces between components. A wide variety of applications are involved: those associated with signal detection (pressure or temperature changes, chemical composition and, of course, electrical and optical signals); those associated with data processing, including computer interfacing and data storage; and those associated with communications systems and data transmission. Some components are required to operate in hostile environments: molten steel, outer space or chemical reactors; while for others the primary requirement is reliability at minimum cost (most notably consumer products).

Photolithographic imaging dominates the microprocessing of semiconductor circuitry, and the minimum dimensions of the active elements are limited by the wavelength of light used. This limit is, practically speaking, about 0.7 μm, and the drive to find alternative technologies must consider the microstructural stability of the device and, most especially, the stability of the interfaces between the active components. Microstructural stability may be affected by both kinetic and thermodynamic considerations. The size and shape of the grains and particles may be determined by the surface energy and anisotropy of surface energy (thermodynamic considerations), but also by kinetic mobility of the surfaces and the dependence of mobility on orientation (kinetic processes associated with surface, boundary and bulk diffusion mechanisms). Two major considerations are the effective diffusion distance in the system and the mobility of the microstructural features.

The packaging of microelectronic components must satisfy specific engineering needs which cover four major areas: electrical contact, thermal dissipation, environmental isolation and mechanical protection. Two basic strategies have been developed. In the first the active components are isolated in a hermetically sealed can, and contact wires make the connection to the outside world, while in the second a potting compound is cast around the assembly, leaving the electrical contacts exposed. It is a characteristic of both assemblies that the active chip is far smaller than the finished assembly, which, in addition to the strictly engineering requirements, must also be large enough to be handled and manipulated in the construction of the system (computer, television, heart monitor or whatever).

Electrical contacts may be made by soldering, typically in the automated soldering of a printed circuit board, but for microelectronic devices ultrasonic welding of fine wires (gold or, sometimes, aluminium) is a common process. The wires, which may be less than 20 μm in diameter, are fed through a ceramic capillary guide and the ultrasonic energy for bonding is channelled down the guide from a piezoelectric source. Good adhesion between the semiconductor chip and the substrate on which it is mounted can be obtained with a ceramic substrate, most commonly alumina. The adhesion depends on the dimensions of the chip and the thermoelastic mismatch between the chip and the substrate. However, adhesion is not the only consideration: successful performance at high frequencies places severe limitations on the dielectric properties of the substrate and its thermal conductivity.

Electrical insulation of the components from each other and from the environment introduces another major component into the design of the chip itself. Historically, silica has been the most important insulator, typically prepared by the controlled oxidation of the silicon crystal. The oxidation process depends on the orientation of the silicon surface and the nature and concentration of the dopant species. Both silica and silicon nitride insulating films have been prepared by chemical vapour deposition. Polymer insulating films are also important, both in the processing of the chip by photolithography and, in some cases, in the final product.

The range of joining processes which have been adapted to the semiconductor and microelectronics industries covers the full range of processes which have been considered in this text: soldering and welding of electrical connections, the bonding of chips to their substrates, the growth of a variety of adherent films, and the use of polymeric adhesives as both potting compounds and sealing agents, as well as in the form of photoresist films. The difference is in the scale of the join and the engineering tolerances associated with this scale. For the system to be reliable all the components must fulfil their intended function. In VLSI (very large scale integrated circuits) this can only be achieved by accepting a certain level of redundancy in the circuit, but even so the reliability requirements are comparable to those in a jumbo jet or nuclear reactor, with the added requirements for mass production, not only in the case of consumer products but for many other devices as well.

It will be interesting to see how the classical modes of failure – loss of adhesion, decohesion, stress corrosion, fatigue and brittle failure – are addressed in the next century, when the components have dimensions below the grain size of today's engineering materials and the devices are, for the most part, assemblies of single crystals approaching submicron dimensions.

Further Reading

1. C. R. M. Grover, *Microelectronic Materials*, Adam Hilger, Bristol, 1989.
2. J. Bargon (ed.), *Methods and Materials in Microelectronic Technology*, Plenum Press, 1982.
3. G. E. McGuire (ed.), *Semiconductor Materials and Process Technology Handbook*, Noyes Publications, 1988.
4. P. J. Peterson (ed.), *Corrosion of Electronic and Magnetic Materials*, ASTM STP1148, 1990.
5. L. T. Manzione, *Plastic Packaging of Microelectronic Devices*, AT&T, 1990.

Problems

11.1 Why should the grain boundaries in a thin film orient themselves perpendicular to the film surface?

11.2 Compare the mechanism of ultrasonic welding to that of fusion welding, and discuss the surface preparation needed for successful ultrasonic welding.

11.3 Joining processes for microelectronic devices generally require extreme measures to prevent contamination. What reasons can you suggest to account for this (at least three!)?

11.4 What is meant by 'electromigration'? Why does electromigration lead to eventual failure in the current-carrying metal runners on a microchip?

11.5 Calculate the current required to melt a 20 μm diameter aluminium connector wire, given the follow: melting temperature = 660°C, resistivity = 2.7×10^{-8} Ω m, specific heat = 2.44 J K^{-1} cm^{-3} and latent heat of melting = 40×10^4 J kg^{-1}.

11.6 What are the engineering requirements which must be satisfied by a microelectronic packaging material?

11.7 Estimate the cross-sectional area of a copper connector strip required to carry a current of 1 mA if the maximum temperature rise is 20°C?

Resistivity = 1.7×10^{-8} Ω m – Specific heat = 0.20 J K^{-1} cm^{-3}

11.8 Compare the thermal stresses associated with power dissipation in microelectronic substrates for silicon chips fabricated from Al_2O_3, AlN, and BeO.

11.9 What are the advantages and disadvantages of a polymeric insulating film for a gallium arsenide detector device, as compared with CVD silica?

11.10 Metal powder-filled resin can be used as a conductive bonding agent to replace solder. What do you see as the possible advantages and disadvantages?

11.11 As the density of active elements in a microcircuit increases, the dimensions of the connecting wires must decrease. Discuss the comparative advantages of beam-lead technology with those of wire bonding as the demand for increased miniaturization escalates.

Index

9 780471 964889